SAKHAROV

A Biography

Also by Richard Lourie

Fiction

The Autobiography of Joseph Stalin
Zero Gravity
First Loyalty
Sagittarius in Warsaw

Nonfiction

Hunting the Devil: The Pursuit, Capture, and Confession of the Most Savage Serial Killer in History
Russia Speaks: An Oral History from the Revolution to the Present
Predicting Russia's Future: How 1,000 Years of History Are Shaping the 1990s
Letters to the Future: An Approach to Sinyavsky-Tertz

Selected Translations

Memoirs by Andrei Sakharov
Visions from San Francisco Bay by Czeslaw Milosz
Goodnight! by Abram Tertz (Andrei Sinyavsky)
The Life and Extraordinary Adventures of Private Ivan Chonkin by Vladimir Voinovich

The publisher gratefully acknowledges the support and generosity of Mr. Ronald S. Lauder and the Mrs. Estee Lauder Philanthropic Fund in making possible the publication of this book.

❖ ❖ ❖ ❖

SAKHAROV

A Biography

by Richard Lourie

BRANDEIS UNIVERSITY PRESS
PUBLISHED BY UNIVERSITY PRESS OF NEW ENGLAND
HANOVER AND LONDON

Brandeis University Press
Published by University Press of New England, Hanover, NH 03755

Printed in the United States of America
5 4 3 2 1
Book design by Dean Bornstein

Library of Congress Cataloging-in-Publication Data

Lourie, Richard, 1940–
Sakharov : a biography / by Richard Lourie.
p. cm.
Includes bibliographical references and index.
ISBN 1–58465–207–1 (cloth : alk. paper)
1. Sakharov, Andrei, 1921– 2. Dissenters—Soviet Union—Biography. 3. Physicists—Soviet Union—Biography. 4. Political prisoners—Soviet Union—Biography. 5. Human rights workers—Soviet Union—Biography. 6. Soviet Union—Politics and government—1953–1985. I. Title.
DK275.S25 L68 2002
323'.092—dc21

2001005246

Acknowledgment for photographs used in this book:
The photographs are housed at the Andrei Sakharov Archives at Brandeis University and are reproduced by permission of Elena Bonner (the copyright owner).

To Ronald Lauder,
who was the first to "see" this book,
and to my mother,
who did not live to see it

Two great things happened in the twentieth century. Physical nature yielded up a very major part of her secrets. . . . Physical nature is now our humble and despised slave: we have nothing to fear from her, though we have everything to fear from the effects of our own socially uncontrolled and perhaps uncontrollable manipulation of her. But when it comes to our understanding of the laws of social development, there has been little if any progress. On the contrary, we have witnessed a total collapse of the most elaborate, best-orchestrated theory of society, born in the nineteenth century—a theory which had also become the state religion, and the object of total or partial belief, for an enormous section of mankind.

The dramatic demise of Marxism is perhaps almost as great an event as the triumph of physics. Hence the life of Andrei Sakharov is probably *the* life of the age. . . .

ERNEST GELLNER,
Encounters with Nationalism

My fate was, in a sense, exceptional. It is not false modesty but the desire to be precise that prompts me to say that my fate proved greater than my personality. I only tried to keep up with it.

ANDREI SAKHAROV

CONTENTS

SAKHAROV

A Biography

PROLOGUE

Reports from KGB Chief Andropov to the Central Committee

13 JUNE 1968

Sakharov, Andrei Dmitrievich (born 1921 in Moscow; Russian; not a party member; holds a doctorate in physics; elected in 1953 to full membership in the USSR Academy of Sciences) is one of the creators of the fatherland's *thermonuclear* weapon. For his special contribution to the defense of our country, he has three times been made a Hero of Socialist Labor and has also received Lenin and State Prizes. . . .

According to information received by the KGB, certain hostile elements, in their demagogic attempts to prove the existence of "the Party's excessive interference in science and the arts," the necessity for "improving socialism," etc., are continuing their efforts to exploit the name of the well-known Soviet scientist, Academician *Sakharov*, by systematically asking him to endorse documents of politically harmful content. . . .

. . . in order to eliminate the opportunities for anti-Soviet and antisocial elements to exploit Academician *Sakharov's* name for their politically harmful acts, we consider it would make sense for one of the secretaries of the Central Committee to receive *Sakharov* and to conduct an appropriate conversation with him.

20 APRIL 1970

We believe that, in order to receive timely information on SAKHAROV's intentions and to discover the contacts inciting him to commit hostile acts, it is advisable *to install secret listening devices in Sakharov's apartment.*

17 APRIL 1971

... the KGB received new information that SAKHAROV had openly switched from a few politically harmful acts of a malcontent to consciously hostile political activity.

Meeting regularly with anti-Soviet individuals, some of whom are mentally ill, SAKHAROV looks at the world around him mainly through their eyes. It seems to him that he is constantly subjected to provocations, surveillance, eavesdropping, etc.

CHAPTER ONE

A DIFFICULT BIRTH

RUSSIA would have a new saint. In mid-July 1903 close to half a million pilgrims—the poor and the ill, the sinful and devout, all hoping for cures, miracles, forgiveness—streamed to the small town of Sarov in the province of Nizhny Novgorod to attend the "glorification" of Serafim. Born in 1759, Serafim had retreated to the Sarov wilderness when he was eighteen, taking a vow of silence to prepare his soul for "the Word of God, the bread of angels." His life in the monastery of Sarov and as a hermit was one of exemplary humility, high spirituality, and the gifts of healing and prophesy; he foresaw Russia passing through an era of ruthless atheism before returning to Christ. In legend and icon, wild bear ate from his hand. However, shortly before his death in 1833, rumors reached the local police that Serafim's relations with the local nuns were not entirely ethereal. The case was dismissed for lack of evidence, but as the subsequent history of Rasputin would demonstrate, a mix of lust and sanctity in holy men was not at all alien to the Russian taste.

Serafim's glorification, as the canonization of a saint in the Russian Orthodox Church is called, was to a considerable extent stage-managed by Tsar Nicholas and Tsarina Alexandra, each of whom had compelling reasons for creating a new point of contact between heaven and Russia. Nicholas, disturbed by his nation's rush to industrialization and its shadow of revolution, wished to establish a direct, mystical bond between himself and his people. The glorification of Serafim was part of the tsar's "mystical activism," Eternal Russia counteracting the poisons of modernity. Pietistic, exalted, willful, Alexandra had given her husband four daughters and was now despairing of ever producing the male heir who would ensure the continuity of the Romanov dynasty, which had ruled Russia for over three hundred years. She too had come to believe that direct contact with "Holy Russia" through the spirit of a holy man was the key to the grace of an heir.

The royal couple arrived in the city of Arzamas on the imperial train and completed their journey by carriage, greeted at every step by adoring throngs, which allowed them the experience, or the illusion, of a personal relationship with the masses that circumvented the bureaucracy of St. Petersburg. At the glorification ceremony, the candle flames in front of the icons flickered as the huge cathedral bells tolled and the voices intoned:

> We magnify
> We magnify thee
> Holy blessed Serafim
> And we honor thy holy memory
> For you to pray for us
> To Christ, our God.

Then Tsar Nicholas and the grand dukes personally shouldered Serafim's coffin out to the square in front of the cathedral where a "deep silence fell on the crowd, broken only by the sounds of women weeping and lamenting. Peasants scattered bits of linen and skeins of thread along the path in front of the pallbearers, so that afterward they might gather up these precious tokens filled with the grace of the saint and take them home."

The three days of ceremonies were solemn, exultant, but the royal couple found time for their own needs, both private and dynastic. As Edvard Radzinsky wrote in *The Last Tsar:* "At night the empress bathed in the holy pond, imploring Serafim for the birth of a son, while Nicholas sat on the bank. Her body was white in the silver water."

Those were days of peace and prayer, hope and faith for Nicholas and Alexandra, a moment of serene equilibrium in the Russian summer. It was all to be swallowed up, their Russia, their lives, which were also to end on a mid-July night fifteen years later. Even the town of Sarov would be incorporated lock, stock, and barrel into Arzamas 16, the secret Soviet nuclear weapons lab where Andrei Sakharov would spend nearly twenty years of his life.

The imperial prayers were answered a year later, on July 30, 1904, when Alexandra gave birth to a son, Alexis. The dynasty was assured. But by then Russia was embroiled in hostilities with Japan that was originally touted to be just what the country needed, a "short, victorious war" that would deflect the nation's attention from domestic crises. Short it was. In less than two years Japan, which had gone from samurai swords to dreadnoughts in half a century, humiliated Russia on land and sea. Russia lost its fleet, billions in gold, nearly half a million men, not to mention territory and authority. A peace brokered by the American president Teddy Roosevelt in Portsmouth, New Hampshire, was concluded on September 5, 1905.

From the very start 1905 was a year of presagement. Many of the themes that would dominate the history of Russia, and the world, were first sounded then: revolution, racism, scientific breakthrough.

Once again a religious procession was to play a pivotal role in the fate of the Romanovs. In January 1905 a priest named Father Gapon led a peaceful demonstration of workers to the Winter Palace in St. Petersburg in the age-old belief that if only the tsar could hear their grievances, they would soon be redressed. Carrying icons and singing patriotic songs, the demonstrators were mowed down on the snow by rifles and sabers. The demonstrators had been doubly wrong about the tsar: he was neither in his palace nor was he concerned about their grievances despite his professed policy of direct contact between ruler and people. And they were also wrong about the priest who led them: Father Gapon was an agent of the tsarist secret police, which had decided to control the workers' movement by co-opting it.

Street fighting erupted in Petersburg, Moscow, and a dozen other cities. Mutinous sailors turned the guns of the battleship *Potemkin* on the city of Odessa. A new meaning was given to an old word, *soviet*—spontaneously formed councils of workers and soldiers. Though a failure, the Revolution of 1905 was termed by Lenin a "dress rehearsal" for the Bolshevik seizure of power in 1917.

The politics of racial hatred in its modern virulent strain also first erupted at this time. The Russian press routinely referred to the Japanese enemy as "yellow devils." Initially, Europe and especially the British, who directly aided the Japanese, were on the side of the "brave little Nippers" against the Russian Goliath. Nevertheless, they were shocked by the very outcome they desired—a Western nation defeated by an Eastern one, whites by Asians.

It was also in 1905, in the great pogrom of Odessa, that over a thousand Jews were slaughtered. That same year a Jew who would himself become a refugee from racism, Albert Einstein, published three papers—on the photoelectric effect, Brownian motion, and relativity—the latter of which changed our image of the universe and would in time result in the unleashing of atomic energy.

Andrei Sakharov's grandparents took part in the events of those heady and violent days, though on opposite sides of the barricades. Ivan Sakharov was a gadfly lawyer who had made his name and fortune by defending victims of pogroms, of negligence in steamship accident cases, and in the political trials of the day. He was known to the police as a "constant lawyer for strikes." In 1905 he formed his own political group, represented the teachers' union, and was a founding member of the Constitutional Democratic Party, known for short as the Kadets.

Ivan Sakharov was very much a man of his times both in his mentality and his clean break with his own family's past. The Sakharovs had been priests for generations, the nonmonastic Russian clergy being allowed to marry. The Sakharovs first surface in history in the eighteenth century in the province of Nizhny Novgorod, which was to figure often in the life of Andrei Sakharov. At that time the family had no last name. Family legend has it that when Sakharov's great-grandfather Ioann arrived on foot to enter the seminary in Nizhny Novgorod, a teacher looked him over and pronounced: "Since you're as pure and white as sugar, we'll call you Sakharov" (*sakhar,* accent on the first syllable, being the Russian for sugar). Looks were deceiving. Ioann, who headed the church of Arzamas between 1845 and 1864, was stern and demanding, zealous in his efforts to convert Catholics and Jews, and not above reporting to the higher ecclesiastical authorities that a drunken brawl had broken out in a church, a deacon having apparently punched a peasant.

Ioann's son Nikolai, also a graduate of the Nizhny Novgorod seminary, married the daughter of a priest, with whom he had something like a dozen children. A simple, gentle, and modest man, Nikolai always carried his prayer book with him, on whose first page he had inscribed his credo: "Do not hurt anyone." Of Nikolai and his wife it was said: "They lived long and died the same day," she a short while after him on February 1, 1916.

None of Nikolai's sons entered the priesthood. All the children were well educated, even the daughters, one of whom, Maria, was schooled in Geneva and became a physician in Moscow. Others became teachers, agronomists. Andrei Sakharov's grandfather, the gadfly attorney Ivan, had entered the University of Moscow in 1879 to study law. There in 1881 he met his future wife Maria Domukhovsky, a descendant of the old Russian nobility (with connections to the aristocracy of Poland) whose records of service at court went back to 1655.

In her youth Maria had the sort of boyish features that can make a woman's face somehow seem even more feminine and appealing. Dark-haired, dark-eyed, high-browed, her face in old photographs is intense, direct, and emanates the sort of dangerous purity that led many women of her generation into selfless acts of political assassination. In fact, after abruptly abandoning her first husband after six months of marriage, Maria Petrovna fled to St. Petersburg, where she stayed with a friend, Sofia Usova, who was connected with a terrorist organization, Narodnaya Volya (The People's Will or The People's Freedom, the word *volya* carrying both meanings). Maria caught the eye of the police early on. In 1884, when Sofia Usova was arrested and exiled, Maria's apartment was searched, and she was kept under surveillance for years, leaving a long paper trail in the archives of the tsarist police.

Maria had met Ivan Sakharov in 1881, a bad year. Tsar Alexander II, who had liberated the serfs in 1861, two years before the slaves were freed in America, now met the same fate as Lincoln. He was killed by a terrorist bomb hurled under his carriage, a document introducing a form of constitutional monarchy left unsigned on his desk. The years between that assassination and the Bolshevik Revolution were essentially one long night of reaction and the struggle against it by people like Ivan and Maria.

Ivan and Maria were very much of a kind. Both had broken with their backgrounds, both were highly critical of their society, both insisted on personal freedom—they did not marry until 1899, by which time they already had six children though, in a sign of the latitude of Russian manners, neither suffered any disapproval from the Sakharov family. And in the best tradition of the Russian intelligentsia, which required commitment and engagement, Ivan and Maria were active opponents of the injustices of tsarist Russia. Ivan provided exiles with legal counsel and helped unite husbands and wives in exile. Maria collected money for exiles, sent them packages, visited them in prison. Their apartment was searched because of Maria's correspondence with exiled members of Narodnaya Volya, and their baggage was always combed through upon their return from frequent trips to western Europe. It was an era of open borders, and the family went abroad two or three times a year, visiting France, Italy, Switzerland, taking the waters in Germany. Ivan found himself in a position not unknown to other lawyers in other countries and times: he prospered by attacking the system. After moving frequently, the family finally settled in a six-room apartment on Granatny Lane in the heart of old Moscow.

After the Revolution of 1905 was beaten back, Tsar Nicholas's government unleashed a wave of exile and execution, as many as fifteen thousand men and women hung by the neck until dead. Ivan Sakharov protested by editing a collection of essays calling for the abolition of capital punishment, which included Tolstoy's famous "I Cannot Keep Silent." His book influenced the thinking of the time and of his grandson Andrei.

To say that Andrei Sakharov's maternal relatives were on the other side of the barricades in 1905 does not mean in the least that they partook of the reactionary spirit with its jingoism and pogroms. The Sofianos were military men, staunch patriots, solid citizens. Andrei Sakharov's maternal grandfather, Alexei Sofiano, commanded an artillery brigade in the war with Japan, was much decorated, and afterward always insisted that "Russia would never have been defeated had it not been for the anti-patriotic actions of the Bolsheviks and the other revolutionaries." A fine rugged figure of a man with the close-cropped hair, beard, and upturned mustaches fashionable at the time, Alexei Sofiano retired from the army as a major general after the debacle with Japan. When hostilities broke out with Germany in 1914, he returned to active duty at the age of sixty, demanding to be sent to the front. Again he fought gallantly; again he was decorated (photos show his chest bemedaled from epaulet to epaulet). In October 1917 he retired again with the rank of lieutenant general and was supposed to receive a pension but never saw a kopeck of it—that same month the Bolsheviks seized power and were not about to support tsarist generals in their old age.

General Sofiano married twice, the second time to Zinaida Mukhanova, whose last name indicates Tartar origins. Tartar and Mongol are often used interchangeably by Russians to refer to the many peoples that comprised Genghis Khan's Golden Horde, which overran Russia in 1237 and ruled the country for some 250 years. Their legacy included the Russian word for money, *diengi,* a model of autocratic political structure, and an emotional ambivalence: the Russians were particularly xenophobic about Asians (Nicholas was playing on that with his expression "yellow devils"), but they also took pride in descent from Tartar princes. Zinaida was of the landed gentry/military class, and her lineage was quite ancient. Andrei Sakharov would be in the eleventh generation of that line of the family, from which he would inherit the "Mongol cast" of his eyes and a combination of "obstinacy" and "awkwardness in dealing with people."

The Sofianos were of aristocratic Greek descent, originally from the island of Kea, which Plutarch described as small in size and population but a great source of illustrious people. Historically, Greeks and Russians were united by a common religion and a common enemy, the Turks. But the relationship between the two countries predated the Christian era: parts of southern Russia and Ukraine had been Greek colonies as early as 8 B.C., and there were enough Greeks, both as residents and refugees, in the Caucasus region for Stalin to exile one hundred thousand of them en masse to Central Asia in 1949. It was natural for military men such as the Sofianos to switch their allegiance to Russia, the great power in the region; besides it was nothing unusual for foreigners, especially soldiers, merchants, and instructors of various sorts to seek their fortune in Russia; the great Russian poet Mikhail Lermontov was a descendant of George Learmonth, a Scottish mercenary who served the tsar in the early seventeenth century. From the end of the eighteenth century through the First World War, Sakharov's Russified Greek ancestors served Russia bravely and well, taking part in all the major wars and campaigns, serving in the cavalry, artillery, navy. Nearly all displayed exceptional valor and initiative, winning medals, promotions, and grants of land (not all of which were actually delivered upon, the communists not the only renegers). The Sofianos fought both in the unambiguously glorious war against Napoleon and in ignominious actions like the quelling of Polish uprisings or the "pacification" of the Crimean Tartars, for whose rights Andrei Sakharov would battle nearly two centuries later, not without some consciousness of recompense.

* * * *

The years between the aborted Revolution of 1905 and the outbreak of World War I were turgid, schizoid, a mix of apathy, prosperity, and terror. The workers slipped out of politics back into a life of labor and vodka; the intelligentsia turned to eroticism and mysticism, the consolations of the bedroom and the beyond. Deprived of the right to publish, assemble, and organize strikes, the revolutionaries responded to execution with assassination (three thousand in 1907 alone). Shocked into a semblance of reform, the tsar tolerated the existence of a quasi parliament, the Duma, but used all his powers to thwart it at every turn. The ambivalence of the government was reflected in the step-on-the-gas-and-brake policies of its prime minister, Stolypin, who both launched vigorous economic reforms helping capitalism to boom and had so many revolutionaries executed that the hangman's noose was nicknamed a "Stolypin necktie." Stolypin himself was shot dead in 1911 by an agent provocateur of the secret police while attending an opera in Kiev.

Though Ivan Sakharov continued to battle for justice in the courts and the press, his wife, Maria, now dedicated herself to the raising of their six children, five boys and a girl, for whose legitimacy they had successfully petitioned in 1901. A more dramatic change had transpired within her: the friend and ally of the bomb throwers had become a devout Christian. Andrei Sakharov would know her only as a churchgoing believer who lit candles before the icons in her room, though her dedication to European culture would never fade. The family life in the handsome, substantial two-story building on Granatny Lane was one of abundance and generosity, high purpose and high spirits. A niece of Ivan's from the "sticks" of Nizhny Novgorod was overawed by the sumptuous spaciousness of the Sakharov home, the oriental rugs, the piano, and delighted by the knowledge that any gift she requested for New Year's would be granted. But what she loved most was the warmth of the Sakharov family, Maria at its center dispensing wisdom and comfort. "What a beautiful family! It's not often you see one."

Andrei Sakharov's father, Dmitri, the fourth child, was born in 1889. He entered Moscow University in 1907, studied medicine for a short while, then switched to physics. He was briefly expelled for taking part in an illegal student demonstration in 1911 but otherwise did not come into any sharp conflict with his society. Dmitri Sakharov's main conflict sprang from his other great passion—music. An accomplished pianist with perfect pitch, before entering the university he had attended the renowned Gnessin Conservatory and received their gold medal upon graduation. His tastes ran to Beethoven, Bach, Mozart, Chopin, and the Russian composer Scriabin. Dmitri was also enamored of the poetry of Alexander Blok, with its blend of nobility and decadence, and set to music one of Blok's poems, "You walked away into the field." His connection to music was keen and deep. Dmitri suffered "almost physically" from shifts in pitch and "identified tones and semi-tones with particular colors." (The novelist Vladimir Nabokov was gifted with a similar form of synesthesia, though in his case it was letters that had colors. Nabokov's mother was, however, "optically affected by musical notes," a phenomenon known as *audition colorée,* which Nabokov remarks was first discussed in 1812 by an albino physician.) Anyone so gifted and sensitive whose talents had already been recognized with a gold medal would have deliberated before choosing the sobriety of physics and pedagogy over the life of a concert pianist or composer, chancy anywhere and any time, but never more so than in that slowly collapsing Russia which the poet Blok heard as actual sounds—a creaking, rending, ripping.

Something did break in 1912. In April workers in the Lena Gold Mines in Siberia went out on strike demanding an eight-hour workday. The army was called in and 150 were shot dead, causing a new wave of strikes after seven years of sullen quiescence by the workers. An unknown revolutionary, later remembered as a "gray blur," by the name of Joseph Stalin, briefly in St. Petersburg that same April to launch the newspaper *Pravda,* wrote that the "shots broke the ice of silence and the river of popular resentment was set in motion. It has begun!"

The revolutionaries and the poets both hated arrogant capitalism and a rotten social order that needed to be swept away by elemental forces. That sentiment even moved Dmitri's favorite poet, Alexander Blok, to note in his diary for April 5, 1912: "Unspeakably overjoyed by the destruction of the Titanic (there is still an ocean)."

The elemental catastrophe that both poets and revolutionaries assumed would cleanse the world was not long in coming. In August 1914, as Anthony West wrote in a biography of his father H. G. Wells, "the whole cast of befurred and befeathered, breastplated and helmeted crown princes, grand dukes, archdukes and princelings who stood for Europe's divisions" and who lusted after new markets stumbled into war and were supported by the bloodlust enthusiasms of their peoples.

Early that same summer Ivan and Maria Sakharov had made one of their frequent trips abroad, to take the "waters" in Bad Hamburg and to show the children the landscape and monuments. As Andrei Sakharov was to hear the story, Dmitri was on a "Belgian beach when he learned that war had been declared; he immediately got on his bicycle and pedaled through the night nearly fifty miles to rejoin the family. Space was found on the deck of a coal steamer, and the family set off for home. The weather was stormy and everyone . . . suffered from seasickness. The ship sailed silently through fog without lights, for fear of German warships. And indeed, . . . at one point a massive silhouette with gun turrets loomed up through the fog."

Home in Moscow, Dmitri Sakharov was immediately conscripted as a medic and served six months at the front in the Masurian Lakes region of Poland. Russia was once again ill prepared for war, and it wasn't long before one-quarter of its soldiers were sent into battle unarmed, instructed to take weapons from the dead. Though he remained closemouthed about what he had seen, the one story Dmitri would tell was of that same sort: during a poison gas attack, an officer refused to put on the single gas mask his platoon had been issued and died along with his men. Dmitri also kept a long steel dart as a memento of the lethal ingenuities of war—dropped by Germans from airplanes at the beginning of the war, the darts were supposedly able to pierce both a rider and his horse.

In a final act of folly, in 1916 Nicholas II appointed himself commander in chief of the army, believing that the mystical bond between tsar and people could overcome such mundane details as a lack of ammunition. Surrounded by brass bands, kneeling soldiers, and icons, Nicholas went off to the field leaving the country in the hands of his wife, Alexandra. By then she was under the spell of Rasputin, who could staunch her son Alexis's bouts of hemophilia. An odd mix of libido and charisma, Rasputin had supposedly walked from his native Siberia to Jerusalem. He had started out in a sect known as the Khlysty, the Whippers, whose service was based on sexual orgies followed by whippings and tearful repentance, then another cycle of the same. (Rasputin's daughter also ended up with whip in hand but as a lion tamer in an American circus. She wrote an adoring biography of her father, stressing his prowess and endowments, shortly before her death in Los Angeles.) Rasputin was assassinated in December 1916 by aristocrats—it took poison, knives, bullets, and drowning to put an end to that creature of preternatural vitality.

Now the soldiers of the Russian army were training the weapons they took from the dead against their own officers. They mutinied and deserted by the tens of thousands. Peace and land were all the peasants wanted, and the country was still three-quarters peasants. The cities were freezing, the bread ration down to a few ounces a day.

After his military service Dmitri Sakharov found employment teaching physics in private schools, where he met Katya Sofiano.

She had been raised in quite a different tradition. Daughter of the redoubtable, mustachioed general, she grew up on horseback and loved to sing army songs and Ukrainian folk songs. She was sent to the Institute for the Daughters of the Nobility in Moscow which, her son remarked, "provided more finish than education; it was neither up-to-date nor practical, and offered no real training for a profession." But she was not a superficial young lady at all. Katya had the Sofiano virtues: integrity, loyalty, piety. And she was reasonably modern and self-starting as well, finding herself employment teaching gymnastics in a Moscow school after graduation.

By early 1917 she and Dmitri were in love. The daughter of the Sakharovs' neighbors in the house on Granatny Lane recorded in her diary for February 25, 1917: "Today I met some physics teacher, Dmitri Sakharov, who's inexpressibly ugly, clumsy. Only his eyes are nice—dear, kind, pure. Katya's in love with him and he with her, so much so that they can't, and, it seems, don't wish to hide it. It's a joy and a pleasure to look on them."

For all the turmoil and deprivation, it wasn't at all a bad time to fall in love. Shortly after the diarist made her entry, the tsar abdicated and Russia was seized by euphoria, becoming, if ever so briefly, the "freest country in the world," as Trotsky put it. Control of Russia's cities was soon divided between the provisional government and the spontaneously formed councils of workers and soldiers known as soviets. In fact, no one was in control. The new provisional government, in a surfeit of liberalism, emptied the prisons. Citizens were routinely robbed on their way to attend revolutionary rallies. It was the hour when the Russian liberals, Kadets such as Ivan Sakharov, would be tested and found wanting, men of moderation unable to cope in a season of extremes. In his magisterial study *A People's Tragedy,* Orlando Figes says of the Kadets that they increasingly "abandoned their liberal self-image as a party of the nation as a whole and began to portray themselves as a party for the defense of bourgeois class interests, property rights, law and order and the Russian Empire." Ivan Sakharov had in fact recently purchased a home and considerable land near the fashionable spa town of Kislovodsk in the south of Russia and may well himself have been susceptible to the need for order and property rights; the friend of revolution in court, in practice he proved a constitutional liberal, a political philosophy that would have a poor career in the short run but would prove remarkably hardy over time.

In his book *The Story of a Life,* Konstantin Paustovsky, who as a young man had a knack for being in the places where history happens (his high school class had been invited to the opera the night when Prime Minister Stolypin was assassinated in Kiev), gives a vivid account of Moscow in the days between the February and October Revolutions. "It was a cold spring in 1917, and the young grass along Moscow's boulevards was often covered by hailstones like frozen grain. . . . The cold spring of 1917 was followed by a sweltering summer. A hot wind blew piles of crumpled, torn newspapers along the streets. . . . On the walls of buildings the wind ruffled dozens of posters. The air was filled with the kersonelike smell of printer's ink, and the smell of rye bread. . . . The city was filled with soldiers pouring back from the front. . . . Moscow was transformed into a turbulent military camp. The soldiers settled in around the railroad stations. The squares in front of them were wreathed in smoke like the ruins of a conquered city. This was the smoke not of gunpowder but of cheap tobacco. The breeze blew up little waterspouts of sunflower shells. . . .

". . . A shroud of dust hung over the city. . . . Yellow street lamps burned day and night. People forgot to turn them off. To save electricity the government ordered all clocks turned ahead. The sun went down late in the evening. . . . The whole city was on its feet. Apartments were empty. . . . Strangers thrown together at a meeting became either friends or enemies in a split second. Four months had gone by since the [February] Revolution, but the excitement had not died down. Anxiety still filled people's hearts."

But not only anxiety, and not every heart. Dmitri and Katya continued their courtship despite all the lurch and violent drift of the year 1917. But things took a definite turn for the worse for the Sakharov family after the Bolshevik coup in October 1917, relatively bloodless in Petersburg, the battles more pitched and prolonged in Moscow. In late November the Kadets, the party of which Dmitri's father, Ivan, was a founding member (as was Nabokov's father), were banned by the Bolsheviks, their members branded "counterrevolutionary" and "enemies of the people." Several of the party's leaders were arrested, an act that the writer Maxim Gorky called a "disgrace to democracy." In banning the Kadets, the Bolsheviks had more in mind than temporarily paralyzing a political opponent—they were intent on wiping out the bourgeois as a class. Trotsky was quite explicit on the virtue of doing precisely that: "There is nothing immoral in the proletariat finishing off a class that is collapsing: that is its right." Days later, Lenin founded the political secret police, the Cheka (as an acronym it stood for "All-Russian Emergency Commission for Struggle against Counter-Revolution and Sabotage; as a word it meant "linchpin"), one of whose tasks would be the extermination of people such as Ivan Sakharov, a liberal, a lawyer, a Kadet.

Aware that the currents of history had turned hard against him, Ivan left Moscow in late 1917 with his wife, Maria, and their twenty-year-old son, Georgy. Possibly contemplating fleeing the country for America, they traveled to the family property in Kislovodsk, far from the center of political unrest. Food was more easily available in the sunny abundant south, the ideal place to lie low until things had sorted themselves out. Any outcome was necessarily unclear in late 1917—Russia was still at war with Germany, and several of Russia's armies were still loyal, if not to the tsar then to tsarism.

"The year 1918," Paustovsky's account continues, "began with short thaws, gray snow, and such a hazy sky that smoke from factory chimneys climbed up to the clouds, stopped, and spread out under them in heavy wreaths of black. . . . The Moscow streets still smelled of printer's ink. Gray scraps of paper and posters hung on the walls. Decrees of the new Soviet government were pasted on top of them. They were printed on crumbly gray paper."

It was a cold and hungry city in which those decrees were posted. The intelligentsia got by the best they could, playing Grieg on the piano, reading the works of the great nineteenth-century revolutionary humanist Alexander Herzen. The same young woman who had confided to her diary that she found Katya's beloved Dmitri Sakharov "inexpressibly ugly, clumsy," noted on February 24, 1918: "The temperature in the apartment is again 8 degrees, there's nothing to eat, the bread ration is 1/8 pound per person. We've already eaten all the groats. There's a little rice and potatoes left. What's going to be? One consolation—Herzen."

"Bag men" brought in food from the countryside, which they swapped for jewels, clocks, dresses, silverware—money had no meaning. The latest decrees and developments were discussed passionately in every freezing apartment. Events were speeding up, time and space shifting. On February 8, 1918, the Russian calendar, lagging thirteen days behind the West's, was brought into sync with the European standard, one of the few instances of the Soviet drive to catch up with the West that yielded any practical results. (Nearly two weeks evaporated into the ether, catapulting the celebration of the October Revolution into November.) In early March, hard pressed by the advancing German army, Lenin shifted the capital from Petersburg to Moscow. The Japanese invaded eastern Russia. By the spring, Civil War had broken out.

None of this could prevent the marriage of Dmitri Sakharov and Katya Sofiano, which took place on July 1, 1918. The diarist noted: "Today at two o'clock Katya married Dmitri Ivanovich Sakharov. . . . Now it's two o'clock at night. . . . I think of her with longing, fear, how is she doing?! . . . Everything went well, even better and more beautifully than I had expected. Marvellous weather, bright sun, everyone in white, we all went on foot to the Assumption on the Graves Church. A beautiful long table decorated with wild flowers, darling Katya looked so beautiful, everything passed in the blink of an eye, like a dream. . . . The wedding headdress and veil looked so lovely on her but she did not wear it long. Someone spilled hot tea with milk on her, her whole arm and shoulder. They bathed and bandaged her arm, she changed her clothes and then left to visit the groom's relatives and from there to the station, to the dacha. It was the first time in my life that I'd attended a wedding and was so keyed up that several times I almost started bawling, I felt so sorry for her! . . . The young couple was escorted to the station and now she's been cut off from us."

On the same day as the wedding, British and French troops landed in the northern seaport of Murmansk. Their immediate task was to protect military stores sent by the Allies to Russia, but quickly that task would widen—to induce Soviet Russia to stay in the war against Germany. Churchill and the British went even further, wishing to "strangle Bolshevism in its cradle." In practical terms that meant supplying the White Army with advisers and a new weapon of war, the tank, which the writer Ilya Ehrenburg described with fresh eyes as "majestic and nauseating . . . a combination of something archaic and ultra-American, Noah's ark and a bus from the twenty-first century."

But advisers and tanks did not ultimately avail the cause of the Whites, who were fighting for everything the Russian people had so emphatically rejected: the return of the old ruling class. Still, any coherent description of the Russian Civil War is almost necessarily a fiction. The Whites were not a single army but several, fighting without any coordination in different parts of vast Russia. The foreign invaders included French, British, Japanese, Americans, and the thirty-five-thousand-man Czech Legion fighting its way eastward to Siberia; anarchist "bands" controlled, if briefly, cities and swatches of territory. Cities changed hands a dozen times. People went out in the morning after a night of gunfire to see whose government they were living under—they could tell by the soldiers' caps.

Three days after the wedding of Dmitri and Katya, Lenin ordered that Nicholas, Alexandra, and the royal family, currently under house arrest, now be guarded by the secret police, the Cheka. This was a prelude to their execution on the night of July 16, Lenin having decided that he did not want the Whites to have a "live banner" to rally around. Trotsky provided another explanation for the killing of the "crowned executioner" and his family: "The execution of the tsar's family was needed not only in order to frighten, horrify and dishearten the enemy, but also in order to shake up our own ranks to show them there was no turning back, that ahead lay either complete victory or complete ruin." A servant returning later to the deserted house found the empress's icons in the trash, including one of her favorites, Saint Serafim of Sarov.

But the great killer during the Civil War was disease: typhus, influenza, smallpox, cholera. More soldiers died from bacteria than bullets (by 1920, 30 percent of the Red Army, more than a million men, had contracted typhus). And it would be hunger and typhus that would decimate the Sakharov family in those years.

In November 1918 Ivan Sakharov made the long, arduous, and risky journey from Kislovodsk to Moscow to attend the christening of a grandson and to attempt to put the family's financial affairs in some order. On his way back to rejoin his wife and son, Ivan Sakharov came down with typhus. His strength gave out in Kharkov, Ukraine's second-largest city. Ivan, the great lover of Europe, spent his last days in a city that Lloyd George in conversation had mistaken for a general. An unknown samaritan brought Ivan to a hospital where he died in early December.

Around that same time the newly married Dmitri and Katya fled Moscow for the Black Sea town of Tuapse, where they were joined by Dmitri's mother, Maria, and his brother Georgy. Dmitri eked out a living teaching science, an income that he, winner of the gold medal at the conservatory, supplemented by playing the piano to accompany silent movies. But prices had gone sky-high. Maria was so desperate to get out of Tuapse that she even wrote in April 1919, noting that it was the second day of Easter, that "maybe things will become better if Soviet troops occupy Tuapse. And so we here are awaiting their arrival so to be reunited with Russia!"

Then Maria left Tuapse with Georgy to join another of her sons, Ivan, along with his wife and three children. In 1920 Ivan's two sons starved to death. Georgy, who himself had come down with typhus, rose from his sickbed to comfort the boys' mother, then lay back down and died. Maria also came down with typhus, but she survived. There was nothing in the least unique in that decimation—another great physicist, Peter Kapitsa, lost his father, his son, his wife, and his newborn daughter in a matter of months between late 1919 and early 1920.

Though some sporadic fighting continued until November, by April 1920 the Civil War was essentially over. For whatever reason—lack of ideology, lack of discipline and coordination, lack of any real help from the Allies—the Whites had been defeated, and their final battles were for space on the ships departing Black Sea ports for Constantinople. And sometime in 1920 Dmitri and Katya Sakharov made their way back to Moscow from Tuapse, bearing with them stories about a "night they spent in an enormous barn crowded with Red Army soldiers delirious from typhoid fever, about the machine-gunning of Kalmyk families, men, women, and children, trying to escape from famine, and about starving people frozen to death on the steppe."

It was a time of hunger and music. The letters and diaries of the Sakharovs' friends and relatives speak of bread, butter, potatoes, Bach, and Beethoven, with little else in between. Life had been reduced to survival and higher consolation. Dmitri and Katya were lucky: they found a dank basement apartment in Moscow, and he was soon employed again teaching physics. He also proved adept at scouring the countryside for produce, which he would lug back to the city. Dmitri was now thirty-one, Katya twenty-seven. Neither of them could have welcomed the new Russia with enthusiasm. They had already seen too much, lost too much. Enthusiasm was for the young communists armed with ideas, zeal, and cruelty. Dmitri and Katya were loyal to their family, their city, their country. And they had already resisted the temptations inherent in geography—leaving the country would have been relatively easy from Tuapse on the Black Sea, only hours away from Turkey and easternmost Europe. They had already voted by returning to Moscow. And they would vote again that some sort of life was possible in that stunned and devastated land where Civil War had devoured ten million lives. By September 1920 Katya Sakharov was pregnant. In late May 1921 Katya's brother writes to his godmother: "Something wonderful has happened in Moscow! On May 21 at 5 o'clock in the morning Katya Sakharov gave birth to a son whom she has decided to name Andrei. It was rather a difficult birth."

CHAPTER TWO

A SOVIET ZODIAC

THE signs were mixed. At first, the birth of Andrei Sakharov was a cause for jubilation: Katya was radiant and, as the diarist noted, Dmitri, the new father, always the soul of moderation, was "terribly excited, not at all his usual self." But the labor had been exhausting, and the child was weak. If Katya had milk, the child would live. But she was malnourished like everyone else. Yet general's daughter, horseback rider, teacher of gymnastics—Katya had good vital reserves. Her milk flowed; the boy would live.

Andrusha, as they called their son, was also lucky to be born in springtime, not having to survive his first months of life in a Moscow winter. Historically, his timing was also excellent. In 1921, the year of his birth, peasants were slaughtering Bolsheviks who went to the countryside to requisition food for the hungry cities, their power base among the working class. But Lenin found a solution: a temporary return to capitalism. Small manufacturers and distributors would be allowed to provide the peasants with the goods they wanted in exchange for produce that, though now taxed, was no longer taken at gunpoint. For the ideologues, this was a compromise brushing on treason. But Lenin found the perfect slogan: Two Steps Forward, One Step Back. It both appealed to common sense and had the higher justification of the dialectic, the "holy trinity" of Marxism. History was driven by the clash of thesis and antithesis resolving itself into the temporary harmony of synthesis. The forward steps of revolution and civil war had met a counterforce, the reluctance of the peasants, wily, sullen, violent. Lenin's New Economic Policy (NEP) worked fast, as it had to. Sausage, ham, eggs, butter, and milk suddenly materialized in stores. Gaudy prostitutes of both sexes put the most primitive form of capitalism—flesh for cash—into effect; cocaine, popular among the high-living decadent upper classes before the revolution, made a comeback.

The "step back" was doubly lucky for the newborn city dweller Andrei Sakharov because famine struck the heartland Volga region in the year of his birth. One of the idealistic young people who volunteered to work on the hospital trains was Ruth Bonner, whose daughter Elena, yet to be born, would be Sakharov's second wife: "The train would stop and our job was to go out . . . carrying stretchers. . . . We'd go around and pick up the people who could no longer walk by themselves. . . . Sometimes we'd bring back a corpse instead of a live one. You couldn't tell who was alive and who was dead, and every so often you'd make a mistake."

Another five million died, the numbers of the victims rounded off to the nearest million. Cannibalism was rampant. Nearly every Russian family has a story to tell—an aunt arrested when someone noticed fingers floating in the soup, a little girl going out her front gate and seeing people dead of starvation in the streets, the last of the meat on them, the buttocks, sliced off. In a sweeping humanitarian gesture, now largely forgotten by both sides, America, which had invaded a few years before, now sent in the American Relief Association, which fed ten million people a day during the height of the famine.

Two other important decisions the Bolsheviks made in early 1921 affected the young Sakharov less immediately, more contextually. A goodly proportion of the Russian workers, peasants, and military men now believed that Lenin had usurped the revolution (a disillusioned revolutionary had already seriously wounded him in a 1918 assassination attempt). The Civil War against such an obvious enemy as the Whites was one thing, but the uprising of the "pride and glory of the revolution," as Trotsky called the mutinous sailors on the island of Kronstadt, off St. Petersburg, was a contradiction that could be resolved only by the dialectic of force. Trotsky, creator of the Red Army, hero of the revolution, led the attack over the ice against Kronstadt to extinguish the flame where it burned reddest. Kronstadt would come to signify the moment when a believing communist finally lost his faith. In the joke on the subject, three old communists are arguing. The first one says, "The Moscow trials were my Kronstadt." The second one goes even further back and says, "Collectivization was my Kronstadt." The third, outdoing both of them, says proudly, "Kronstadt was my Kronstadt!"

At that same party congress where NEP and the resolution to suppress the Kronstadt rebellion were passed, another motion also carried, one that was kept secret from the rank and file, a purely procedural, inner-party matter. Since the party's hold on power was so obviously tenuous, Lenin insisted that none of the typically leftist tendency to split into factions could be tolerated. Forming anything like an opposition meant excommunication from communism. Lenin had granted Stalin the means to seize power, the inclination already present.

One star disappeared from the firmament in the year of Sakharov's birth. Alexander Blok, loved by Sakharov's father and later by Sakharov himself, died in poverty in 1921, having lost his faith in the elemental redemptions of the revolution. In 1918 he had been in tune with the music of history, writing *The Twelve,* in which an apostolic dozen Red Guards are led through a blizzard by Jesus Christ. But by 1921 the poet could detect only a terrible silence: "All sounds have stopped."

Sakharov spent his earliest years in what has been called a "time of hope and expectation . . . one of the calmest periods in Soviet history," though some wise and severe witnesses like Nadezhda Mandelstam dissent: "it was in the twenties in which all the foundations were laid for our future: the casuistical dialectic, the dismissal of older values, the longing for unanimity and self-abasement." Yet there also was, after the violence and chaos, some stability. Sakharov's grandmother, no supporter of the regime, put it simply: "Still, the Bolsheviks have managed to get things in order; they've strengthened both Russia and their own hold on power. Let's hope things will be easier for ordinary people now."

Things weren't easy for anyone, including the Sakharovs. Amenities were utterly lacking. They didn't have a baby carriage for Andrusha; the production of baby carriages, not to mention babies, falls drastically during a civil war. And any prams of prerevolutionary manufacture that had survived would have been beyond Dmitri Sakharov's means. The days when people of the educated professional class, including Dmitri's father, could rent six-room apartments, buy oriental rugs, and travel to Europe to imbibe the culture and take the waters were not particularly distant in time, but so many events had piled up in the interim—world war, the fall of tsarism, revolution, civil war—that it was difficult to much regret those days. It was a new Russia. A new socialist poverty, a democracy of the poor. In any case Dmitri Sakharov solved the baby carriage problem quite handily by taking his newborn son for outings in the fresh air by placing him in a large folder designed to hold musical scores. Dmitri had gotten through the Civil War by working as a silent-movie accompanist, and now his music was again sacrificed to practical service.

Andrei, the first Sakharov born in the new dispensation—the Soviet Union would be officially founded in 1922—was especially valued as a child because of the losses the family had so recently suffered. A cousin says he was his mother's "little prince." And his father's as well. Dmitri kept a diary chronicling his son's development, a little slow at first because his weakness prevented him from lifting his head, resulting, Sakharov later thought, in a flattening of the base of his skull. The father not only recorded his son's first words but even made up little entries in which Andrusha spoke for himself: "Today I spent the whole morning crying, Mama was very upset, then I calmed down and looked out the window. Very interesting." Since Dmitri was more the observer than the fantasist, the account no doubt derived from actually watching his son regain his equilibrium, charmed by the world past the window.

After a year, Dmitri, Katya, and Andrei moved back to the old family home on Granatny Lane (from *granata,* grenade or shell; in sixteenth-century Moscow the cannonball manufacturers had their workshops there, a suspiciously propitious address for the future "father of the H-bomb"). In the meantime that grand dwelling had shared the fate of all of Moscow's larger apartments: it had been transformed into a communal apartment, meaning that each of the six rooms was now supposed to house an entire family, kitchen and bathroom shared. It was a mathematical equation for the impoverishment of the Sakharovs, and of Russia: a six-room apartment for one family became six apartments for six families. Things were simply six times worse.

But even in this the Sakharovs had what was, by the standards of the time, some additional good luck. Four of the six rooms were occupied by various members or groupings of the Sakharov family, which meant that there was much less chance of running into a stranger yawning in the kitchen. The feeling was, one Russian woman put it, like "stepping from your bedroom into Red Square."

Communal apartments could be cozy and neighborly or cauldrons of spite and fury. The kitchen was the place the trouble started, too many women and children, too many boiling pots, too much dripping laundry hanging from the ceiling, something bound to go wrong any second.

Dmitri, Katya, and Andrei were the only family to be allotted two rooms, totaling a bit over three hundred square feet. One room served as a bedroom, living room, and nursery; the other, little more than a section of hallway, served as Dmitri's library and study and as a dish and linen closet. Such conditions enforced concentration, self-discipline, immersion in the inner life. Quarters were even further cramped by the Dutch tile stove that heated their rooms and that had to be fed wood shouldered in from the courtyard, also the site of the icehouse where food was stored, a great outdoor communal refrigerator.

"The house was very old," Andrei would remember, "and the ceilings leaked constantly. The kitchen was used by all six families, and in that terrible cramped space six Primus stoves would sometimes be roaring. But the building still had its magnificent doors inlaid with Karelian birch, a broad staircase, and exquisite banisters." Those solid stone houses were built with thick walls to keep winter out, and the light inside was either pale and faint or had a brown-gold Rembrandtesque glow. The house on Granatny Lane faced another building that had gone through a somewhat different form of Sovietization. An elegant private mansion set in considerable grounds, it was now the Bureau of Weights and Measures of the All-Union Institute of Standards. Twice a year, on May Day and on November 7, the anniversary of the revolution, Andrei would see the same slogan proclaiming: The Comintern Is the Gravedigger of Capitalism!

As he had in the Civil War, Dmitri Sakharov successfully adapted to circumstances. He found employment at the Institute for Red Professors, which was founded in 1921 to supply institutions of higher learning with qualified teachers in economics, sociology, and science. Later on he secured a position where he would remain for most of his working life—at the Bubnov Pedagogical Institute, which had to be renamed after Bubnov was arrested, at which point the safest possible choice was made and it became the Lenin Pedagogical. But again Dmitri Sakharov felt the need to supplement his income. Lacking the fire and self-will of the great achievers, he was a natural and excellent explainer, both in the classroom and, he now discovered, on the page. Andrei Sakharov said his father's style was "crisp, precise, and lucid, but it cost him a great deal of time and effort. He agonized over every word, and copied out sentences again and again in his elegant hand. I used to watch him, and perhaps it was this more than anything else that taught me what it is to truly work."

Dmitri Sakharov's popularizations proved popular. His first book, *The Struggle for Light,* which elucidated the history and physics of lighting devices from ancient times to the present, quickly sold out its first edition of twenty-five thousand copies, a great success, given that paper had recently been in such short supply that the poets would meet in cafés and recite their poems to each other from memory. And his other books, whose titles—*The Physics of the Tramcar, Experiments with a Light Bulb*—give a sense of the mass audience they targeted, also did well. The masses were hungry for scientific knowledge, an appetite stimulated by the new Soviet state, which clearly assessed the value of science to both industry and armament. "You must give our barbarians one thing: they understand the value of science," said Ivan Pavlov, whose experiments with conditioned reflexes had made him the first Russian to win a Nobel prize.

Dmitri Sakharov's success meant that his family could now afford a bit more than the bare necessities—a crystal radio set, books, and later on an imported wooden scooter for Andrei, a room rented in a little dacha outside the city—which placed the

Sakharov family "virtually at the top of the social ladder" such as it was in those times, in that building. But as Soviet society did in some ways become classless, those fine distinctions took on even greater meaning: the children at least always know which girl has the nice dress, which boy the patched pants.

Minor economic differentiators aside, Andrei Sakharov was unambiguously born into a specific social grouping, something between a class and a clan: the intelligentsia, educated people whose sense of honor and duty compels them to take action against injustice. By temperament and mentality, they tend toward secular skepticism, creative doubt, loyal opposition. But they can also take extreme form: Lenin and some of the other Bolsheviks were of the intelligentsia, its crude and jagged cutting edge. And there were also spiritual extremists. Nikolai Fyodorov, a philosopher who had a great influence on Konstantin Tsiolkovsky, the pioneer of Russian rocketry, and who was the only man in whose presence the great Leo Tolstoy felt humble, sought to reconcile science and Christianity; the greatest possible good that science could do for humanity would be the resurrection of all those who have ever lived. Staggering, in a sense inarguable, the idea was anything but moderate.

Moderation was Dmitri Sakharov's highest value because it recognized the true complexity of reality. His favorite proverb was: To live a life is not to cross a field. He never spoke of politics with Andrei, because he himself was not greatly interested in such matters and also for the standard Soviet reasons: why burden a child with things he cannot understand and, worse, might repeat? Dmitri's only spoken criticism of the Bolsheviks—and given his values it was a withering one—was that they lacked balance. Moderation was a value that Dmitri strove to instill in his son by precept and example, and one that Andrei wished to emulate but felt he could not: "There was a ferment inside me, an inner conflict, and moderation was something I could achieve with great effort, if at all."

The intelligentsia into which Andrei Sakharov was born had its own elaborate system of signs and symbols, manners and mores. Ideas were discussed with fervor at the kitchen table over tea late into the night. People were expected to understate their own achievements, but some of their exchanges on points of principle could be hair-raisingly blunt. As in a Chekhov play, the bookcase, usually glass-fronted, was the essential item of furniture. Thick tomes, slim volumes, bound in dark blue, deep maroon. Though the intelligentsia could be divided into scientific and humanistic, any engineer was expected to (and could) quote the lines of poetry that were for him both beauty and words to live by. The intelligentsia had its own complex code of behavior, tending toward the bohemian in matters of romance but coupled with a very strong sense of family obligation—someone's mother was usually living with the family. Their sense of humor favored the witty pun with historical resonance or the pungent, well-crafted joke, which was always the best way to vent the otherwise inexpressible. They liked long philosophical strolls, punctuated by frequent stops to impress a point, an index finger jabbing a shoulder. Any attention paid to clothes was considered vulgar or simply silly.

A bust of Tolstoy, a print of Raphael's *Madonna,* views of Rome and Venice, a romantic lithograph of Beethoven adorned Andrei Sakharov's grandmother's room in the communal apartment, though in her case these cultural icons were supplemented by an actual one, a devotional candle constantly flickering in front of it. Her room was a shrine to the values of the Sakharov family —Russian and European culture, Christianity, patriotism, hard work, high ideals, modesty, courtesy, a quiet but implacable independence.

Two major events occurred when Andrei was four. His younger brother Yura was born. (No detailed book was kept for recording Yura's development or imaginary thoughts. Whatever the confluence of reasons, he was not to feel the light of full favor and full love.) At that same time, Andrei began to teach himself to read, "spelling out the words on signboards and the names of steamships." His mother worked with him to "perfect" his skill.

Sakharov's early reading was fairly typical of any intelligent boy of the time. Pushkin was a pleasure and a passion—his poems did not have to be memorized; the words just leapt into your mind. He and Yura would contest each other by humming only the rhythm of a Pushkin poem, the other required to come up with the words. Andrei was very fond of *The Three Musketeers, Les*

Misérables, David Copperfield, Uncle Tom's Cabin, The Adventures of Huckleberry Finn, War of the Worlds. Uncle Ivan read Gogol aloud "brilliantly capturing the intonations and mimicking the gestures"; Andrei discussed Tolstoy's *Childhood, Boyhood and Youth* in great detail with his grandmother. Like many other young Russians of that era, he also read an American writer, Mayne Reid (1818–1883), a purveyor of Western yarns and gothic tales that were also devoured by the young Nabokov, who devotes a few pages of his autobiography, *Speak, Memory,* to Reid. (By an odd coincidence of precisely the sort that would amuse Nabokov, two other Americans with similarly pronounced last names —John Reed, who wrote a vivid eyewitness account of the October Revolution, *Ten Days That Shook the World*, and whose ashes were emplaced in the Kremlin wall, and Dean Read, an American rock singer who was popular later under Brezhnev, achieved renown in Russia and much less, if any, at home.)

Uncle Ivan, who had lost two sons to starvation, looked with added affection on that earnest yet merry and curious boy. Andrei liked to watch Uncle Ivan, who could fix anything, tinker with a motorcycle of which he had use—to actually own one in those days was exceedingly rare. But zooming away on that motorcycle lent Ivan a certain dash, something in which he outdid all his brothers. Convinced by Nikolai Bukharin, who would become the author of Stalin's 1936 Constitution before becoming one of his victims, to "serve the people," Ivan abandoned his first love, engineering, for the law and rose to prominence in Soviet financial circles. In time that inner circle would prove one of the most dangerous places on earth. It was close to a certainty that Ivan's high intractability would land him in trouble.

But there was a greater world outside the Sakharov communal apartment, the rectangle of the courtyard that enclosed the family circle. Like any boy, Andrusha went out to play what he nicely calls childhood's "eternal games": Cossacks and robbers, hopscotch, hide-and-seek, and mumblety-peg, in which a circle drawn in the dirt is divided into smaller and smaller triangular sections by the toss of a pocketknife, a game which early on revealed to Andrei Sakharov his lack of physical grace and adroitness—so prized in the world of children—and leaving him with a scar on his foot that would grow, along with the rest of him, "something like three times larger."

He was a fantasizer, not a knife tosser. And the courtyard—one large tree, some grass, a scattering of dandelions—was spare enough to turn into jungles, castles, other planets. But that courtyard was also a place where Andrei Sakharov learned some first lessons about himself. Not quick on his feet or with replies, he was also difficult to intimidate, unbullyable, and proud of it. And he had a streak of the devil in him. At age six Andrei began his great friendship with Grisha Umansky by punching him in the nose and making it bleed. Grisha was from a poor Jewish family who lived on the ground floor of the building, their one window looking out on the garbage heap. Grisha was glad to escape that stifling room, which he shared with his father, "a morose sickly looking shoemaker," his mother, "stout and loud," and his elder brother Izzy. Grisha's "enormous pale blue eyes" glowed with the light of a fervent imagination, which is what attracted and bound Andrei to him: "We would walk around the courtyard for hours on end, describing our fantasies and entertaining each other with yarns that were a cross between science fiction and fairy tales."

Ethnic distinctions were made among the children in the courtyard but more as fuel for teasing than from any real malice. The one exception was a young Polish boy who served as a lightning rod for the traditional Russian distrust of Poles as treacherous and too clever by half. However, Sakharov and the other children of the courtyard reserved their "fiercest animosity" for the stuck-up children who attended the nearby elite Kremlin school. His later clashes with the Kremlin had their early roots.

Though Russia has a well-deserved reputation for anti-Semitism, philo-Semitism, rarest of words, was often to be encountered among the intelligentsia. What drew Andrei to Grisha were qualities that even then he sensed as "Jewish intelligence. Or perhaps 'rich, inner life' might be a better way of putting that quality found among even the poorest of Jewish families. I don't mean to imply that other peoples lack that richness, but there is a special intensity to the Jewish spirit."

The intensity of Grisha's spirit is evident in one photograph that survives of the children of 3 Granatny Lane. Thirteen of them gathered in the courtyard in 1928 when Andrei was seven. Andrei is wearing short pants and a sweater; one little girl is decked out in a jacket with a fur collar and fur sleeves; the weather cool, but the afternoons still good for playing outside. Arranged by height, only one of the thirteen is smiling; the rest look solemn and strikingly poor. Even some of the littlest boys wear the gray, snap-brim caps favored by Russian workers. The background is a wall of well-worn wood, possibly the icehouse, in whose open door stands a boy with the features of a Renassiance prince holding an orb, the courtyard kickball. Andrei, hair brushed forward, is the only one not looking at the camera. Biting his nails, Grisha stares directly into the lens, but his eyes seem also deeply distracted. In the final days of the Second World War Grisha would be killed when a bomb hit his truck. Valya, the boy holding the ball, would die during one of his first combat missions as a fighter pilot.

Valya was the son of a housepainter, Grisha the son of a shoemaker, and Andrei the son of a teacher of physics. In the new Soviet Moscow, all classes had been squeezed down into a rough democracy of poverty, tenement equality. Those socialist slums bred crime and gangs. A seventeen-year-old boy who lived across the street from Sakharov was known as Mishka Chugalug because he gulped vodka straight from the bottle. It may have been drunkenness or just recklessness that cost Grisha a leg when hitching a ride on the back of a streetcar, an irresistible temptation to the nerviest of city kids. Sakharov knew Mishka was in a gang and may have even played an inadvertent role in his murder. Having faced down a bunch of street toughs who demanded five kopecks, Sakharov walked away without surrendering a kopeck or a kopeck's worth of dignity. Pleased as Punch with himself, he continued on his way until he was accosted by a "tall fellow of about twenty-five, with pale, mean features and a cap pulled down over his eyes." This was more robbery than bullying, and Sakharov gave him ten kopecks but protested when the hoodlum still wouldn't let him go.

"Leave me alone. I live here."

"You do? You know Mishka Chugalug?"

"Yes, I do."

"Tell me where he lives."

"In No. 6."

"Alright, now beat it while you're still in one piece."

A few weeks later Mishka's body was found on the steps of a neighborhood church. His eyes had been gouged out, his tongue cut off, both traditional woundings among the criminal element. Apparently, Mishka had been careless with his tongue, whereas the gouging of the eyes derives from the superstition that the last thing a murder victim sees, his killer, is somehow imprinted on the victim's eyes. To cover that supernatural trail, the eyes must be damaged or removed. Once Sakharov went out his front door it was no longer a sheltered life.

Sakharov's schooling did not take him further out into the world. Most likely on the urging of the outspoken and more worldly-wise Uncle Ivan, the family decided that Andrei should be educated at home by private tutors. Even before the revolution, home study had been something of a tradition in the Sakharov family but now had the added impetus of avoiding Soviet schools, with their second-rate instruction and insidious political indoctrination. Still, as Andrei later realized, there was a price to pay—avoiding socialism meant avoiding socialization: "home study impeded our psychological and social development, and for myself, this kind of education reinforced the innate lack of ability to communicate with people that has troubled me for much of my life."

In the fall of 1927, when Andrei was six, the family hired a religious young woman by the name of Zinaida to instruct him in reading, writing, and arithmetic. After classes, she'd take him with her to the Cathedral of Christ the Savior, telling Andrei tales from history and the Bible as they walked the back lanes of Moscow. Both Zinaida and her favorite cathedral would soon come to a bad end. She taught Andrei only until the next spring but continued to appear at the Sakharov kitchen, increasingly frightened by the country's official militant atheism. Sakharov's mother would feed her, give her a little money. Then, seeking religious freedom, Zinaida was arrested attempting to cross the Soviet border, sentenced to ten years, and never heard from again. Built to commemorate the victory over Napoleon, the Cathedral of Christ the Savior, with its five gold domes and fifty tons of bells, was, after being looted, razed by dynamite in December 1931. Stalin intended to replace it with the largest building in the world, the Palace of Soviets, topped by a statue of Lenin pointing the way to the future with an index finger fifteen feet long. But the project had no future—the ground beneath the cathedral proved too spongy to support any such building; after decades the gaping pit became the world's largest outdoor swimming pool.

For the next four years, between the ages of seven and eleven, Andrei was privately tutored in a little group that included his cousin Irina, his younger brother Yura, and a boy named Oleg, who was to become, along with Grisha, the other great friend of Sakharov's youth. Cheerful, "slightly absent-minded . . . [with] something courtly and old-fashioned, even comical in his manner," Oleg had a passion for history and literature. With an amazing ability to memorize (which, like an aptitude for chess and ballet, seems intrinsic to the Russian psyche), Oleg could quote long passages from the *Iliad* and the *Odyssey* in Russian translation, but only he mastered them in the original. Oleg had early on decided to become a historian and never wavered in that intention, eventually writing his dissertation on second-century-A.D. Roman foreign policy. Sakharov credits Oleg with "opening up . . . whole vistas of knowledge and art, the entire field of the humanities." Under his influence Andrei discovered Shakespeare's *Hamlet* and *Othello,* and especially Goethe's *Faust,* which would leave a lifelong impression.

Sakharov was often at Oleg's house admiring the portraits of Descartes, Newton, Ampère in the book-lined room that served as a study, and losing himself for hours in their edition of the *Brockhaus and Efron Encyclopedia,* a staple of educated homes in those days. But Andrei never invited friends to his house. Precisely because of its self-contained intimacy, the Sakharov family did not easily admit outsiders.

The little study group broke up in 1932, and Andrei at age eleven was now educated alone by two elderly sisters, a dreary, lonely patch of his childhood highlighted only by being bullied by the son of a Red Army commander who lived in the same building as the tutors and "who considered himself a superior breed to the likes of me."

Observing that the year of solitary study was not helping Andrei much scholastically or socially, Dmitri and Katya decided to enroll him in the fifth grade in Public School 110 in the fall of 1933, when he was twelve. But it quickly became clear that public school was also not improving matters greatly in either department. The work was too easy and, as Sakharov puts it, "I made no friends there: nor, for that matter enemies." He felt that he suffered from "incontactability"; no sparks flew at all, neither negative nor positive. He was invisible to the other children, except, as bad luck would have it, in carpentry class, usually dreaded by young intellectuals like Sakharov who had already learned they would never be good with their hands. "During one of my first shop lessons, a couple of older boys decided to see whether I was a crybaby and jammed my fingers in a vise. Somehow or other, I managed to hold back the tears, and the next time, one of the bullies offered to help me with my carpentry: I was having a terrible time constructing a stool." Since a great deal of childhood's victories and defeats hinge on whether you stand up to bullying or are subjugated, that alone may have made that spell in school worth the pain and the time lost.

In that school Sakharov was also first subjected to socialist indoctrination, although it took the relatively harmless form of the teachers' boasting about spectacular Soviet achievements. All the exploits the children were told of—the Moscow–Kara Kum–Moscow automobile race, the rescue by Soviet pilots of the crew of the ship *Chelushkin* trapped in the Arctic ice—had a dimension that was airbrushed out of all reports. The route of the road race had been lined with armed guards to prevent escapes, and the USSR grandly turned down offers of American assistance for the trapped ship, not out of national pride but out of fear that they would discover another Soviet ship trapped in the ice, this one with thousands of prisoners dying inside. The teachers were as ignorant of the truth as the students they instructed.

Triumphs in industrializing the economy and collectivizing agriculture were also trumpeted in school, again of course with no mention that those who resisted those forcible policies ended up in the snows, mines, or death ships. The children were also encouraged to cheer on the brave Spaniards in their war against fascism. Oranges from Spain, bright, exotic, nutritious, began to appear in dull, undernourished Moscow, a perfect gift.

The Sakharov family did not discuss politics with the exception of Uncle Ivan, who called socialism efficient for consolidating power but not for satisfying human needs. He even drew a "caricature of Stalin with fanglike teeth and a sinister grin behind the moustache." A systematic attempt was made to shield Andrei from the spirit of the times, brutal and exultant, but at best the atmosphere could only be filtered. It was there in the silences that occurred as he entered the room, in the frightened face of Zinaida appearing in the Sakharov kitchen, in the arrogance of the son of the Red Army commander.

Ignoring politics, the Sakharovs tried to live a private life of work and family celebrations—children's birthdays and name days, Christmas and New Year's. They would gather for homemade ice cream, games, charades, and magic tricks, usually performed by Sakharov's father, Dmitri: a coin that couldn't be brushed off his hand, matches snapped into pieces inside a handkerchief only to emerge unbroken. All the children joined in the charades playing beggars, pirates, millionaires. Dmitri's specialty was "An American Reading His Newspaper," a skit whose props, a coat stand and cane, are known but whose humorous content is lost. For one birthday Uncle Ivan gave Andrei a tricycle, which he learned to ride in the long, wide corridors of the communal apartment. In winter there was sledding down Moscow's white and carless streets, and in summer trips out to the dacha, where Sakharov learned to love the Russian countryside of meadows, reedlined winding rivers, dense pine, graceful birch. He never really learned how to swim, summer's chief pleasure, preferring macro- and microcontemplation: "I can think of no greater pleasure than lying on my back by the edge of the woods and looking up at the sky and trees and listening to the buzz of insects in summer or turning over on my stomach and watching the ants scurrying to and fro over the grass."

As he grew older, Andrei was also observing his family, invisible to them when they were discussing serious matters and "don't realize how attentively children are listening." He heard stories about teenagers fleeing famine, hiding in "tool compartments under freight cars, often dead when finally pulled out." His aunt adopted one who survived and grew up to be an electrician who worked on the Soviet Union's main particle accelerators.

"I began to hear the words 'arrest' and 'search' more often," he says of the early thirties. It would have been surprising if he hadn't. Forced industrialization and forced collectivization of agriculture, the two prongs of Stalin's policy, impaled people by the millions. Uncle Ivan finally fell afoul of the system. Arrested in May 1930 along with fourteen others, some of whom received three-year terms of exile, Ivan himself, after being held a few months, was released.

An attentive boy would learn more about what was actually happening in his country by eavesdropping on the adults than from listening to his teachers' solemn boasts about ice-locked polar expeditions rescued by Soviet pilots. In fact, apart from propaganda and a little carpentry, Andrei did not seem to be learning much at all in his first foray into public education. After a few months in P.S. 110, Andrei's parents yanked him from school, having decided that a crash course of private study would better prepare him for the entrance exams to seventh grade. His education in mathematics and physics would be handled exclusively by that master elucidator of science, the teacher, writer, and occasional magician, his father, Dmitri Ivanovich Sakharov.

CHAPTER THREE

THE WORLD AGLOW

"PAPA made me a physicist, otherwise God knows what would have become of me," said Sakharov in a statement as ardent as it is incomplete, for only God knows what makes a physicist. Those hours alone with his father were more than a course of instruction. They were a rite of passage; Sakharov was initiated into the order of science. Imbued with its worldview, naturally internationalist. "There is no national science just as there is no national multiplication table," was Chekhov's classic formulation. Andrei learned the hierophantic language of mathematics. The knowledge that Dmitri imparted was necessarily of the classical physics that would soon be outdated, celestial mechanics ousted by quantum mechanics, interlocking regularities by the futuristic magic of relativity. But classical physics was good physics and the best way to impress the scientific mentality—observation, logic, experiment, doubt—on his son's all too boundless fancy. Andrei was still closer to the rhapsodies of the courtyard than to the razor-sharp exigencies of reason. Now he was lost in the worlds of Jules Verne and H. G. Wells, which dizzied him as they did Edward Teller with their "unlimited possibilities of man's improvement." Every science-minded boy of the time devoured the works of the only two masters, and in some way impossible to measure the history of the twentieth century would take some shape and tone from those authors' fantasies.

The intimacy of father and son was not limited to science and initiation. Dmitri began to confide his own disappointments in life, confidences containing implicit expectations. He was aware, said Sakharov, "that he had never fully realized his own potential; he liked to talk to me about this."

It was clear to Dmitri that his son had all the attributes for exceeding him and becoming a true scientist, not just a teacher, a popularizer, but one who exults in discovery. His son had the imagination and intellect, the awe and the appetite. Almost too quick to understand, ill at ease in explanation.

A vision completes an initiation. Andrei now had his: "I was fascinated by the possibility of being able to reduce the whole gamut of natural phenomena to the comparatively simple laws of interactions between atoms, as expressed by mathematical formulas. I did not yet fully appreciate all the subtleties of differential equations, but I sensed and delighted in their power."

"Fascinated," "delighted" are words that Sakharov used rarely, only when they were exactly right.

He breezed through the seventh-grade entrance exams, impressing the teachers less by his knowledge than by his "free and confident manner." But that manner evaporated as soon as Andrei had to deal with his classmates. In that high school, which went from seventh through tenth grade, students shared a bench and a desk, a side-by-side proximity that often resulted in lifelong friendships. In the courtyard Sakharov had been able to form close ties with other boys with whom he could live the life of the imagination. But even though his deskmate, Misha Shveitser, was of that sort, a devotee of art and literature, a future film director, Sakharov could not connect with him: "during the year and a half we shared a desk we never once had a heart-to-heart talk, and the only ten-minute conversation we ever had on a street corner seemed a major event to me."

And it was even worse with the girls. "To our right sat a row of girls, and their life was even more of a mystery to me. I would steal a timid glance in their direction, but never spoke a word to any of them. When the year came to an end, Misha moved and sat next to the girl whom I found most intriguing. I was far too bashful to give her the slightest hint of my interest."

His friendships were few and strictly intellectual. In late 1935, early 1936, the twin brothers Akiva and Isaac Yaglom heard from a mutual friend about a "very strong mathematician" by the name of Andrei Sakharov. The brothers arranged for a meeting and invited Sakharov to join them at Moscow State University's special evening classes and Saturday lectures for high school students. In his memoir, "Close Friend, Distant Friend," Akiva Yaglom recalls most vividly the lightning eccentricity of Sakharov's mind, saying that he was "often the first to see the correct answer, but his explanation of how he came upon it was not often particularly clear to the other students and sometimes put the instructors in an embarrassing position." In another classic formulation: "There are two kinds of geniuses, the 'ordinary' and the 'magicians.' An ordinary genius is a fellow that you and I would be just as good as, if we were only many times better. There is no mystery as to how his mind works. Once we understand what they have done, we feel certain that we, too, could have done it. It is different with the magicians . . . the working of their minds is for all intents and purposes incomprehensible. Even after we understand what they have done, the process by which they have done it is completely dark."

Sakharov himself was not slow to realize his own exceptional intelligence. All the outward signs were there—the approval of his teachers, the rank of honor student, the admiration of his classmates. Ease with the girls or in the games of boys were not his strong suit; reason was strength and knowledge his pleasure. They alone fed his self-confidence. Imbalanced by intellect, sorrowed by all lack of the common touch, Sakharov now developed a hidden pride. It had to be hidden because the credo of the Russian intelligentsia to which Sakharov by birth and nature belonged was not to flaunt one's gifts or achievements. On the contrary, they were the one class habitually to employ that most un-Russian form of expression: understatement. A world-class mathematician would say only, Yes, I do a little math. Woefully undersocialized and paralyzingly sensitive, Sakharov had few defenses. He assumed a pose of solitary pride, though stances have a way of seeping into psyche.

Sakharov entered physics at its great hour. The neutron was discovered only in 1932, when Sakharov was eleven, by James Chadwick, a protégé of Ernest Rutherford, who had himself discovered the atomic nucleus and had roared at the annual meeting of the British Association for the Advancement of Science in 1923: "We are living in the heroic age of physics!" It did not seem that way to the Yaglom twins and the other young hotshots in their little group. "Before the war mathematics was much more popular among Moscow high-school students than physics," recalls Akiva Yaglom. Sakharov attended the math group in eighth and ninth grade but declined Akiva's invitation to continue in the tenth: "If the university had a physics group, I wouldn't miss it for anything," he told Akiva, "but I don't feel like going to the math group any more." That struck Yaglom as odd. He wondered if Andrei had been bored by the amount of time the group devoted to proving theorems that to him "seemed perfectly obvious and in no need of proof." But there was no asking. "He greatly disliked speaking about himself," says Yaglom. "We never called him Andrusha."

Yaglom misread Sakharov's decision. It was not what he was rejecting that mattered but what he was embracing. "In the stories modern physicists have made of their own lives, a fateful moment is often the one in which they realize that their interest no longer lies in mathematics," observed the science writer James Gleick. "Mathematics is where they always begin, for no other school course shows off their gifts so clearly. Yet a crisis comes: they experience an epiphany, or endure a slowly building disgruntlement, and plunge or drift into this other, hybrid field."

For Sakharov it was epiphany. He was in that state of exalted excitement that another physicist, I. I. Rabi, described as "rockets all the time. The world was aglow."

Nothing could be clearer than what to do next after high school. "It was taken for granted that I would study physics at the university," he says. "I entered the physics department almost automatically." But in choosing physics he did more than lead with a strong genetic bent assiduously shaped by his father. His father's disclosure of his own disappointments—something his son noticed he "liked to do"—was also a challenge to outdo his father, who had not been able to pursue either of his passions, music or science. Andrei understood that and admired his father for his sacrifice, even more for being a person who had achieved both a classic sense of moderation and the ability to "enjoy life to the full (a rare talent indeed)." But within, Andrei felt neither moderate nor a great enjoyer of life. Physics was the world in which he could leap with daring assurance. Physics delivered him from what Einstein called the "merely personal . . . an existence which is dominated by wishes, hopes and primitive feelings. Out yonder there was this huge world, which exists independently of us as human beings and which stands before us like a great eternal riddle, at least partially accessible to our inspection and thinking. The contemplation of this world beckoned like a liberation, and I soon noticed than many a man whom I had learned to esteem and to admire had found inner freedom and security in devoted occupation with it."

In part for reasons that were subjective and personal, Sakharov was attracted to the objective and the impersonal. He shared that view with Einstein, T. S. Eliot, and Soviet foreign minister Andrei Gromyko, who once replied to journalists: "I am not interested in my own personality."

For many young physicists, death was the great discord. Roald Sagdeev, who would head the Soviet Space Agency, was a Tartar named after Roald Amundsen, the Norwegian explorer, first to reach the South Pole in 1911. His father's imminent death accelerated Sagdeev's thinking and development: "Somehow it entered my mind that life has a finite span. I was terrified, really shaken by this revelation . . . philosophical discussions with my father may have given me the first impulse to become a scientist."

Something similar occurred with Enrico Fermi. "Both Fermi's biographers—his wife Laura and his protégé and fellow Nobel laureate Emilio Segrè—assign the beginning of his commitment to physics to the period of psychological trauma following the death of his older brother Giulio when Fermi was fourteen years old," wrote Richard Rhodes in *The Making of the Atomic Bomb*. And when Andrei Sakharov's grandfather, the bemedaled general, suddenly died in 1929 at eighty-four without having been sick a day in his life, the eight-year-old Sakharov had taken that as an affront to harmony and a breach of right order: "This was my first experience with the death of a relative. I'd already been puzzling about death for some time—it seemed to me a monstrous injustice of nature."

The exhilarations of science and the loss of religious faith coincided for Sakharov. His mother and grandmother had been the ones who had taken him to church, where he was entranced by "the chanting, the exaltation of worshippers at prayer, the flickering candles, the dark faces of the icons." But it was his father who, when Andrei was twelve, took him to his laboratory and showed him "dazzling 'miracles,' but miracles I could understand."

By age thirteen, "moved by the spirit of the times" and by his father's "subtle influence," Sakharov ceased to consider himself a believer and to say the prayers which his mother had taught him, attending church only rarely. Hurt, his mother made no outward objection, able to radiate disappointment and disapproval with silence. But she had to be resigned to the force of Dmitri's influence and of Soviet society, modern life. The Soviet Union was a "militantly atheistic" society, one dogma of Marxism that did not need to be much imposed. The shift from faith to science happened at every level of society. Sakharov, descendant of a long line of priests and a child of the Moscow intelligentsia, was in this respect no different from a poor Belorussian Jew who in his youth worked in a bakery and kept himself awake at nights by reading Darwin: "One of the books had color illustrations. Pictures of hands—a monkey's, a primate's, prehistoric man's, modern man's. They had clearly evolved over the generations. This proved that God did not create man. I ceased believing in God." The studious young baker became a Stalinist, seeing class struggle as a continuation of the struggle for existence. Though Sakharov took a different route, the process was much the same.

It's been said that the physicists who grew up in the 1920s and 1930s can be divided into two types: those who played with chemistry sets and those who tinkered with radios. Though never good with his hands, Sakharov was able to build a crystal radio of his own based on a design by his father. Later on he began doing physics experiments at home, using his father's book *Experiments with a Light Bulb* to get going but only until he could "finally" work on his own. That he was not a natural-born hands-on experimentalist he learned from some carelessness with a battery and small motor. "The electric shock I received was a memorable lesson in induction."

He supplemented his experiments by reading popular science books: *Physics for Fun*, *The Universe around Us*, *Space Travel as a Technical Possibility.* He studied calculus and was so taken by Paul de Kruif's *Microbe Hunters* that he even considered a career in microbiology, one of those dreams that with time "simply faded away," though it never did quite disappear entirely.

Whatever made Sakharov choose physics, it had all come down to a simple binary yes or no when Akiva Yaglom invited him to participate in the math club in tenth grade. Sakharov declined, but they continued to discuss science, Sakharov always avoiding both the personal and the political. Yaglom had no inkling that Sakharov had been educated at home and, as he says, "I have absolutely no memory of discussing anything but mathematics and physics with Sakharov during our years in high-school. I am, however, certain that my brother and I never had any conversations on political subjects with him," not because they didn't trust him but because they sensed such issues were of "no interest whatsoever" to him.

Politics seemed drab and distant until January 1, 1934, when Uncle Ivan was arrested again.

Case No. N-9086 accused Ivan as belonging to a group of eight people involved in "counterrevolutionary activities and preparing for one of its members to escape abroad in order to contact the Menshevik Center." The charge of counterrevolutionary activities was now so common that the state decided to economize on ink by referring to such acts simply as "c.r. activity," which is how the charge against Ivan actually reads. Oddly enough, the charges were largely true except for the fiction of the "Menshevik Center" (the Bolshevik Center could not help but assume the existence of a Menshevik Center). Uncle Ivan's offense was both real and serious. The group of which he was a part was indeed conspiring to help two people to flee Russia: Uncle Ivan's high school friend Prince Obolensky and his wife. Obolensky, realizing what the future held in store for princes who had managed thus far to survive, also feared that his illustrious name on his own passport would set off alarm signals at any Soviet border point. Ivan gave his old friend his passport, and the prince was carrying it when he and his wife were picked up near Tashkent, with an eye to traveling to China, where many Russian aristocrats had already fled.

Under interrogation, Ivan admitted the act and was outspoken in his reasons for it: "I was and I remain someone who thinks politically and relates critically to everything. . . . I am in disagreement on matters of principle and politics with the policy conducted by Soviet Power and the Bolshevik Party. . . . I have come to the conclusion that we must have the freedom to form political groups and by that means arrive at the real truth in the political battle going on in our country. . . . The economy needs a planned decentralization, with widespread local initiative—we don't have anything of the sort. . . . The basic goal of the state is to affirm the individual's creative capacity in all areas, both in economic and cultural life. . . . In this country the individual as such is not free. Finally, freedom of speech and freedom of the press should be allowed, every person should have the chance to express their opinions, again we have nothing of the sort whatsoever."

In Ivan's mug shots, the classic profile and full face, his head is held high; he looks shocked, shrewd, determined.

The exact texture of reality in Stalin's Russia cannot be quite appreciated without a feel for what happened next. At that time the head of the secret police was Henryk Yagoda, a former pharmacist with hangdog jowls and a postage-stamp mustache, who continued to experiment with drugs that induced confession or death. The secret police, at least its upper echelons, were a perverse parody of the intelligentsia. Felix Dzierzhinsky, the organization's founding father, found time to write verse in his native Polish, Lavrenty Beria affected a pince-nez, and Yuri Andropov kept a statue of Don Quixote on his desk. The intelligentsia and the secret police even socialized together. Isaac Babel once suddenly found himself alone with Yagoda at a party at the home of Maxim Gorky, the very person Yagoda would later be accused of poisoning (and indeed may well have). As Babel put it: "to break an unbearable silence, I asked: ' . . . tell me, how should someone act if he falls into your men's paws?' He quickly replied: 'Deny everything, whatever the charges, just say *no* and keep on saying no. If one denies everything, we are powerless.' "

In 1934 Yagoda was one of the most powerful men in the Soviet Union. Ivan's wife had a brother who knew Yagoda from high school in Nizhny Novgorod. Using that family connection, Ivan's wife got through to Yagoda, who apparently looked with favor on the matter, because the sentence was, as one commentator put it, "by our fatherland's standards, surprisingly mild." One of the group was even released while the others were given three years' exile to remote Kazakhstan, which Ivan managed to get switched to the more amenable Russian city of Kazan on the Volga.

The year began and ended with shock and alarm. Sakharov had been in seventh grade only a few months when, on December 1, 1934, Sergei Kirov was assassinated. Kirov, the boss of Leningrad, was the number two man in the party, and in some people's eyes—Stalin's included—the only man in the country who could possibly replace Stalin. Kirov had several advantages. He was a Russian, unlike Stalin, who was from Georgia, a country Russians both enjoyed and distrusted. Kirov had the right look, tall, blond, blue-eyed; a real Bolshevik, he was all business but affable. It has never been exactly clear which of Stalin's two main principles—kill all rivals, exploit all situations—was operative in Kirov's assassination. In any case Stalin now had an ideal pretext for a campaign of terror against his enemies, real, potential, imaginary.

Andrei Sakharov learned of the Kirov assassination at a school assembly: "The whole school gathered in the auditorium, and the principal, an Old Bolshevik, told us what happened, fighting back her tears." His father's reaction began with a cognitive error from which, once corrected, he drew an erroneous conclusion. Riding a streetcar, Dmitri "caught a glimpse of a black-bordered portrait in someone's newspaper and, thinking it was [Defense Minister] Voroshilov who had died, he rushed home in a panic, anticipating a revival of the Red Terror of 1918. He calmed down when he learned that it was Kirov instead. The name meant nothing to him."

But soon enough everyone would know who Kirov had been and why, officially, he had been killed. The next day's papers contained decrees on combating terrorism (meaning anyone with the remotest connection to Trotsky or who could otherwise be construed as an enemy) and a large photograph of Stalin beside Kirov's coffin. A ballet company, bridges, theaters, cities were named after Kirov, exalting his status in order to make the crime all the more heinous or to cover its trail. Dmitri Sakharov had offered himself ersatz comfort, for the terror that ensued surpassed that of 1918—or anything previously known—by several orders of magnitude. Ivan was lucky to be arrested on January 1, 1934, instead of after December 1, 1934, when it would have been futile to approach Yagoda and switching one's place of exile an impossible luxury.

In common parlance, Russians conjur the Terror by referring to the worst year in an absolutely minimal formulation—not even 1937, but '37. That was the only way to indicate horrors that beggar description. But this was a different terror from that of the early thirties, when it had been the workers and peasants who came under the hammer. Now it was the party, the intelligentsia, the military. Stalin was true to his theory of governance that he once confided to Winston Churchill: "I prefer fear to convictions, because convictions change."

In sad amazement, the poet Osip Mandelstam said that people think life is normal because the streetcars are still running. Not only were the streetcars running during the terror of the late thirties, Moscow's grandiose subway system, overseen for a time by a tough comrade from the mines of Ukraine, Nikita Khrushchev, was under construction. But it was a tense and poisonous era, with Hitler's rise to power, the war in Spain, and the Moscow show trials where Bukharin, once the "darling of the party" and the person who had inspired Uncle Ivan to "serve the people," admitted that he had wanted to kill Lenin and destroy the revolution. But the streetcars ran; sports stadium were filled with enthusiastic youth performing mass calisthenics (even Stalin did his morning knee bends before signing the lists). The year 1937, marking the one-hundredth anniversary of the death of Russia's greatest poet, was proclaimed the Year of Pushkin, the public declamation of his verse a more refined variation of the truck motors that were sometimes run to cover the sound of firing squads. The joke of the day, necessarily repeated to very few, was: If Pushkin had lived in our times, he still would have died in '37.

But Sakharov was oblivious to most of that when entering the physics department of Moscow University in 1938 at the age of seventeen. He was still living at home, still a loner by inclination and inability: "I didn't make a single close friend during my first three years at the university." But he did impress his fellow students. The physics department at Moscow State University was relatively small before the war, and the students entering in 1938 were all soon acquainted, one way or the other. The inevitable club formed, and Sakharov was one of its twenty to twenty-five members. As he had in the math circle, Sakharov immediately distinguished himself by the eccentric brilliance of his mind, which flashed from point to point, omitting tiresome obvious linkages. He sat in the back of the classroom and didn't take notes but reconstructed the lectures from memory at home. Sakharov struck one classmate, Leon Bell, as "tall, skinny, invariably wearing black, narrow, too-short pants, and a black jacket with too-short sleeves. Everyone's arguing, excited, only Andrei, a pack of notebooks and books, under his arm, remains silent and seems to feel too shy to express his own opinions. I thought to myself—the poor kid, he's so inhibited and awkward, what'll ever become of him?" One of the few women students, Sofia Shapiro, watched the lanky young man pacing the corridors alone and caught a glimpse of his intensity and pride, his "head thrown slightly back, bright eyes, and a very concentrated gaze." But when speaking with him, she found him approachable, glad to help, free of conceit. "One thing I'm quite sure of, girls didn't interest him at all," she says, never recalling him at any of the student get-togethers in the evening. "And of course he didn't dance."

He did, however, strike up an intellectual friendship with a student his age, Misha Levin, who lived nearby. They often walked home together, discussing physics and math, but since they were junior members of the Russian intelligentsia who suffered no split between the two cultures, their conversations were often laced with references to poetry and especially to Pushkin, almost a mania of Andrei's and still very much in the air after the official Pushkin Year of 1937. At the time the students of the Moscow University physics department were divided into two furious camps: those who wanted to master the new physics of relativity and quantum mechanics immediately, and those who advocated a through grounding in classical physics first. Sakharov took the side of Kalashnikov, the teacher, who argued that "Einstein and Bohr loved classical physics and knew all its fine points and it was precisely for that reason they realized the necessity of rejecting it. An understanding of the new physics can be reduced to rules and formulas, but you have to reach the new physics by experiencing the classical physics and struggling your own way free of it."

Sakharov drove his point home with a reference to Pushkin's "little tragedy," *Mozart and Salieri:* "Kalashnikov's right. You shouldn't imitate Salieri."

"What does Salieri have to do with it?" said Levin.

"Remember," said Sakharov and quoted:

> When the great Gluck appeared
> and revealed new mysteries to us
> (mysteries deep and captivating),
> did I not abandon all I'd known before,
> all I'd so loved and fervently believed,
> and did I not follow him with good cheer
> and no complaint, like a person who'd gotten lost
> and by a chance encounter is shown a different way?

Levin agreed: "One must not just abandon and then follow with good cheer and without complaint. The break with the old should be agonizing." A debate on physics settled by poetry quoted by heart.

Sakharov was such an admirer of Pushkin that he had even adopted one of Pushkin's mannerisms. Whenever the poet had astonished even himself with some great lines, he would exclaim "O Pushkin, you son of a bitch you!" At moments of insight, rubbing his hands in delight, Sakharov would repeat those words aloud.

What surprised Levin was not only Sakharov's detailed knowledge of Pushkin's work, life, and era but his uncharacteristic emotionality for anyone connected with Pushkin's death. Pushkin was killed in a duel in 1837 by D'Anthès, a debauched Frenchman who had humiliated Pushkin by the public attentions he paid to his wife, a society beauty.

Sakharov spoke with "malice" and "enmity passing into hatred" about D'Anthès and others involved in the duel, and confessed to even being unable to read *The Count of Monte Cristo* as a boy because one hero of the novel had a similar name. But Sakharov was never an uncritical admirer of Pushkin either; he had a lifelong dislike of Pushkin's infamous poem "To the Slanderers of Russia," which justified the quelling of the Polish Uprising of 1831 by presenting it as a family dispute, the "arrogant" Poles refusing to flow into the "sea" of Russia.

Levin caught other glimpses of the flint behind the mild exterior. That Sakharov was more in the know politically than he seemed became clear in an incident involving an assistant instructor named Turovsky. Terrified of his students, he always avoided conflict with them, handing out passing grades to all. Sakharov explained why. Before the revolution the teacher's last name had been Troitsky which, connected with the word for trinity, sounded priestly in origin. To be on the safe side after the revolution, the family changed their name to Trotsky. When that hero of the revolution was transformed into the archenemy of the people, the family was in immediate need of yet another name, and this time they settled on the more neutral Turovsky. But any trouble requiring paperwork might reveal both of Turovsky's previous last names, the last thing in the world he wanted when people with the slightest connection with Trotsky were disappearing, not to mention those who had even once taken his name.

Confiding this information to Levin, which he had apparently gained from his father, Sakharov swore him to silence: "But not a word to anyone. God forbid that turns out to be the pebble that starts a terrible landslide."

Levin kept his promise for fifty years.

* * * *

Sakharov was connected with Pushkin and his era not only by knowledge and emotion but by family history as well. In 1827 a relative on his mother's side, Pavel Mukhanov, was asked by Pushkin to serve as a second in a duel, but though part of family history, the incident itself belongs to some parallel universe of non-events because the parties effected a reconciliation. More important, Pavel's brother, Piotr, was one of the Decembrists, a loose confederation of aristocrats who took advantage of the interval between the death of Alexander I and the ascension of Nicholas I to stage an insurrection in St. Petersburg on December 14, 1825, as well as in the south of Russia. Russia had always known revolts and regicide (tsarism having been defined as absolute autocracy tempered by assassination), but this uprising, though swiftly crushed, had inaugurated a new era of active, organized, ideological revolt. The Decembrists wanted a republican form of government and a free-trade economy, though one of its more radical members did call for the Jews either to be assimilated or to be "deported en masse to Asia Minor to establish their own independent state," an idea Stalin toyed with at the end of his days. The Decembrists were either hanged or exiled, having first been personally interrogated by the new tsar, Nicholas I. Pushkin, though close to the conspirators, had not been recruited either because they feared his penchant for loose talk or because they wished to spare the poet the dangers of a chancy enterprise. When questioned by the tsar, Pushkin gallantly admitted that had he not been in exile at the time he would certainly have joined the ranks of the rebels. In any case the Decembrists acquired the tragic charisma of martyrs of liberty. Though they were aristocrats, they still had a place in the Soviet iconostasis as progressive heroes.

Whatever Sakharov might have known about his rebellious forebears or his Uncle Ivan's principled resistance to interrogation, it had no discernible effect on him in his university days. His only problems were with the obligatory courses in Marxism-Leninism, not that he had any stirrings of independent political thought: "It never entered my head to question Marxism as the ideology best suited to liberate mankind, and materialism too seemed a reasonable enough philosophy." What stuck in his craw was the "attempt to carry over the outmoded concepts of natural philosophy into the twentieth century (the age of exact science) without amendment. . . . I was unable to absorb words devoid of meaning."

Neither was Sakharov particularly troubled by the precipitous shifts in Soviet policy or their justification in the seminars on Marxism-Leninism. In August 1939 the Soviet Union signed a nonaggression pact with Nazi Germany, which was immediately reflected in the catechism. The teacher would ask: "Do the treaty and the rapprochement between the USSR and Germany demonstrate opportunism or principle?" All a student had to do was give the right answer: "Principle. They reflect the congruity of our positions." These classes were something to be endured, not questioned. Active intellect was better saved for the theory of probability, the calculus of variations, group theory, topology, even though they were taught in a "cursory manner" and all his life Sakharov had a sense of "inadequate grounding."

On September 1, 1939, Nazi Germany invaded Poland and, on the seventeenth, according to the secret clauses of the Nonaggression Pact, the Red Army occupied the eastern half of Poland. As Poland's allies, England and France, were now at war with Germany. Stalin had bought himself some time, though he seemed to fail to notice that, with the division of Poland, Nazi Germany now directly bordered on the USSR.

In early 1940 the Sakharov family had more immediate concerns than the fighting in France, the Battle of Britain, or even the USSR's Winter War with Finland: in failing health, Grandmother had a stroke at the end of the year that deprived her of speech. The radical-turned-believer braved dying with equanimity. Sakharov's father moved into his mother's room to be able to help her at any moment. Sakharov's mother asked Andrei not to go see his dying grandmother. "I can't understand why she made that request—or, for that matter, why I agreed so meekly. After all, my relationship with grandmother should have taken priority over my mother's desire to shield me from unpleasantness, especially since I was then already grown (a fact my mother probably failed to notice.)" He did, however, disobey his mother on a few occasions, once holding a glass of rose hip tea to his grandmother's lips, the last nourishment she ever took. When she died in late March 1941, she was given a religious funeral, illegally attended by Uncle Ivan, who was still serving his term of exile. It was the last time Andrei would ever see his favorite uncle, his face "contorted with grief" at his mother's death. With her passing, "the soul seemed to depart from the house on Granatny Lane."

Three months later, on June 22, 1941, Sakharov was at the university for a review session before the last exam of his third year. That session was interrupted by a sudden summons to the auditorium where along with the rest of the country Sakharov heard the voice of Foreign Minister Molotov announce that the Soviet Union had been "perfidiously" invaded by Germany. The words that lodged in Sakharov's mind were those of Molotov's concluding vow: "Our cause is just. The enemy will be defeated. Victory will be ours!"

Russia had received a violent but restorative shock. The country was now delivered from an alliance with Hitler that, despite the barrage of propaganda, had never sat well with a populace conditioned to view fascism as the natural enemy. In a shelter during a German bombing raid on Moscow, Sakharov ran into his Aunt Valya, whose husband had been executed in the thirties. She said what everyone felt: "For the first time in years, I feel like a Russian again!"

Stalin, who trusted no one, had trusted Hitler, a lapse of principle that cost Russia dearly. The first days were especially disastrous. German tanks penetrated thirty miles into Russia. The Germans had total control of the air, having destroyed 1,200 Soviet aircraft in a day, most of them on the ground. In less than three weeks, thirty Soviet divisions had been essentially wiped out. Communications were so disrupted that for a time the Kremlin had no idea of how bad things were. The German army—4.6 million soldiers supported by 5,000 planes and 3,700 tanks—drove unstoppably toward industrial Leningrad, Kiev (capital of Ukraine, rich in grain and coal), and the great prize, Moscow.

In early July, after nearly two weeks of silence, Stalin finally addressed the nation, casting the war in patriotic terms and referring to Soviet citizens as his "brothers and sisters," a sign of just how desperate the situation had become. In that same month Sakharov's classmates who were members of the Komsomol (Young Communist League) were sent to dig antitank ditches outside Moscow, but Sakharov was not among them, never having joined the organization, "for reasons of inertia, not ideology." Sakharov was, however, among the students summoned to the Air Force Academy for a physical examination, which he failed due to a chronic heart condition discovered at the time. It would have been shameful to simply go on studying. He found a position repairing radio equipment at a university workshop, then attempted to devise a magnetic probe for locating shrapnel in wounded horses. The experiment wasn't successful, but he gained valuable knowledge that came in handy during his later war work and at the time experienced a "psychological boost" from doing independent research. Andrei also joined the volunteer air defense units at the university and in his apartment building, "helping to extinguish incendiary bombs when the air raids on Moscow began. . . . Almost every night from the end of July onwards, I stood on the roof watching as searchlights, tracer bullets, and Junker bombers crisscrossed the uneasy skies over Moscow."

One night, hearing the air-raid siren while in a public bath-house, he decided to ignore regulations and began walking home through the deserted streets of Moscow when a fragment from an antiaircraft shell ricocheted off a wall and, in a bit of lucky physics, only nicked his boot. Then he continued homeward, "with the glare from burning buildings lighting my way." The world aglow.

CHAPTER FOUR

WAR AND LOVE

Moscow prepared for a battle to the death. In the summer and early fall of 1941 anyone fit for combat was rushed to the front line, closer all the time. Everyone else dug antitank ditches and trenches. Andrei Sakharov was there shovel in hand, engaged in the simple physics of war and survival. He also unloaded explosive components from trains, chemical residues being absorbed into his skin, his saliva bitter for days.

Dying at a rate of twenty to one, the Russians had slowed the German advance, but only slowed it. Moscow and Leningrad were well protected by antiaircraft artillery, meaning they would have to be taken on the ground. And it was on the ground that the Red Army was going to defend them, because the air force had essentially been obliterated in the first days of the war. There were instances of Soviet pilots running out of ammunition and simply ramming German planes, the kind of action that immediately became legendary and made the due impression on the enemy's assessment of Soviet morale. And matériel.

By September 8 Leningrad was encircled; the nine-hundred-day siege had begun. On September 17 Kiev fell to the Germans. Hitler could now concentrate his forces on Moscow. On September 24, 1941, the Germans launched Operation Typhoon, the drive on Moscow. They had to hurry. By October the rains were turning Russia's roads and fields into viscous mud, slowing infantry and tanks. And, depending on the year, November could be either fairly mild or frostbite-cold. German supply lines were already stretched thin, and the army had come uniformed for victory, not winter.

Moscow was on edge. Rumors surged into the void of hard

news: the Germans were on the outskirts of the city, the Germans were parachuting spies into Moscow. A future friend of Sakharov's, Natalya Gesse, a little sparrow of a woman, was hobbling through Moscow on crutches after a recent operation. "More than once," she says, "I was surrounded by mobs suspicious that I had broken my legs parachuting in from a plane." High officials fled the city in cars packed with food to eat and goods to sell. Some were stopped, robbed, beaten to death. On October 15 the secret police mined the city, and most of the Soviet government and Lenin's mummy were evacuated to Kuibyshev, four hundred miles southeast of Moscow. Foreign embassies and foreign correspondents had already been shifted there. Dread and suspicion kept mounting until on October 16 the city flared in a full-blown panic. The mobs at railroad stations were "human whirlpools" that tore children from their parents' hands. Stores were looted. Orgies broke out. Officials tore up their party cards; the police simply evaporated.

Sakharov was in the midst of it. "As office after office set fire to their files, clouds of soot swirled through streets clogged with trucks, carts, and people on foot carrying household possessions, baggage, and young children. Somehow or other I made my way to the university, where a crowd of students had gathered, eager to make themselves useful, but nobody said a word to us. I went with a few others to the Party committee office, where we found the Party secretary at his desk; when we asked whether there was anything useful we could do, he stared at us wildly and blurted out: 'It's every man for himself!' "

But order was gradually restored in the following days. Having decided that Moscow must be defended at all costs, Stalin remained in the capital, a symbolic act that stiffened morale. One week after the panic, the government ordered the university evacuated. Amid the crowds at the Kazan railroad station, which was adorned with 151 statues and portraits of Stalin, Leon Bell spotted his classmate Andrei Sakharov. Watched over by his parents, Sakharov was mechanically munching on some food, utterly immersed in the journal *Successes of the Physical Sciences*. Bell walked over and asked: "What're you reading?" Without saying a word, Sakharov showed him the article, a survey of colorometry. "To my evidently superfluous question 'why,' I received the brief and exhaustive reply—'It's interesting.'" Bell found him "goofy."

But then, after the usual delays and confusion, it was suddenly time for him to say farewell to his father, mother, and twenty years of life at home. The adventures came almost at once. He covered part of the first leg of the journey—to Murom, a city 175 miles east of Moscow—on a platform car carrying tanks to a repair shop. The tank crews told tales of rout and slaughter that did not match the official coverage in the least. In Murom, when not studying quantum mechanics and relativity and achieving "new insights into those subjects," Sakharov was achieving insights of quite another sort. He stood and watched wives and mothers searching for loved ones among the wounded left on stretchers on the platforms until a train could take them. Waiting for his own train to arrive, Sakharov was billeted with a mother and daughter who turned the needs and disorder of war to their own uses: by day the daughter stole sugar from the grocery store where she worked, by night the mother was "visited by a succession of soldiers." Lessons not learned in the homes of the intelligentsia.

It was in Murom on November 7 that Sakharov heard Stalin again address the nation at the traditional parade for the anniversary of the revolution. Luck favored him—it was too overcast a day for German aircraft to attack. Stalin continued to beat the patriotic drum: Pushkin, Tolstoy, Chekhov, Tchaikovsky, military heroes like Alexander Nevsky, who had defeated the Teutonic knights in 1242, and Kutuzov, the general who, by retreating, let Napoleon's hubris cause his own downfall. Stalin also used the speech to send an emotional signal to the collective psyche: "The German invaders want a war of extermination. . . . Very well then! . . . They shall have it. . . . No mercy for the German invaders! Death to the German invaders!"

Listening to that speech and to a description of the parade on Red Square, Sakharov "realized that it was all carefully staged; nevertheless it had a powerful impact on me."

Shortly after Stalin's speech, Andrei Sakharov and the other science students of Moscow University were assembled to begin their month-long journey that would take them east past the Ural

Mountains, which divide Europe and Asia, then south through the steppes of Kazakhstan toward their final destination, Ashkhabad, capital of the Turkmen Republic, just at the border with Iran. Forty students were packed into a freight car, fitted out with double-decker bunks and a stove. Food and fuel were scarce. After crossing the Urals the temperature dropped to twenty below zero, compelling Sakharov and the other students to engage in that most Soviet of practices, filching from the state—in this case stealing chunks of the coal piled at the stations to fire the steam locomotives or dismantling fences and any other "working wood." The students of Sakharov's car had nailed an evergreen tree to the roof of their car, both out of high spirits and—as one of them, Joseph Shklovsky, put it—to identify their car in the chaos of "junctions clogged with convoys as we made our way back to our freight car with vats of kasha or pails of boiling water, diving under cars and across tracks." But leaving the train, whether to swap clothes for food with the peasants or steal fuel, involved dangers of its own. When Sakharov's train left without him, he had to catch up by traveling on an open coal car, keeping his head low to avoid decapitation by sudden bridges. Finally making his way down to the platform of the parlor car, he ran into an acquaintance of his father's coming out for a smoke. He informed Sakharov that on the very day he had left Moscow his home on Granatny Lane had taken a direct hit from a German bomb and been largely destroyed. Several people were killed but no members of the Sakharov family. What had seemed coddling to Bell was in fact providence.

Eventually, Sakharov caught up to his train, rejoining his car where, as in a prison cell, people's essential nature is revealed. "During the month-long journey, the cars became separate communities, as it were, with their own leaders, their talkative and silent types, their panic-mongers, go-getters, big eaters, the slothful and the hard-working. I suppose I fitted into the silent category," said Sakharov. He was definitely excluded from the "big eater" category. Having stuffed his knapsack with books on quantum mechanics and relativity rather than clothing, he went hungrier than most—the peasants weren't about to swap bread and eggs for the latest breakthroughs in physics.

For obvious reasons the stove was the heart of the car's social life. Sometimes the mood was somber, pensive—the battle for Moscow had been raging when they left in late October, but since then they'd been mostly cut off from radios or newspapers and did not know if the capital had fallen. "To distract ourselves from those bitter reflections, we grains of sand caught up in the whirlwind of war found preposterous ways to amuse ourselves," says Shklovsky. Every group has its joker, and in theirs it was Zhenya, whose red stubble beard grew almost up to his eyes. He regaled his mates with tales of his experiments in crossbreeding the various sorts of lice with which they were all infested. "Every evening," recalls Shklovsky, "he gave us new details of the bold experiment, embroidering his account with fantastic minutiae. The group roared with laughter." And in every group there is a hero: when the chimney pipe for the stove blew off during a blizzard on the steppes of Kazakhstan, one first-year student won immortal fame by climbing up onto the roof of the freight car and repairing the damage. But no group had anyone like Leon Bell. Slight and feisty, Bell was a red-headed Jew from Texas who enjoyed tormenting his mates with tales of succulent barbecues. Bell's father had emigrated to Texas from Ukraine but, disillusioned with America, he had returned to Russia to build socialism and was of course promptly arrested.

They sang endlessly—folk songs, revolutionary tunes, and Soviet ballads of the prewar years: ". . . train follows train, year follows year. . . ." The students made up their own songs, some to the tune of current favorites such as "Into battle for our native land, into battle for Stalin," which in their rendition ran:

> With a vacuum in our bellies and callused hands,
> And driving rain to soak us to the skin—
> Our teeth are honed on the granite of science,
> And after granite—clay can never win!

Bell had become Russian enough to enjoy all those songs of the song-loving Russians but was still American enough to introduce a little jazz into the proceedings. And so it was that their train plowed through the snowy steppes of Kazakhstan to the tune of "I ask you confidentially, ain't she sweet?"

Joseph Shklovsky, in his colorful memoir *Five Billion Vodka Bottles to the Moon*, written later when he was already the country's leading astronomer, paints himself as the leader of the freight car's collective, because of his mastery of cursing, an art he had learned as a construction foreman on BAM, the alternate northern line of the Trans-Siberian Railroad. Sakharov had the bunk to the left of Shklovsky, who saw him close-up: "Tall, skinny, and seedily dressed, with deep set eyes and bushy hair, he almost never took part in our schoolboy pastimes. He spoke almost inaudibly and was diligent in doing the menial, dirty work so prevalent in convoy life. Everything indicated that the boy had been ripped by the whirlwind of war out of an intellectual family before his skin had time to thicken. One day he asked me a preposterous favor: 'Do you have anything I can read on physics?' . . . My first impulse was to send this mama's boy and his ridiculous request straight to hell."

Sakharov may have stood apart from the group, but when they looked out, they all saw the same thing he did, their country "wounded by war. The trains moving east with us carried evacuees, damaged equipment, and wounded men; those racing past us toward the west carried combat troops: their faces were tense as they peered from the train windows, and they all looked somehow alike."

Far from the "confines" of home and family, that journey roused first stirrings of the erotic. Male and female students traveled in separate cars but socialized whenever they could. "One of them took an interest in me, and some of my fellow travelers enjoyed a few jokes at my expense when they saw I wasn't completely indifferent to her charms."

The Andrei Sakharov who stepped off the train in Ashkhabad on December 6, 1941, was not quite the same person who sat on his knapsack watched over by his parents a month ago in Kazan Station. He had now witnessed war, grief, and corruption, had now experienced camaraderie, self-reliance, a taste of flirtation.

Hungry for news, Sakharov devoured the latest communiqués about the battle of Moscow, the Red Army having launched its first offensive against the German invader the day he arrived. The October rains had reduced the advance of Operation Typhoon from ten miles a day to four. The November cold froze German machine-gun oil. The cruder Soviet Pepesha—wood stock, fat drum—continued to shoot their eight hundred bullets a minute. It was not only weather. The Red Army was reinforced by large numbers of fresh Siberian troops used to the cold and by the T-34 tank, which could take and deliver more punishment than its German counterpart and ran well in mud and snow.

The Germans were perilously close, their main tank force less than eighteen miles from Moscow. Some German reconnaissance tank units broke into the western suburbs of Moscow but were soon thrown back. Some critical point of agony for humans and machines was reached. For the first time since the outbreak of the war, Hitler's army had been stopped dead. The German general Guderian noted in his diary: "The offensive for Moscow has ended. All the sacrifices and efforts of our brilliant troops have failed. We have suffered a serious defeat."

On December 7 the Japanese attacked Pearl Harbor, obliging Hitler to declare war on the United States. On December 8 Hitler signed orders placing his troops outside Moscow on the defensive. Within six weeks those troops were beaten back from sixty to two hundred miles away. Moscow had been saved and was never in mortal danger again. As Alex de Jonge says in his biography of Stalin: "the Red Army had turned a Blitzkrieg, which they never could have won, into a war of attrition, which they could not lose."

Disappointed to be separated from the girls, gladdened by the news of Soviet success in arms, Sakharov and his fellow students were soon settled in the center of Ashkhabad and taking their classes in a suburb known as Keshi. Punning on Vichy, they dubbed themselves the "Keshi government." The students were devoted to two pursuits: physics and food. In principle, as the Soviets liked to say, there should have been plenty of food in Ashkhabad long known for cattle and sheep, grapes and melons, dates and yams. But the savagery of the Civil War and collectivization had caused many Turkmen either to slaughter their herds before they were requisitioned or to drive them into neighboring Iran or Afghanistan. Cultivated lands were abandoned, irrigation canals fell into disrepair. And so, by the time the contingent from

Moscow University arrived, the Turkmen Republic was a lean and hungry place.

Since they were students, finding food was always part lark. Some went out into the Kara Kum Desert, whose black sand hills reached heights of sixty feet, and caught turtles for soup—the "soup tropics." Others ate mulberry leaves straight from the trees, horrifying the locals. The biology students caught and cooked stray dogs, occasionally inviting the physicists for dinner. A scandal erupted when the city fathers found out, and intramural dog dinners were at once a thing of the past. To supplement their ration of less than a pound of bread a day, the students forged ration coupons for noodle soup, scoured the local markets for scallions and potatoes, though cooking oil was never to be found. Sakharov achieved a measure of renown by discovering that castor oil, available in the drugstores, was, once you got used to the smell, not bad at all for frying potatoes.

But after a time heat and hunger were no lark. The scorching, dust-laden wind the locals called "the Afghani" and the desert expeditions for food brought on heat prostration, causing at least one death. Sakharov came down with severe dysentery, which in those conditions was life-threatening.

In his first three years at Moscow University, Sakharov had not made a single friend. In Ashkhabad he made two: Piotr Kunin, with whom he founded a little group to study the general theory of relativity, and Yasha Tseitlin, a Ukrainian Jew whose imagination and "melancholy empathy" reminded Sakharov of Grisha from the courtyard. But the keenest moments still occurred in solitude: "As spring approached, I abandoned my stuffy bedroom for the flat, clay-covered roof of our dormitory. At night I gazed up at the star-filled southern sky, and at dawn I watched the sun's first rays light up the Koppet Dagh Mountains. The red-tinted peaks seemed to be transparent."

Contact with some of the locals was considerably more prosaic. It was in Ashkhabad that Sakharov first experienced two phenomena that he found puzzling and distasteful: worker hostility toward the intelligentsia for having it easy, and anti-Semitism. Both were directed against him because he was clearly a student and because he pronounced his *r*s like some Russian Jews, at the back of the mouth with a clearing-of-the-throat sound as opposed to the usual hard trill of tongue on palate. But otherwise he was focused on physics, the original five-year program having already been shortened to four in the double-time spirit of the era and now even further truncated by war and dislocation. Teachers were scarce. The Terror of the 1930s had maimed every class and profession. Though its importance was yet to be recognized, physics had not fared as badly as some other fields, especially biology, where Lysenko's crackpot theory—genes didn't matter, acquired characteristics could be transmitted directly—had won Stalin's approval. To challenge Lysenko was to challenge Stalin. Many biologists had been purged and arrested, even some of Lysenko's followers, a signal from Stalin that his general approval in no way limited his specific will.

The situation may not have been so drastic in physics, but physicists were arrested, exiled, executed (at least 20 percent of Soviet astronomers simply disappeared). Some of the leading professors at Moscow University were forced to resign, including the illustrious Igor Tamm, who would play a major role in Andrei Sakharov's life. Among their sins was the espousal of the theory of relativity, which was officially deemed "antimaterialist," deistic elements having supposedly slipped into Einstein's cosmology; his articles "Science and Religion" and "What I Believe" were sharply attacked in the USSR. In the twenties "bourgeois" scientists had been actively recruited, but by the thirties Soviet science underwent its own Stalinization. Genetics and relativity were bourgeois sciences and, as the Marxist thinker Plekhanov had put it, "Bourgeois scientists make sure that their theories are not dangerous to God or to capital."

Of the few teachers Sakharov encountered in Ashkhabad, the most important for him was Anatoly Vlasov, a former pupil of Tamm's. Vlasov taught Sakharov basic quantum mechanics, which governed interactions on the particle level and which supplemented but did not quite complement relativity. As a teacher, Vlasov "varied from brilliant to incomprehensible. He had a very odd manner of lecturing—covering his face with his hands and droning on in a monotone without looking at anyone." This later proved a symptom of mental illness rather than eccentricity.

But Vlasov was sane enough to spot Sakharov's potential. And for Sakharov that put him in the best of company: "Aside from my father, Vlasov was the first person to see the makings of a theoretical physicist in me." In July 1942, because of the tremendous heat in Ashkhabad, Vlasov conducted the final examinations in theoretical physics outdoors in a public garden. Officially, Sakharov and the others in his class were to graduate with a degree in "defense metallurgy," though they knew little about the subject in general and even less about its application to defense. After a few perfunctory questions, Vlasov gave Sakharov his A, then asked if he would like to continue on as a graduate student in theoretical physics.

"I had been expecting this offer, and turned it down politely. I felt it would be wrong to continue studying when I could be making a contribution to the war effort, although I had no clear idea what that might be."

Sakharov used the excuse that he had already been assigned to a munitions factory in the city of Kovrov, though both teacher and student were well aware that changing that assignment would simply be a matter of paperwork. His first choice between duty to science and duty to country would both have specific consequences and initiate a lasting pattern.

Once again Sakharov traveled across war-ravaged Russia on the journey from Ashkhabad to Kovrov, some 150 miles northeast of Moscow, but this time he was alone, sleeping on his suitcase in the teeming stations and packed trains where people "talked endlessly, as if compelled to share the horrors that were haunting them." His lucky shrapnel-nicked boots were stolen in one of the many delousing stations he was subjected to en route, an institution described by Alexander Wat, a Polish poet in a Russian prison: "they took your clothes, which took a beating from the heat and came out wet. Then everything—all the underwear, clothing, coats—was thrown out, everything jumbled together. You had to be quick."

After ten days of waiting and appointments in Kovrov, he was informed there was no work for him there and he should report to the Ministry of Armaments in Moscow. The pretext he had used for refusing graduate work had proved entirely empty.

After open, sunlit Ashkhabad, Moscow was blacked out, grim. His parents, whom he hadn't seen in ten months, looked careworn, exhausted, aged. His brother Yura had just graduated high school and was expecting to be drafted. The Red Army needed men by the millions. Though it had stopped the Germans at the gates of Moscow, the Red Army had failed to relieve the siege of Leningrad in the north, and now the Germans had broken through in the south of Russia, Hitler bifurcating his forces, sending his strongest troops to seize the oil fields of the Caucasus, the remainder to the Volga city of Stalingrad, which was of both strategic and symbolic value.

Sakharov was assigned to work in a munitions factory in the Volga town of Ulyanovsk, some 450 miles southeast of Moscow but well north of the fighting around Stalingrad. Known before the revolution as Simbirsk, Ulyanovsk was originally a fortress city built on the five-hundred-foot-high western bank of the river in 1648 to defend the Russian frontier against raids by Tartars and Mongols. With the Russian conquest of Siberia, the city lost its military importance and became a shipping center; chanting, ropes around their chests, Volga boatman hauled barges of grain and fish along the river. The local peasants became adept at an odd assortment of industries, including the making of tombstones, saddles, and guitars. By the nineteenth century Simbirsk had become a sleepy provincial river town with a reputation for excess and eccentricity. The journalist Louis Fischer unearthed some illustrative local lore: "There is a story of two Simbirsk landlords who used to meet regularly for hunting, card games, and vodka-drinking bouts. Each owned a large-caliber cannon. If one wished to invite the other, he fired a cannon ball onto his estate. If the second accepted he fired back a shot. If he fired two shots he was inviting the first to come to his manor house. If each insisted on the other's visiting him, they continued the cannonade until their ammunition gave out, when they met halfway to decide what to do."

The main street of the city was named for Goncharov, native son and creator of the fictional but very Simbirskian character of Oblomov, whose name became synonymous with epic sloth. Good-natured, philosophical, requiring hours just to rise from his bed if he rose from it at all, Oblomov had an innate distaste for action, which he viewed as vain commotion. In the lexicon of the intelligentsia "Oblomovitis" diagnosed the illness that kept the best of the nation from rising against tsarist injustice. And yet, as if in dialectical defiance of local tradition, Simbirsk was also the birthplace of that furiously purposeful, unceasingly active, and absolutely un-Oblomovian Vladimir Lenin, whose real last name, Ulyanov, was bestowed upon Simbirsk after the leader's death in 1924.

By the time Sakharov arrived there on September 2, 1942, Ulyanovsk was split between a provincial town on the west bank and the vast munitions plant on the far shore. Past the plant were the workers' barracks and, out past them, huts, the peasantry blurring with the proletariat. On his own again, a young traveler with his bag slung over his shoulder, arriving in a new city at daybreak, Sakharov had just missed the morning labor train that shuttled workers over the bridge. So he decided to walk the tracks to the ferry. On the way he stopped into a local library and, in the mood for a novel, took out Steinbeck's *Grapes of Wrath,* which he read, enjoyed, and lost, involving him in morally painful negotiations with the librarian.

His first assignment, to the chief mechanic's department, struck Sakharov as "thoroughly absurd," since he knew nothing about the manufacturing process. The chief mechanic was of a similar opinion and, "without so much as a glance" in Sakharov's direction, sent him off to the woods to fell trees. Sakharov accepted the assignment with cheerful fatalism though the work was deeply exhausting and hardly the best use of his talents. At the end of the day the workers sat around a fire and roasted potatoes gleaned from nearby fields. By that fire Sakharov first heard Stalin held to some account. A worker grieving for a son just killed at the front said bitterly of Stalin: "If he was Russian, he'd take more pity on the people."

Even though he took pride in withstanding the rigors of the work, Sakharov, who had been unable to construct a stool in carpentry class, inevitably injured himself felling timber. When the wound became infected, he had to return to Ulyanovsk, walking more than ten miles back to the tracks where he hopped a freight train the rest of the way.

This time the personnel department assigned him to the drafting department as a junior engineer, not quite as absurd as his first posting since, with some effort, Sakharov could summon the little he knew about the subject. His new work had the advantage of taking him to nearly every department in the munitions factory, and so he learned much about both production and the working class. In vast, dim rooms, women, mostly peasants from the nearby villages, operated large, deafening punch presses, sitting cross-legged on their stools to keep their feet, shod in crude wooden clogs, off cold floors slick with water and lubricants. The shifts were eleven hours. "Malingering"—failure to report for work, or even lateness—could cost a worker a five-year stretch in the Gulag that would make the factory seem cushy. For the women workers, pregnancy was their only ticket back home, though even that required standing in line all night in the hopes of seeing the deputy personnel director, who was only available for twenty minutes before disappearing to "more important Party business."

Lunch at the munitions factory was a few spoonfuls of porridge mixed with American powdered eggs "served on sheets of paper and eaten on the spot, washed down with ersatz tea from a tin cup." Sakharov's factory later began manufacturing its own spoons so the gruel would not have to be lapped straight from the paper. That plus the bread ration—sullen lines unforgiving of a moment's absence—and the little milk, carrots, and cucumbers Sakharov could buy from the peasants was what kept him alive.

From September 1942 until July 1943, the time of Stalingrad, Sakharov lived in a workers' barracks where in each room up to twelve people slept on three-tiered plank bunks. The toilet was in the courtyard, too far for the exhausted men: "there were always frozen puddles of urine outside the door." At least that same exhaustion ensured peace in the barracks except for the occasional incident, as when one worker, a "giant of a man . . . with fair hair and the light blue eyes of a child" drank methyl alcohol, went berserk, and was taken away by ambulance never to be seen again.

In early November Andrei Sakharov did something quite un-Soviet: he took the initiative. Ordered to process a batch of rusty metal that any fool could see was unsuitable for shell casings and which was only damaging the dies, Sakharov ordered the process halted then and there. At a meeting the next morning he was raked over the coals by the foreman: "Comrade Stalin has issued an order—not one step back! Soviet soldiers are fulfilling that order and fighting the enemy at the cost of their lives, but Engineer Sakharov abandoned his battle station without having completed a vital task. At the front, deserters are shot on the spot. We cannot tolerate such behavior in our plant!"

The tirade was met with silence. Even Sakharov voiced no indignation that an action dictated only by intelligent patriotism would be called treasonous. In any case there were no repercussions, someone with greater authority than the foreman having decided, after one look at the metal, that the Soviet Union had not been in the least betrayed by Engineer Sakharov. Though the incident left him with a bitter taste, it did not prevent him from taking the initiative again. Sakharov now presented himself to the head of the central laboratory, who happened to need someone to invent an instrument to check the cores of armor-piercing bullets. Finally, some work that suited and challenged him. On his first day at the laboratory on November 10, 1942, Sakharov was introduced to his new colleagues. One of them, a good-looking young woman with sandy blonde hair and green eyes, took particular notice of him. He was not bad-looking, but she was already being courted by some of the young bravos of the town, and he would have fared poorly in any comparison with them. What drew her eye was his expression of civilized alertness peculiar to the intelligentsia, which she had come to know during the four years she had spent in Leningrad studying chemistry and glass production. But she had not finished her fifth year and graduated because her father had ordered her home at the outbreak of war. Being a dutiful daughter had proved fortunate, for many of her friends were losing their lives in Leningrad as the death toll rose to a peak of ten thousand a day.

Her name was Klavdiya Vikhereva, Klava for short. Born in 1919, she was a year and a half older than Andrei, a fact that she did not discover until much later, having assumed they were the same age because they had entered university at the same time. Klava was a member of what the Russians called the "first-generation intelligentsia," a term that was scornful on some lips but proud on others—under the new system the children of peasants and workers could attain the highest levels of education. Klava's parents were of the peasantry, and she had grown up helping out with the planting and the harvesting. Her father, who was mechanically inclined, now worked in industry as did she. She belonged to the educated class by inclination—devouring the Russian classics in her youth—and by her scientific training. Strong and healthy, in her teens she regularly swam the considerable width of the Volga and once disguised herself as her sister when her sister's boyfriend came, out of both mischief and rivalry.

Andrei disappeared after that first view, engrossed by his assignment: to devise a "nondestructive" means of testing whether the "cores of armor-piercing bullets had been sufficiently hardened." Production problems left many of the shells too soft to pierce armor, of which the Germans had a formidable amount. The old testing method was cumbersome, labor-intensive, and inefficient. Samples were taken at random, placed halfway through a hole in a steel plate, and broken open with a piece of steel pipe. Within a month Sakharov replaced this crude method with one closer to the physics he had studied. He would magnetize the core of a shell then demagnetize it; a soft core would require less coercive force to demagnetize. His first real effort at independent applied science and at contributing to the war effort resulted in success, with all the attendant perks and pleasures. His device was immediately approved for use, winning him an award of three thousand rubles (compared to his monthly salary of eight hundred) and some status and greater freedom at work.

And so it was a more confident Andrei who, along with the other young men, began visiting the chemical lab, which was staffed largely by young women, most of them locals, Klava among them. In a blend of hospitality, compassion, and courtship, the young women treated the young men to baked potatoes from their own vegetable gardens, potatoes often being the main part of a meal, if not a meal in itself. Fairly quickly, Klava and Andrei began separating off from the group, going to the theater and the movies where, along with the usual patriotic war films, they saw foreign movies such as *That Hamilton Woman*. Klava's several suitors were taken aback that she had forsaken them for "that scarecrow" even though their relationship was still only a friendship, though one intensified by the dramatic uncertainties of wartime.

In early 1943 Sakharov's laboratory acquired a "costly optical device, a spectroscope for the semiquantitative analysis of steel and other metal alloys." The problem was that no one knew how it worked. Since Sakharov was the only person in the entire laboratory with any knowledge of optics—and even that exclusively theoretical—he was assigned the task of figuring out how this "miracle of technology" actually operated to distinguish among various grades of steel that were "constantly being mixed up in the storeroom," a euphemism for chaos. He solved the problem quite quickly, and to verify his results he sent some samples to the chemical lab. "Klava was assigned to analyze a few of them. Either from carelessness or because the exhaust hood was defective, she came down with hydrogen sulfide poisoning. That incident was one of the things that brought us closer together."

Now, quickened by compassion, friendship shaded into courtship, though its accoutrements—shoes and potatoes—were decidedly unromantic. Rowing with Klava on the Volga, Sakharov managed to drop one of her shoes in the water, his clumsiness both endearing and irritating. Klava, more nimble and resourceful than Andrei, had found him a pair of sturdy shoes to replace those stolen from him at the delousing station—he had gone through the first winter in Ulyanovsk in a pair of thin summer shoes, all he had left. And potatoes, which had brought them together in the first place, now marked a new progress. In May of 1943 Sakharov went with Klava to see her parents at their home, a one-room wooden hut. He offered to help in the spring planting of potatoes, having already seeded some of his own by the munitions factory wall.

Right from the start Andrei got along with Klava's father, Alexei, who called him Andrusha without thinking twice about it. Cheerful, sociable, Alexei Vikhirev was a man of the people and a jack-of-all-trades; adept at farming, shoemaking, and operating a lathe, he also liked to compose songs, crying when singing the lament he had written for his native parts flooded by a gigantic dam and reservoir project. And he would weep bitterly when recalling how a "handsome colt" of his had been requisitioned by the local collective farm and died soon after. "The loudmouths are bragging again" was his usual response to radio reports of the latest achievements. Though fiercely patriotic, Alexei put his trust in the folklore of rumor—Stalin was responsible for his wife's death, lavish banquets were held in the Kremlin, millions of rubles were spent on Trotsky's assassination—which was in fact closer to reality than anything in the paper. Fond of vodka, he kept icons in the house and read the Gospels.

Reserved, aloof, brooding, Klava's mother, Matryona, had never recovered from the loss of her true love, a young man who had been killed in World War I. But he still had nearly all Klava's mother's love, and what little was left went to Klava's younger sister Zina.

In the spring of 1943 Andrei Sakharov suffered a recurrence of the severe dysentery he contracted in Ashkhabad. Klava nursed him back to health, the family sacrificing precious rice, goat's milk, eggs, potatoes, and chicken to save his life; to sustain themselves the family had to "dig up frozen, half-rotten tubers to make pancakes by using the rather complicated 'technology' developed by generations of starving peasants." Once again illness and compassion were agents of intimacy. Physical proximity, the atmosphere of wartime, the first days of spring, and suddenly they crossed the line to love.

Aware that in matters of feeling as well as physics his mind had a tendency to skip important steps, Andrei proposed to Klava in the manner of a bygone era. "He didn't offer me his heart and his hand by speaking, but by writing. It wasn't from shyness or bashfulness but so that I would understand everything right. I may have been the only woman in wartime Russia who was proposed to just as in the old-fashioned novels!"

On July 10, 1943, Klava's father, Alexei, blessed the young couple with his most prized icon, a blue, orange, and gold depiction of the Holy Mother of the Burning Bush, who protected wooden huts like his from fire. Alexei made the sign of the cross over them and offered a few words of advice. Then, after home brew and a humble repast, Andrei and Klava had to leave for town and the nearest office where quick drab ceremonies were performed and marriages registered. Holding hands, they ran off through the fields.

CHAPTER FIVE

"A CHANGE IN EVERYTHING"

"GREAT events and changes are taking place in my life," wrote Andrei Sakharov in a letter to his parents on June 25, 1943, "and the time has come to inform you of them. I am marrying for love." But before speaking of Klava, Sakharov voices a complicated wish: he wants his parents to "rejoice with him" and "feel no bitterness or much misgivings, i.e. that you have sufficient trust in my judgment."

Only then does he introduce Klava. "She is Russian," he tells them first, a natural point of interest in a multiethnic society where, especially among the intelligentsia, intermarriage was common. Sakharov proceeds to what he finds most "touchingly attractive" about Klava: her blend of the "impetuously childish," the womanly, and the adult, which he defines as her "serious attitude toward everything concerning the life of the mind."

It took almost a month for the letter to travel the less than five hundred miles from Ulyanovsk to Moscow, by which time Sakharov had already been married for nearly two weeks. Returning home to find the letter propped on the doorknob of their apartment, his mother responded at once without waiting for Sakharov's father, who was away visiting their other son, Yura. Her letter was warm, effusive and gave no evidence of misgiving, though the salutation—"Our dear little and beloved boy"—was of the sort to cause a grin or wince.

> With all my heart I embrace and congratulate you and Klava and invite you both to come visit us, my only fear is that this cannot happen any time soon. I am so glad that you aren't alone any more, you so loved family and suffered so much from loneliness. As soon as possible I want to get to know and love the

woman my darling boy loves. . . . I kiss and bless you and Klava.

Mama

He sent his parents a photograph of him and Klava taken that July. They could not have found their son's appearance particularly reassuring. Andrei's close-cropped hair, already receding to a widow's peak, and too large shirt make his face and neck look gaunt and frail, the result of his recent recurrence of dysentery. The shirt may not even have been his—much of his clothing had been infested with lice in the dormitory and burned. Taller than Klava by several inches, Andrei inclines his head toward hers, his eyes melancholy. Klava looks healthier, more serene; her blouse, open at the neck, reveals the corded muscles of the girl who regularly swam across the Volga, now married to a man who could barely dog-paddle. Her expression is serious, open yet uncertain, as if she is aware of how carefully that photograph will be scrutinized in Moscow.

Sakharov had now moved from the squalid solitude of the workers' barracks to his in-laws' home on the outskirts of town. Their house, one large room—about 180 square feet—was built around a Russian stove used for both cooking and heating. Water from the well. An outhouse in the yard. Klava and Andrei were given a corner of their own, screened off by a blanket strung from a rope, the custom in such close quarters. Hardly ideal honeymoon conditions. But Soviet couples were adept at finding times and places where they could be alone. They were young; it was July.

"Great events and changes" were also taking place in the country at large. Even Stalin had modified his ways. Placing self-interest over self-importance, Stalin realized that he could not win the war against Germany unless he relinquished command to his top military men, especially Zhukov, his newly appointed deputy. (That way, they could be blamed for defeats while he could always take credit for the victories). Political restraints were also lifted from the officers in the field, whose orders no longer had to be countersigned by the political commissars attached to their units. Improvements were made in technology and strategy. The cabins of the T-34 tanks were enlarged and equipped with radio communications, which they had lacked before, fighting alone and blind. Taking a lesson from the Germans' panzer division, the Red Army adopted the concept of the tank corps, rapid, motorized, mobile, a concentrated force as opposed to the former use of tanks to support infantry strung out over a broad front. Lend-Lease from the United States and Britain, which would total over $11 billion, supplied field telephones, jeeps, trucks, locomotives, planes, medicine, Spam, and the powdered eggs that Sakharov spooned off sheets of paper in the munitions factory. That took some of the pressure off Soviet industry and allowed it to focus on armaments. In 1943 the Germans produced 17,000 tanks; the Soviets produced 24,000.

The change in Soviet strategy was nowhere more apparent than in the epic tank battle at Kursk, some three hundred miles south of Moscow. By the spring of 1943 the front between the German and Soviet armies had stabilized all the way from Leningrad to the Black Sea. In January of 1943 the blockade of Leningrad had been partially breached, but that city would not be delivered from starvation and bombardment for another year. The victory over German general Paulus's army at Stalingrad in February 1943 was of both symbolic and strategic significance but hardly decisive. The Germans could still win the war, and they had a plan, code-named Citadel, to do precisely that. That plan, developed by Field Marshal von Manstein (Hitler too having relinquished total control of the war effort), targeted an area 120 miles wide and 60 miles long where the main force of the Red Army—40 percent of its manpower, 75 percent of its armored forces—bulged into German-held territory (the Kursk salient). As Richard Overy writes in *Russia's War:* "Manstein planned to envelop the bulge with two heavily armored pincers that would cut the neck of the salient from north and south. The aim was to destroy a large part of the Red Army at a critical juncture of the front, allowing German forces to recapture the southern area, or to swing northeast behind Moscow."

Edgy, Stalin called for an immediate "preemptive offensive." Zhukov rejected his plan. Stalin did not object. The "deep battle" plan ultimately adopted was more in keeping with traditional Russian military strategy, and a portion of it was even code-named

Operation Kutuzov, after the general celebrated by Tolstoy in *War and Peace* for his "passive" strategy of allowing Napoleon to overextend his forces. The Soviet plan was to draw a German attack, meet and parry, then suddenly counterattack with an immense and well-concealed reserve force.

Goaded by feigned Soviet offensives, the Germans attacked on July 5, 1943, in what has been called the "largest set-piece battle in history." The Germans had 900,000 soldiers including the formidable SS panzer divisions "Death's Head," "Reich," and "Adolph Hitler Guards," 2,700 tanks, 2,000 aircraft, 10,000 artillery pieces. The Red Army had the advantage in men and matériel—1,337,000 men, 3,444 tanks, 2.900 aircraft, and 19,000 big guns. The Germans still controlled the sky, but that would matter less as the tanks began to engage at close range.

For once the Russians were formidably well organized, and Stalingrad had given them a taste for victory. Zhukov, who was at the front, called the roar of artillery, bombers, and tank cannon a "symphony from hell." In the largest tank battle in history, 850 Soviet T-34s clashed with 600 German tanks in an area some three miles square. "The air itself seemed scorched," says Overy. After the Germans had been stopped, Zhukov toured the battlefield with Nikita Khrushchev, the party representative at the front, and was awed by the hideous vistas. Khrushchev took a more exuberant and vindictive view: "Our detractors used to say that the only reason we were able to defeat Paulus's colossal army at Stalingrad was that we had the Russian winter on our side. They had said the same thing about our defeat of the Germans outside Moscow in 1941. Ever since Russia turned back Napoleon's invasion, people claimed that winter was our main ally. However, the Germans couldn't use this excuse to explain their defeat at the Battle of the Kursk Salient in 1943. They fired the first shot; they chose the time, place, and form of the battle. All the cards were in the hands of Hitler and his cutthroats. It was high summer."

On August 3 the Red Army launched its surprise counterattack; ferociously successful, it sent the Germans reeling back. Some of the shells piercing German armor at the Battle of the Kursk Salient may well have been tested by the magnetic methods Sakharov had devised. Having already been "recognized as an 'expert' on magnetic quality-control methods," in mid-1943 Sakharov switched from armor-piercing shells to the bullets used in the TT automatic pistol. Again the task was to find a quality-control method that was more reliable and less wasteful of precious resources than the one currently in practice, and again he succeeded; though his device was "not exactly a triumph of design," it was, after proving its merits, accepted for use in the war effort.

Some of the larger issues raised by this very practical piece of work inspired his initial independent forays into theoretical physics, "gaining confidence in my ability as a theoretical physicist, something vital for any beginner." He began writing papers summarizing his approach, insights, and results, none of which were published. Sakharov sent two of those papers to the renowned physicist and professor Igor Tamm, who did not, however, respond. While Tamm "sensed the high promise" of Sakharov's early work, he "couldn't quite figure out" what the young physicist was trying to say.

* * * *

> This spring there is a change in everything.
> More lively is the sparrows' riot. . . .

wrote Boris Pasternak in his poem "Spring 1944" about a Russia coming back to life after three years of German invasion. Leningrad had finally been liberated from the nine-hundred-day siege on January 27, 1944. By spring the Germans had been driven from nearly all Soviet territory, and on June 6, 1944, the Normandy invasion opened the long-awaited second front.

That spring Klava became pregnant and Sakharov helped his father-in-law, Alexei, plow and plant millet in virgin land a dozen miles from their house. Alexei harvested the millet that fall but had no horse to transport the grain back home where some of it would be eaten, some sold at the market for cash. But a solution was found, says Sakharov: "the two of us hitched ourselves up to a cart and pulled it after us through the night, until we finally reached the house just before daybreak."

In 1944, emboldened by his recent successes, Sakharov began an "intensive study of theoretical physics," which led to a grotesquely comic incident typical of the times. Choosing the "warm and well lit" Party Educational Center for his research, Sakharov soon came to the attention of the woman who ran the center, one of those Stalinist busybodies who played a low-level but essential role in surveillance everywhere from streetcars to hotel corridors. Seeing that the young man "was not reading Lenin or Stalin or even Marx or Engels, but some completely mysterious books," she reported Sakharov to the head of the laboratory who, embarrassed by the absurdity of the situation but obliged to do something, reprimanded him "so politely it hardly sounded like a reproach."

But munitions was still his main work in 1944, theory a luxury. He was proud of his contributions to solving problems in testing the 14.5 mm shells used in antitank guns. Defective shells exploded in the bores of the guns, ruining valuable weapons and endangering the lives of the men operating them. Previous testing measures, "hellish and unreliable," would result in an entire batch of armor-piercing shells being scrapped—"all fifty thousand of them!"

Sometime in the second half of 1944, Sakharov's father, Dmitri, taking advantage of his prestige as a popularizer and his old acquaintance with Igor Tamm, asked the physicist to consider Andrei for the graduate program at FIAN (Physics Institute of the Academy of Sciences), supposedly saying, with intelligentsia understatement, that his son might make a "passable physicist." Tamm may also have remembered the two eccentrically expressed papers he had received from the young scientist working at the munitions plant in Ulyanovsk, or his respect for Dmitri Sakharov's opinion may have been sufficient. In any case, in December 1944 Andrei Sakharov received an invitation from FIAN to take an examination for admittance to the graduate program.

In Ashkhabad, in the spring of 1942, Sakharov had turned down the chance to do graduate work, considering it "wrong." This time the conflict of loyalties took a different form. "I had been ready to shift to pure science for some time. I was sorry to abandon my work as an inventor just when I was starting to have some success, but my craving for science outweighed all other considerations." On the other hand, Klava was due to give birth in little more than a month. Accepting the invitation would mean separation at a critical moment, especially since Klava had already lost a previous pregnancy, miscarrying while lugging a heavy sack of potatoes. And their separation would necessarily be of unknown length, given the housing crisis caused by the war. But both Klava and her father insisted that Andrei accept the invitation. There was no future for a man of his abilities in Ulyanovsk. Klava's years in Leningrad had given her a taste for urban civilization, and Moscow, as the capital, was always the best provisioned of Soviet cities. It was a sacrifice, but in a time when so many had sacrificed their lives it could not have seemed terribly great, painful as it was. And hardship in the name of future opportunity is always easier to bear. On a mid-January night of heavy snow, Andrei said good-bye to Klava and her father at the train station in Ulyanovsk where he had arrived sixteen months earlier, a lone young man, his bag slung over his shoulder.

Sakharov's mother and father met him at the station in Moscow, where they had plenty of time to talk since they couldn't leave the station during the curfew. Russia may have been winning the war by then, but it was still wartime. His parents looked drained, and that touched and concerned him. Their biggest problem was housing. After the house on Granatny Lane was bombed, they had lived in a few places, including a kindergarten, and their current situation was tenuous because the original occupant had returned and was going to court to get his apartment back.

The very next day Sakharov went to see Igor Tamm. Squinting in a blue haze of cigarette smoke, Tamm—"with the same surprising liveliness and ease that marked all his actions"—was busily moving little flags on a map to chart the progress of the great January Offensive. In his tumultuous youth, Tamm had made his own credo—"truthfulness, cheerfulness, health and work"—and was still living by it at fifty. But, as Sakharov could not help noticing, the cigarette Tamm preferred—the Russian *papirosa,* an inch of rich and murderous black tobacco inhaled through a cardboard tube, a mouthpiece not a filter—sent him into fits of coughing. But Tamm could not think or work without a *papirosa* clenched between his teeth. Work over health. The credo had shifted slightly over time.

Tamm asked after Sakharov's father, then got right down to business. Questioning his prospective student "calmly and tactfully," Tamm "quickly penetrated to the bottom" of Sakharov's knowledge of science, which Sakharov, always realistic about his own worth, considered "rather modest, though solid, and, I think, not superficial."

Toward the end of their three-and-a-half-hour conversation, Sakharov found tacit approbation in Tamm's increasingly exacting questions. Tamm told Sakharov then and there that he was accepting him as a graduate student, even though Sakharov had appalled him by confessing to a lack of any English.

"You've got to learn English right away. First, get to the level where you can read the *Physical Review* with a dictionary, you'll have to do this very quickly, on your own, regardless of any graduate requirements." Tamm also gave him two books to read, the main focus of his study: *Theory of Relativity* and *Quantum Mechanics,* both by Wolfgang Pauli, both in German, which Sakharov could read. One encounter had been enough to transform the tactfully probing examiner into an injunction-issuing mentor.

After Sakharov left, happy to be officially a FIAN graduate student, Tamm returned to his desk, covered with intricate piles of papers—notes, calculations, diagrams, equations—which he allowed his wife to dust only once a year. (As Jeremy Bernstein remarks in *Cranks, Quarks, and the Cosmos:* "Theoretical physicists are notorious scribblers; no piece of paper, from the back of an envelope to a place-mat, is safe.") Though Tamm enjoyed regaling his family and frequent guests with stories at dinner table, the silver-haired, craggy man (his grandson thought he resembled Spencer Tracy) was not the sort to write about himself or what he thought of his newest student, who must have seemed an unformed youth, not only because Tamm was old enough to be his father but because of all Tamm had been through as a person and scientist.

Tamm's grandfather, an engineer, had emigrated to Russia from Germany in the 1860s (the original name was Tham). His father, also an engineer, was a man of great principle and courage who had once dispersed a pogrom with his walking stick and during World War II hid people from the Gestapo. On his mother's side, Tamm inherited not only Russian blood but that of a Cossack chieftain and a khan of the Crimean Tartars. Tamm himself was such a hot-headed radical Menshevik in his youth that in 1913 his parents sent their eighteen-year-old son to Edinburgh, Scotland, to study and cool off. Returning to Russia in 1914, Tamm won fame as an antiwar orator, speaking in machine-gun-like bursts. He did, however, serve for a time as a battlefield medic, learning that even under bombardment, "it's entirely possible to keep a grip on yourself." He was very politically active in the tense and heady days between the abdication of the tsar in February 1917 and the Bolshevik seizure of power in October. He left the Mensheviks when he saw they were no match for Lenin's Bolsheviks (Lenin once shouted "Bravo, Tamm!" when he voted with the Bolsheviks on an issue). But Tamm had few illusions about the Bolsheviks, thinking their leaders "fanatics" blinded by their own great but limiting truth.

During the Civil War, Tamm was arrested by both the Whites and the Reds and came close to being executed by both, surviving through pluck and pure chance. The Red Guard who arrested Tamm for having no papers was a former student himself. He gave Tamm one night to solve a difficult mathematical problem to prove that he was what he said he was, a physicist. "Solve it and you'll be released, otherwise you go up against the wall." The next day Tamm still hadn't solved the problem, but the Red Guard judged that the calculations he had written out proved he was telling the truth. But Tamm was not released. Instead, he was transported to the headquarters of the local secret police. En route he remembered that the editor of the local newspaper was an acquaintance, tricked his guard into taking a route that went by the newspaper's office, and managed to alert the editor. That in turn resulted in a series of phone calls reaching all the way up to Felix Dzierzhinsky, the head of the secret police. Part of a group of prisoners who were being executed in small batches each day, Tamm was freed at the last moment.

His younger brother Leonid was not so lucky during the Great Terror: arrested in 1936 for sabotage, he was sentenced "without the right of correspondence," usually a euphemism for execution. In the twenties, the Soviet government's task had been to win the "bourgeois" scientists over to the new system, but the thirties were marked by the Stalinization of science, which was re-created in the image of the party—pyramidal, hierarchical, centralized—and the party intervened in all scientific matters from the selection of personnel to the endorsement of one theory over another. And purges were carried out as assiduously in science as they were in the party.

"I can vouch for my brother," said Tamm, a very brave act at a time when brother was expected to renounce brother. Tamm, who consistently refused to join in any of the feeding frenzies of denunciation, did not himself suffer to any great extent. His Department of Theoretical Physics was closed, but his seminars continued to meet. Though Tamm had lost his taste for politics during the revolution, he never lost his faith in socialism. "I, as an extreme optimist, always think that the new, which will utterly engross me, is just around the corner." For all Tamm's ebullience and vigor, he was also a cool and reserved man, his daughter never having once seen him cry.

* * * *

Andrei Sakharov returned to pure science with ardor, but when he was not immersed in study, his life could not have been more ordinary. If he was not chasing documents to prove that he had a right to that most prized of possessions, a Moscow residence permit, he was planting potatoes in the family's garden an hour outside the city. He had to scout about for people traveling to Ulyanovsk who could be trusted to deliver vitamins and money to Klava, the mails still too unreliable. For that reason as well, Sakharov began numbering all his letters and postcards to Klava with pedantic regularity and convinced her to do the same.

Sakharov's letters are a mix of the tender, the fearful, the practical. Separated from his wife by barriers of geography and bureaucracy, he felt even more apprehensive and helpless than most expectant fathers. On January 24, 1945, he writes: "Yesterday I was seized with terrible anxiety and all sorts of gloomy thoughts about you. Send me a telegram quick to calm me down. Are you drinking tea with milk?" A constant motif of his letters is his wish that they share their thoughts and feelings with each other "as if they were side by side . . . the only way to diminish the pain of being apart." Writing on February 3, he thinks "today or tomorrow will be the day." He was only off by a few. His daughter, Tanya, was born on February 7, 1945, in a cold maternity hospital in Ulyanovsk that had "only paper to burn in the stoves." It was a difficult birth, and Tanya soon suffered a series of ailments, which only increased Sakharov's anxiety and guilt: "how little I'm sacrificing . . . painful to think I was living better than you."

But it was not all fear and affection. He had news, having run into an old acquaintance, the sort who knows everything: "who got married, whose baby can say 'Papa,' who got killed." And there was hope. On April 30, the end of the war with Germany only days away, Sakharov wrote: "Spring is here, the young mothers are out sitting on benches in the little courtyard. . . . You can feel the nearness of victory in the air, everyone is jubilant."

The January Offensive moved at a headlong clip, the Red Army at times advancing fifty miles a day, which would have kept Tamm busy shifting his little flags westward. The German army fought stubbornly in retreat, in part from military discipline, in part from the desire to surrender to any Allied force but the Red Army, whose casualties now rose sharply, no effort being spared to take Berlin first.

Once the Red Army was on German soil, posters went up announcing that the "hour of revenge has struck!" The soldiers needed no reminding. Sakharov's future friend, Natalya Gesse, who had hobbled through the Moscow Panic on crutches and been taken for a German spy parachuted into the city, had in the meantime talked her way into a job as a war correspondent and was with the invading army.

"The Russian soldiers were raping every German female from eight to eighty. It was an army of rapists. Not only because they were crazed with lust, this was also a form of vengeance. Those soldiers had now seen what the Germans had done to their land—the burned-out villages, the partisans hung in the square, people herded off to Germany as slave labor.

"The Russian soldiers robbed every German man and killed most of them too, and they raped all the females. They only knew two expressions in German—'Frau, komm' and 'Uhr' meaning wristwatch. In those days, it was still rare for a Russian to own a wristwatch.

"Stalin knew the value of the medieval custom of giving a captured city over to the soldiers for three days of rape and looting. It's very good for the soldiers' spirit.

"But after those few days it would all be brought to a stop with an iron hand. And the officers would begin having mad, passionate affairs with the German women. So much Slavic blood was mixed with German. . . .

"There were jokes about all that too. A solider is demobilized and sent home from Germany. His wife welcomes her conquering hero home. They have a great meal and they drink and then they go to bed. But he can't get it up. So, he says to her, "Get out of bed.' She gets out of bed. 'Get dressed,' he says, and she gets dressed. 'Now,' he says, 'put up a fight.'

"One time we went into a house in Germany and saw that it had beautiful furniture and a radio and lace napkins. To a Russian it looked incredibly fancy. Then we found out that the house belonged to a shepherd! Now Russians had seen the world a little and had a new standard to measure by. And just because they knew too much, many of those Russians soldiers were sent immediately to the Gulag."

On May 9 Moscow celebrated victory with salvos of fireworks over the Kremlin. Alexander Werth, a war correspondent for the *Sunday Times* and BBC, describes the day in *Russia At War:* "The spontaneous joy of the two or three million people who thronged the Red Square that evening—and the Moscow River embankments . . . was of a quality and depth I had never seen in Moscow before. They danced and sang in the streets; outside the U.S. Embassy the crowds shouted: 'Hurray for Roosevelt!' (even though he had died a month before); they were so happy they did not even have to get drunk, and under the tolerant gaze of the militia, young men even urinated against the walls of the Moskva Hotel, flooding the wide pavements. . . . For once, Moscow had thrown all reserve and restraint to the winds."

Though he rejoiced at the victory, Sakharov was now consumed by physics. Between February and April 1945 he "spent nearly every minute studying Pauli's two books. They were changing my world." Pauli was also a reminder of time lost. Born in Vienna in 1900, Pauli had been a wunderkind, as Sakharov exclaims: *The Theory of Relativity,* "truly the best book" on the subject, was "written by Pauli at the age of twenty-one!" By the time Sakharov finished reading it, he was about to turn twenty-four, the same age Pauli had been when he advanced his exclusion principle (an electron once in orbit excludes any other from occupying that same orbit), for which he received the Nobel prize in 1945. Pauli, known for his exuberantly mordant wit ("Not even wrong" and "So young and already so unknown" were among his favorite put-downs), would die at the age of fifty-eight. As C. P. Snow remarks in *The Physicists: A Generation That Changed the World:* "Extreme longevity was not an occupational feature of major scientists, as it has often been of visual artists and musicians." The probabilities were thus that a scientist would do his most brilliant work in his youth and not live an especially long life. In the real life of a scientist, time was still classical.

And Sakharov's time was consumed by obligations and obstacles. In the spring of 1945, in addition to chasing documents, planting potatoes, and earning a few extra rubles to send to Klava, he was given a new assignment—washing windows. The Academy of Sciences was celebrating its 220th jubilee that spring; important foreign guests would be attending. Curtains had already replaced the blue paper used to black out the windows and of which, after four years, everyone was heartily sick. Sakharov and another student, Lidia Pariskaya, were assigned to wash the tall, grime-encrusted windows in the corridor by the conference hall. Taking advantage of his height, Sakharov would clamber up onto the windowsill, suds the windows with a wash mop, then climb down and "like an artist, inspect his work from a distance, first from one side then another." When he was done, Sakharov said to Pariskaya with cheerful irony: "Well, now I've learned to wash windows—maybe it'll come in handy later on."

Pariskaya liked his smile, "gentle, childlike, trusting," which "practically never left his face," and enjoyed the unselfconscious eccentricity of his gifts—his leaps of mind, his ability to write as easily with his left hand as his right when chalking an equation on the blackboard. But what most impressed her were his spontaneous flashes of realism. She asked him: "So, is your brother talented?"

"Not as talented as I am."

She could not detect a scintilla of braggadocio in his reply. "He was simply stating a fact." Sakharov's passion for exact measurement carried over into his assessment of himself and others, but it failed him where it always fails, in the thick of it.

In mid-July 1945, after a flurry of telegrams to Klava and several snafus—"We waited and waited for you after your telegram but obviously something went wrong on your end again," wrote Sakharov. "Did someone get sick?"—he was finally reunited with his wife and had the father's pleasure of first beholding his child. But the problems began piling up at once: food, apartments, money. And there was another problem that no one had foreseen, an instant antipathy between Klava and Sakharov's mother. The causes were as clear to them as they were opaque to Sakharov, who was baffled and paralyzed by the velocity and intensity of their emotional infighting. "I always tended to avoid confrontations, feeling myself psychologically unable to cope with them, and, as if in self-protection, chose the line of least resistance." Did Klava bridle at some instance of her mother-in-law's forceful nature? Or did Sakharov's mother, who had greeted the news of her son's marriage so effusively ("As soon as possible I want to get to know and love the woman my darling boy loves"), discover that, for her at least, to know Klava was not to love her? Did Sakharov's mother snub Klava as a member of the first-generation intelligentsia, with the education but without the refinement and mores of that class? Was one look enough to see that Klava was a nice, simple provincial girl but unworthy to be a Sakharov, a family that Sakharov's mother considered to be "of a very high spiritual level" and of which she herself had felt unworthy in the past? Or was no woman ever going to be good enough for her "darling boy"?

When Tanya was grown, her mother often told her the story of her first encounter with her mother-in-law, who had said with brutal candor: "You're no match for my son. No matter what, he'll divorce you. You can leave the child with us or take it with you." Those words, says Tanya, left a "deep wound" on Klava's "vulnerable" soul.

Worse, the lack of money meant that for the time being they would all have to live together. To supplement his grad-student stipend, Sakharov lectured, taught at a night school, and wrote up extracts of scientific papers for Soviet journals. But there was never enough. He had to borrow from his father and Tamm. Once, unable even to buy milk, Klava tried to sell some of the candy she had received as part of her rations and was detained on the street by the police for speculation, a potentially serious matter. But, as Sakharov says, "There was nothing exceptional about our hardships, at the time. Nearly everyone's life was difficult in the first post-war years; ours, in fact, was rather easier than most people's. And, most important, everyone in our family had survived."

Victorious, Russia was devastated. Agriculture, still reeling from the violence of collectivization when the war started, was now lamentable. Industry had either been destroyed or converted to military production. Twenty-seven million Soviet citizens had lost their lives during the war. In 1944 the New Family Code began awarding money and medals to mothers of more than five live children, the highest Order, that of Hero Mother, going to those who gave birth to ten or more.

When they were finally able to afford their own housing, Andrei, Klava, and Tanya Sakharov were never in the same rooms for more than two months at a stretch. For a time, they lived in a basement so dank that Tanya developed a rheumatic heart condition and kidney problems. In a dwelling that a colleague recalls as "Dickensian," Sakharov had to run out and buy a chair when the doctor was coming because, apart from the beds, there was nowhere to sit. When they finally found a decent, drier place, it turned out that the landlady had all sorts of tricks to evict tenants and collect double rents. Sakharov had to wedge a sack of potatoes against the door to keep her out at night. The two rooms they later rented in a private home in Pushkino outside Moscow were so cold that Sakharov had to work at his desk with a fur coat over his shoulders. They were kicked out of a house they rented back in Moscow when Klava refused to cooperate with a KGB agent who wanted her to report on anyone her husband met, even though he was not involved in any secret work at the time. Informing was so much a fact of life that Sakharov took this as no more than an "ordinary link in the network of surveillance that enveloped the whole country." In fact, the KGB had already taken an interest in the small number of people working in the arcane but militarily essential field of theoretical physics.

"Each time we were turned out of a room," says Sakharov, "we'd have to go back to live with my parents, and relations between Klava and my mother, bad enough to begin with, would deteriorate further. Klava, my mother, and I were all to blame for aggravating the situation, only my father took a more reasonable attitude."

It was a time of lacerated hopes for the country. United by patriotism, sacrifice, and Stalin's nationalist line during the war, people naturally expected a better life after the cessation of hostilities. As one writer dreamed in 1944: "When the war is over, life in Russia will become very pleasant. A great literature will be produced as a result of our war experience. There will be much coming and going, with a lot of contact with the West. Everybody will be allowed to read anything he likes. There will be exchanges of students, and foreign travel will be made easy."

Wrong on every count. Churchill would soon make his famous "iron curtain" speech. Behind the curtain the poet Anna Akhmatova, the satirist Mikhail Zoshchenko, and the composer Dmitri Shostakovich would soon be savagely attacked, an official signal that any relaxation on expression was a thing of the past in the postwar world of enforced reconstruction and rapidly cooling relations with the former allies. But theoretical physicists were about to become a privileged class.

"On my way to the bakery on the morning of August 7, 1945, I stopped to glance at a newspaper and discovered President Truman's announcement that at eight A.M. the previous day, August 6, an atom bomb of enormous destructive power had been dropped on Hiroshima. I was so stunned that my legs practically gave way. There could be no doubt that my fate and the fate of many others, perhaps of the entire world, had changed overnight. Something new and awesome had entered our lives, a product of the greatest of the sciences, of the discipline I revered.

"*The British Ally* began serial publication of the Smyth report, an account of the development of the atom bomb that contained an abundance of declassified information on isotope separation, nuclear reactors, plutonium, and uranium-235, and a general description of the structure of the atom bomb. I would snatch up each new issue of the *Ally* and scrutinize it minutely with an interest that was purely scientific. . . . After the final installment of the Smyth report appeared, however, I gave little thought to the atom bomb for the next two and a half years."

But the Soviet atomic establishment was more aware of Sakharov than he was of it. Toward the end of 1946 he received a somewhat mysterious invitation to a meeting in the Peking Hotel on Mayakovsky Square in the heart of Moscow. Room 9 looked more like a government office than a hotel room—a portrait of Stalin on the wall, a T-shaped desk from which a man rose and introduced himself as General Zverev. Speaking obliquely, tactfully, without exerting any pressure, the general said: "We have been following your progress in science for quite a while. We'd like you to work with us on state projects of the greatest importance after you complete your graduate studies. You'll have the best of everything for your work—libraries with scientific literature from all over the world, big accelerators, the best pay and living conditions. We know you have a housing problem; if you agree to work for us, you'll be given an apartment in Moscow that will be reserved for you even if you're assigned elsewhere for a while."

Zverev never quite defined who "we" were, what the work was, or where that "elsewhere" might be or how long was the "while" Sakharov might have to spend there. But the general had chosen his temptations wisely: if anything could have swayed Sakharov, it was the promise of better housing. Scientists were now being courted, not commandeered. The *sharashka,* the intellectual forced labor camps described in Solzhenitsyn's *First Circle*, would obviously not do for matters as complex as atomic physics. All the same, Sakharov rejected the offer. "I hadn't left the munitions plant for FIAN and the frontiers of physics, only to abandon everything now." The general hoped he would reconsider.

But there was no walking away from politics in the Soviet Union, as Sakharov discovered when in January 1947 he published his first paper in *The Journal of Experimental and Theoretical Physics*. That paper, in the words of the American physicist Sidney Drell, "estimated the cross section for the production of mesons as the hard components of cosmic rays in a model proposed by Tamm." Sakharov's first taste of ink was, however, doubly tainted, first by his realization that Tamm's model was wrong and second because the title of his article was changed from "Meson Generation" to the vaguer "Generation of the Hard Components of Cosmic Rays." More conscious of security than ever, Stalin had placed Lavrenty Beria, the head of the secret police, in charge of the Soviet atomic weapons program and, as Tamm remarked when explaining the change to Sakharov: "Even Beria knows what mesons are."

In a speech at the Bolshoi Theater in 1946 Stalin had decreed state support for Soviet scientists to overtake and pass the achievements of their colleagues abroad, which hardly indicated any collegial sharing of findings. Every word published in the Soviet Union, from circus posters to the labels on matchboxes, had to pass through the department of censorship known as Glavlit. Scientific articles were of course vetted even more stringently. "Every article," Sakharov the fledgling author discovered, "had to pass through a complex bureaucratic process: references had to be submitted, long questionnaires filled out, and a recommendation procured from a special permanent commission at the institution where the author worked (if the author for one reason or another did not work in a scientific institution, he was absolutely unable to publish). The commission's recommendation had to certify that the article contained no secret information, patented ideas, or ideas and proposals with significant practical application. . . . Glavlit had its own interminable list of subjects forbidden not only in the interests of secrecy, but, in the main, out of political considerations. (For example, it was forbidden to publish information about crimes, alcoholism, health conditions, education, the water supply, suicides, the existing supply and the production of nonferrous metals, precise data on the population's nutrition and income, and movie and theater attendance figures.) . . . It is easy to imagine the resulting delays in publication."

Sakharov may have rejected General Zverev's offer not only out of his dedication to the frontiers of physics but on more practical grounds as well. He did not have much work remaining before defending his dissertation: with a little luck he could wrap up everything by the summer of 1947. He could then count on being appointed to a research position with higher pay. That would take some of the sting out of refusing an offer for better housing. But even the straight and narrow of science had its political obstacles. Before defending his dissertation, Sakharov had to pass a test in Marxist philosophy, a subject that never interested him in the least. That meant actually sitting down and plowing through writers like Lenin's favorite, Nikolai Chernyshevsky, a nineteenth-century utilitarian who held that a shoemaker who made a pair of boots was of more value to humanity than Shakespeare. Tough sledding at any time, it would have been especially difficult for Sakharov in 1947, when he was positively incandescent with his passion for physics. "I felt like the messenger of the gods," he said of reporting on the latest developments in science at Tamm's seminars.

Sakharov's excitement reached its zenith in the summer of 1947: "never before or since have I been so close to the highest level of science—its cutting edge." Yet it was also at this time that Sakharov displayed a failure of the confidence that was always his strong suit. Involved "to the point of obsession" with discrepancies in the measurements of the spectral lines of hydrogen atoms, Sakharov hit on what he felt was a brilliant solution to an important problem. Tamm, however, gave him "neither support, nor approval," saying rather paradoxically that Sakharov's idea was neither original nor workable. Tamm directed Sakharov to an article by the American theoretician Sidney Dancoff, whose calculations would have revealed themselves as "quite simply wrong" had either Sakharov or Tamm taken the trouble to pick them apart. "If our intuition hadn't failed us, we would have questioned Dancoff's work as many times as it took to uncover his error; or, more wisely, we'd have temporarily ignored the contradictions which arose and searched for problems simpler to calculate, whose results could have been compared to the experiment. This was the approach chosen by bolder people with more insight—and more success. But we did not, and so I lost a chance at the most important scientific work of that era."

Sakharov sabotaged himself in another way as well, but this time it was not from insufficient daring but from being, as he put it, "excessively truthful." When questioned by the special examiners in political philosophy, Sakharov admitted to not having read certain of Chernyshevsky's works but said he knew what they were basically about. He received a D, and the need to take makeup delayed the defense of his dissertation until the fall, a "financial blow," since it meant that his family would have to continue scraping by on his grad-student stipend.

In November 1947 Sakharov was finally able to defend his dissertation and did so brilliantly. He then received his candidate degree, the equivalent of a Ph.D., and was appointed a junior researcher in FIAN's Department of Theoretical Physics. Any of the skips of linkage that marked Sakharov's way of speaking about science were entirely absent in the thesis itself, which dealt primarily with electrons knocked out of atomic shells, a subject that had already attracted J. Robert Oppenheimer, the physicist who directed the Manhattan Project. "When you read Sakharov's thesis, which has seventy-one pages," observed the American physicist Sidney Drell, "you are impressed by how carefully all the steps are explained, all the intermediate formulas written down, and numerical coefficients derived. He liked to write as clearly as possible. His friends recall that to handwrite formulas into a typed text was a pleasure for him."

The world of Soviet atomic physics was still quite small, and thus there was nothing unusual about Sakharov's being invited to discuss his thesis at Laboratory No. 2, also known as the Laboratory for Measuring Instruments (LIPAN), the neutrality of both names serving to mask the Soviet atomic energy project run by Igor Kurchatov, who was in LIPAN's small auditorium the day Sakharov spoke. Kurchatov was an attractive yet enigmatic figure, with layer upon layer of personality. Warm, good-humored, a demon for work, loyal to friends, a master organizer, Kurchatov was contradictorily but accurately described as the "liveliest of men, witty, cheerful, always ready for a joke," a "teddy bear" and "first and foremost an 'operator' and what's more, an operator under Stalin—he was like a fish in water." Something of that multiplicity and ambiguity was reflected in his nicknames: "Prince Igor," "The General," and more commonly and affectionately "The Beard," in honor of the luxuriant growth that emerged directly from his chin and which he kept cut straight across the bottom, giving it an oddly pasted-on look. Falling ill with pneumonia in 1942 and too weak to shave, Kurchatov rose from his sickbed with that beard and vowed to keep it until the Soviet Union was victorious. He did not, however, shave it off in 1945, perhaps because by then he was aware that other victories still lay ahead. Kurchatov was in fact one of the first to alert the Soviet government of the military potential of nuclear fission, a goal that he suspected Nazi Germany of pursuing. By 1940 it was also clear that the United States was engaged in an atomic weapons program when the names of leading American scientists suddenly ceased to appear in any of the international journals, a development which Richard Rhodes characterized aphoristically: "secrecy itself gave the secret away."

Aware that secrecy and organization were essential to the success of a Soviet atomic project, Kurchatov had been instrumental in having the program switched from the control of Molotov, the foreign minister, to Beria, the head of the secret police, who could supply the necessary slave labor to construct facilities at high speed and for whom security was second nature. By 1943 Kurchatov was running the atomic program and in 1945 had been given five years to produce a bomb. Stalin told him not to count costs, saying: "If a child doesn't cry, the mother doesn't know what he needs. Ask for whatever you like. You won't be refused."

Failure was not an option. Besides funding and the support of "Mother" Stalin, Kurchatov also needed the best and the brightest for his mission. After Sakharov's talk, Kurchatov invited him to his office for a private chat. Sakharov was impressed by the size of Kurchatov's office and its battery of telephones, each a different color—a sign of importance among Soviet big shots—and one of which was no doubt a direct line to the Kremlin. On the wall behind Kurchatov and thus facing Sakharov was a "larger than life oil portrait of Stalin with his pipe and the Kremlin in the background. The painting, clearly an original by one of the 'court' artists, symbolized Kurchatov's high standing in the state hierarchy." (It remained on the wall even after Stalin was denounced by Khrushchev at the Twentieth Party Congress in 1956. "A present is a present," Kurchatov was said to have explained.) And Sakharov was impressed by Kurchatov's energy, charisma, dedication, but not enough to accept his invitation to join his institute. Sakharov wanted to continue working with Tamm at FIAN; what he wanted now was continuity, stability.

And for a time that wish was granted. In May of 1948 Sakharov was allocated a room in a communal apartment in the heart of Moscow. It wasn't much: 150 square feet for three persons, too small even for a table—they put their plates on stools or windowsills. The kitchen had to be shared with the ten other families and also the toilet, which was on the staircase landing and used by two floors. There was no bath, no shower in that building, which was heated by wood from the yard. But Russians did not need much for happiness in the postwar years. Modest in their material desires, Andrei and Klava were "delighted" with their new home from which they could not be capriciously ejected. Klava and Sakharov's mother had now somehow found a modus vivendi. All the pieces were finally in place. "So began," says Sakharov, "four of the happiest years in our family life."

But in a sense the idyll lasted only two months. Toward the end of June 1948, after the usual Friday seminar, Tamm, acting in an uncharacteristically "furtive" manner, invited Sakharov and another student, Yuri Romanov, to his office, where he broke "startling news." Orders had come down from the very top, the Council of Ministers and the Party Central Committee: Tamm had been appointed to head a group charged with exploring the possibilities of building a thermonuclear weapon. Sakharov and Romanov would be a part of that group.

In Russian fairy tales the hero encounters the same challenge three times before he wins the day, but this version had a distinctly Soviet twist: "in 1946 and 1947, I twice rejected attempts to entice me away from FIAN and the frontiers of theoretical physics. But the third time, in 1948, nobody bothered to ask my consent."

CHAPTER SIX

CHAIN OF COMMAND

OUTWARDLY, Sakharov's life did not change greatly. He still came to work at FIAN wearing the same faded khaki-colored suit he'd had on when arriving from the munitions plant three years before. And he still struck his colleagues as genial and lonely, accessible yet aloof. The more evident changes were occurring in FIAN itself, which had always prided itself on its easygoing democracy among teachers and students. The blackboards that had always been left covered with formulas top to bottom were now thoroughly erased at the end of the day. A well-provisioned, restricted dining hall was opened, a useful privilege in those lean times (in the thirties an actual sign read: "The open canteen is closing, a closed canteen will be opening in its place.") And, most important, a fourth floor was being added to the building where Tamm's group could meet in isolation from the rest of the department, an isolation that would be cordially but firmly enforced by a team force made up of young men who had seen action at the front. Privilege and secrecy were the first signs that FIAN was being absorbed more deeply into Stalin's state.

Though Sakharov was now more abstracted by science than ever, he was still sensitive to the change in atmosphere at FIAN: the passes, the whispered rumors, the long questionnaires that delved deeply into the background of everyone associated with the department. People were said to have good questionnaires, and bad—meaning that they were of suspect social origin, had relatives convicted of crimes or living abroad, or had themselves been associated with political factions no longer in favor. Sakharov had been impressed into atomic weapons research in late June 1948, a particularly tense moment. On June 24 the Soviets closed all rail traffic into Berlin, initiating a crisis that would last a year, the first direct Cold War confrontation between East and West, the final unraveling of the wartime alliance. Sakharov paid only fleeting attention to international developments, probably learning of them as he had learned of the dropping of the atomic bomb on Hiroshima, by reading a wall newspaper on his way to the store; yet the political tensions of the time had already permeated the atmosphere and procedures at FIAN, and it wasn't long before they began to affect him as well. Lidia Pariskaya, his partner in window washing, now observed that Sakharov's smile, though still childish and trusting, appeared more rarely on his lips. She saw him often because unlike the other theoreticians, who sat at their desks and covered every available scrap of paper with calculations and formulas, Sakharov worked nearly everything out in his head. If he wasn't sunk in an armchair, he would either pace the corridors or stand by the window for long stretches of time, utterly still, absent, unapproachable. He would then "squeeze his eyes shut and rub the palm of his hand down his face from his temple as if trying to wipe something away." Pariskaya felt that he was more than pensive—he had become somehow estranged. When she ate lunch with him in the special dining hall, Sakharov chewed his food slowly and in such near total silence that she could just as well have been dining alone.

He worked longer and longer hours, looking more "deathly tired" all the time. Part of the problem was the constant interruptions: The director needs Sakharov! Phone call for Sakharov! Car here for Sakharov!

One day he confessed his problem to a fellow theoretical physicist, Matvei Rabinovich: "Here's the situation. I'm often invited to the Kremlin, for meetings. They usually last until around four o'clock in the morning, then all the participants go off to their cars, but I don't have a car, and no one knows I don't have a car, and I don't tell anyone I don't. And where I live is about seven miles from the Kremlin, maybe even nine."

If he couldn't find a taxi, he'd have to walk home, arriving at daybreak. No wonder he looked exhausted. Out of fear of conspiracies, which are typically hatched at night, Stalin kept vampirishly late hours and expected the same from all those at the top of the Soviet pyramid, Andrei Sakharov now among them. He had yet to realize that he was eligible to request a government vehicle for himself. On grounds of efficiency alone, the state would have preferred such a valuable asset as Sakharov to be home asleep in bed rather than trudging the streets of the city.

Arriving at work late one day, Sakharov looked at Pariskaya with such "ravaged eyes" that she asked: "What's wrong with you?" After a long silence, Sakharov seized his head with both hands and said in a whisper: "You don't understand! It's horrible, horrible! What am I doing?" Then, his voice softer, he added: "Inwardly, I'm hysterical."

"Go home right now and go to bed. Get out of here!" commanded Pariskaya sensibly, and the next day Sakharov "triumphantly" reported that he had slept thirteen straight hours.

The pressure to catch up with and surpass the Americans was so enormous and unrelenting that the work could easily consume every waking hour of a scientist's day. But had Sakharov been suffering from more than mere exhaustion? Was there some buried resentment that he had now been ordered to do what he had twice declined? A fear that he would never do pure science again? He did not have any political doubts or moral qualms about the weapon he was helping to design. Like nearly everyone else involved in the project, Sakharov was, in the words of his colleague Evgeny Feinberg, "completely loyal in regard to the official ideology." Sakharov's inner agitation may in fact have issued from a surfeit of excitement. "Our initial zeal," he says," "was inspired more by emotion than by intellect." That emotion sprang from the sheer magnitude of the challenge: "The physics of atomic and thermonuclear explosions is a genuine theoretician's paradise. A thermonuclear reaction—the mysterious source of the energy of sun and stars, the sustenance of life on Earth but also the potential instrument of its destruction—was within my grasp."

If Claude Bernard's formulation—"Art is I; Science is We"—is correct, it was doubly correct in Soviet science with its penchant for the collective, and so for Sakharov to speak of a thermonuclear reaction "within my grasp" has a whiff of the Faustian about it.

But what Fermi had called the "superb physics" and Oppenheimer the "technically sweet" was not Sakharov's chief motivation. "What was most important for me at the time, and also, I believe, for Tamm and the other members of the group, was the conviction that our work was *essential.*" The work on a thermonuclear weapon was a continuation of the war, the patriotism of survival. Hans Bethe, the German émigré physicist who headed the Theoretical Division at Los Alamos, said of the Manhattan Project: "We all felt that, like the soldiers, we had done our duty." The Russians were more literal. Kurchatov told Sakharov and the other physicists that they *were* "soldiers" and often signed his memos "Soldier Kurchatov." Sakharov accepted the designation unhesitatingly and unequivocally, considering himself a "soldier in this new scientific war."

There were few illusions on either side. By October 1945, less than three months after Hiroshima and Nagasaki, the U.S. Joint Chiefs of Staff Intelligence Committee was already drafting plans for an atomic first strike against the Soviet Union, Moscow being of course the principal target. In a secret report, "Our Army of the Future," Brigadier General Leslie R. Groves, the head of the American atomic bomb project, bluntly stated: "If we were truly realistic instead of idealistic, as we appear to be, we would not permit any foreign power with which we are not firmly allied, and in which we do not have absolute confidence, to make or possess atomic weapons. If such a country started to make atomic weapons we would destroy its capacity to make them before it has progressed far enough to threaten us." The alliance with the USSR was anything but firm, confidence in Stalin anything but absolute. In 1948, when Sakharov went to work on the H-bomb project, the United States had fifty-six atomic weapons, the Soviet Union not a one. Patria and parity were one and the same.

But there was also a personal dimension to thermonuclear research for Sakharov, who saw it as a "possibility to prove what one can do, above all to oneself."

After being recruited into bomb work at the end of June, Sakharov rented a room in a house in the countryside. Klava and Tanya stayed there all summer; he would come out on Sunday and spend a day, sometimes two. Because of his clumsiness—already demonstrated on the Volga when courting Klava—Sakharov was banned from the sailboats. That clumsiness, coupled with the fact that he still couldn't swim, put him in the absurd position of being both the person most likely to cause an accident and the one most in need of rescue. But he loved the "sparkling water, the sun, the lush greenery." Then, at most after two days, he would have to return to Moscow, to FIAN, and again be shocked by the "bizarre and fantastic" world of nuclear weaponry, in such "striking contrast to everyday city and family life, and to normal scientific pursuits." Previously, the professors at FIAN had been shown no special deference in scientific debate; the atmosphere had been free and easy, "cozy" as one colleague called it. All that was now a thing of the past. "From the beginning," says Sakharov, "the Tamm group operated under conditions of strict secrecy, something we weren't accustomed to. Only members of the group could enter the room assigned to us, and the key was kept in the security office. All notes had to be made in special tablets with numbered pages. At the end of the working day, they were placed in a suitcase, sealed, and handed in for safekeeping. I suppose we were flattered by all the rigmarole, but it soon became routine." They may have gotten used to it, but it never sat well. Tamm became increasingly abrupt and short-tempered, though as quick to apology as to anger. Though better rested, Sakharov was still smiling less, repelled by the new heaviness in the atmosphere, though he accepted security as a fact of fate, considering himself "bound for life by a pledge not to divulge state and military secrets, a commitment I undertook of my own free will in 1948."

The Tamm group had a two-part task in the summer of 1948: one quite general, to explore the feasibility of constructing a hydrogen bomb; the other quite specific, to "verify and refine" the calculations on explosions and gas dynamics done at the Institute of Chemical Physics by the Zeldovich group.

An atomic bomb operates by fission, the splitting of the nuclei of heavy elements such as uranium; a hydrogen bomb operates by the fusion of the nuclei of light elements, hydrogen being the lightest element. Under conditions of sufficient pressure and/or heat, the nuclei of hydrogen will fuse, as they do in stars. But stars generate fusion reactions inside their cores, under enormous pressure, which cannot be reproduced on Earth. The alternative is to generate fusion with enormous temperatures, which also cannot normally be produced on Earth. The detonation of an atomic bomb, however, creates enormous temperatures: the interior temperature of the sun is 14 million degrees, while that of an atomic bomb blast is between 50 million and 100 million degrees.

As Sakharov's colleague Yuri Romanov describes the problem at its earliest stage: "It was clear that to sustain a thermonuclear reaction, a temperature on the order of a few tens of millions of degrees was necessary, and that this temperature could be reached only by using an atomic bomb as the ignition. However, if it was simply surrounded by deuterium, an isotope of hydrogen [one proton and one neutron], this would not result in a significant increase in the explosive power. The reason is that the dissipation of matter at high pressure and the decrease in temperature due to the heat transfer would make the fusion rate too slow, and in such an arrangement only a small percentage of the deuterium nuclei would react. The situation would improve significantly if tritium, another isotope of hydrogen [one proton and two neutrons], were used instead. . . . However, tritium is not found in nature, and must be produced in nuclear reactors. . . . Tritium, therefore, is expensive. It is also radioactive with a half-life of 12.6 years, which means that it must be continuously replenished. Thus the construction of the hydrogen bomb was not only a complex technical and industrial problem, but it also required the solution of many basic, purely scientific problems. The group led by Tamm was organized to solve those problems."

It was not long after he began checking the calculations of the Zeldovich group that Sakharov came into direct contact with Zeldovich himself, initiating an unlikely friendship. Short, bald, muscular, Yakov Zeldovich was of the sort who dominate through sheer energy. He was prolific in so many fields—chain reactions, jet propulsion, thermonuclear weapons, elementary particles, astrophysics and cosmology—that Stephen Hawking assumed that "Zeldovich" was an acronym for a collective of authors. Born in 1914, he was seven years Sakharov's senior. Zeldovich's father was a lawyer and his mother a translator who had studied at the Sorbonne in the Europe of open borders before World War I. Though he was of the educated class, Zeldovich's own training was eccentric and largely autodidactic—he had a Ph.D. but no B.A.—partially due to the tumultuous times in which he grew up and partially because he faced discrimination as a Jew. When he and Sakharov grew closer, Zeldovich admitted to having been driven by an "inferiority complex" in his youth, which he had overcome by his own efforts, though Sakharov doubted that Zeldovich had ever truly liberated himself from it. Weight lifter, ferocious tennis player, drinker, dancer, a dervish of libido, Zeldovich impressed some as the "Zorba of cosmology," while others saw a cynic with no values apart from his own success and pleasure. But there was no question about his passion for science and his talent, which is what bound him to Sakharov—who observed that Zeldovich would be "almost childishly delighted when he managed to achieve some important piece of work [and] felt failures and errors keenly."

Working with Yuli Khariton, who would become the administrative director of the Soviet hydrogen bomb project, Zeldovich had published a series of papers in 1939–1940 on the conditions required for producing a chain reaction in natural uranium, causing Tamm to exclaim prophetically: "Do you know what this new discovery means? It means a bomb can be built that will destroy a city out to a radius of maybe ten kilometers." Unlike American physicists, who imposed a "radio silence" as soon as the theoretical possibility of a chain reaction and its potential military uses were discovered, the "behavior of Soviet scientists was quite different," writes David Holloway in *Stalin and the Bomb*. "There is no evidence that they tried to alert their government to the possible implications of nuclear fission before the summer of 1940, and no special organization was created before then to coordinate research on fission. Soviet scientists continued to publish freely on fission in 1940; no effort was made, by the government or the scientists themselves, to restrict publication." Either the Americans were simply smarter, or else they sometimes acted more Soviet than the Soviets.

When Germany invaded the Soviet Union in June 1941, the Institute of Chemical Physics, where Zeldovich worked, was evacuated to Kazan, a Volga city some 430 miles east of Moscow. Fission research had to be dropped for more immediate war needs. Zeldovich and his coauthor Khariton now worked on developing powder fuel for the Katyusha mobile rocket launchers. By that time Klaus Fuchs, the German communist and atomic spy, was informing the Soviet government on British atomic-weapons research. Later he would report directly from Los Alamos. The espionage of Fuchs and that of the American spies and sympathizers —Gold, Greenglass, Rosenberg—alerted the Soviets to actual developments and saved them costly research and development. Hans Bethe called Fuchs "the only physicist I know who truly changed history." But as Richard Rhodes remarks in *Dark Sun,* at the "basic level, there was never any 'secret' of how to make an atomic bomb. Knowledge derived from espionage could only speed up the process, not determine it, and in fact every nation that has attempted to build an atomic weapon in the half century since the discovery of nuclear fission has succeed on the first try." Still, for the Russians, forever behind, to "speed up the process" was worth every ruble it cost.

In 1941–1942 the most urgent question for both the Soviet Union as well as the United States and Britain was whether Nazi Germany was at work on the bomb.

In April 1942 the Soviets had an alarming stroke of luck. Colonel Starinov, a much wounded and much decorated soldier who had fought in four wars—the Russian and Spanish Civil Wars, the Winter War against Finland, and now the Great Patriotic War, the official Soviet designation for World War II—was operating in the south of Russia, engaging in his dangerous specialty: sabotage behind enemy lines. He was working with some of his Spanish comrades who had emigrated to Russia after Franco's victory in the Spanish Civil War and who found the frozen steppes so formidable that they could muster only enough breath to call them "muy frio." Says Starinov: "One of our men found a notebook on a German officer who had been killed in a skirmish and he brought that notebook to me. I couldn't make heads or tails out of it. It looked like some kind of chemical formulas. I sent the notebook to Moscow and it turned out that those were formulas for using uranium for atomic purposes. That officer had been in that recently captured territory looking for uranium deposits."

Alarm bells went off all the way to the top. At that time the Soviets had scant idea of where their own uranium deposits might be located, and here were German intelligence officers scouting for them in captured territory.

On June 14, 1942, a coded radio message went out from Mostow to Soviet intelligence station chiefs in Berlin, London, and New York:

> TOP SECRET
> Reportedly the White House has decided to allocate a large sum to a secret atomic bomb development project. Relevant research and development is already in progress in Great Britain and Germany. In view of the above, please take whatever measures you think fit to obtain information. . . .

followed by a wish list of everything from information on isotope separation and trigger mechanisms to project leaders.

By the end of 1942 Stalin had ordered the resumption of atomic research. In March Kurchatov, The Beard, was appointed head of the atomic project and was already studying the Soviet intelligence materials, declaring them "wonderful . . . they fill in just what we are lacking." By mid-April 1943 Laboratory No. 2 (which Sakharov visited after defending his dissertation in 1947) was set up to run the atomic weapons program. Kurchatov quickly enlisted the services of Zeldovich and Khariton.

The initial Soviet atomic project was modest in scope and largely theoretical in nature. After defeating the Germans at Stalingrad in the winter of 1942–1943 and even more decisively at the Battle of Kursk in the summer of 1943, Stalin did not consider it likely that atomic weapons would prove decisive in the war against Germany, especially since he had intelligence from Fuchs's control that atomic research in Germany had reached a "dead end." Soviet research proceeded slowly for a number of other reasons as well: lack of basic data, of uranium itself, and of graphite for reactors. Recruiting personnel was also problematical. Khariton, for example, did not initially wish to accept the post of administrative director (essentially the role that Oppenheimer played on the Manhattan Project), considering his work on mines and antitank weapons of more immediate practicality. Besides, he had a terrible questionnaire. A gentle, cultured man, the son of a St.

Petersburg journalist who emigrated after the revolution and a Moscow Art Theater actress, and on top of it a Jew, Khariton had spent two years in Cambridge at the Cavendish Laboratory working with Rutherford and Chadwick. He had relatives who had been arrested; his mother was living in Israel. Still, he was a good choice as a leader. A man of "irreproachable, dryish correctness" whose silences could be more communicative than his words, he was a foil to Kurchatov, of whom one subordinate said that he "would sink his teeth into us and drink our blood" until the task was done. Kurchatov, who could also be quite persuasive, prevailed over Khariton's objections by taking the long view, arguing "you can't overlook the time when victory will be behind us and we ought to be concerned about the future security of the country too."

The espionage Kurchatov was receiving helped his research especially after Fuchs arrived in the United States in December 1943. Kurchatov was to all intents and purposes directing Soviet atomic espionage by listing the questions that needed to be answered. He used American materials "for the dual purpose of double-checking the scientific results obtained by members of his team, and for evaluating the possibility that stolen secrets might contain purposely planted disinformation," writes Soviet atomic and space scientist Roald Sagdeev. Kurchatov would examine the work of the theoretical physicists working under him who knew nothing of the espionage material, "then silently open the safe with the precious stolen American secrets to compare the results. 'No, it is not right,' he would say. 'You have to work more and come again.' "

But everything about the Soviet atomic weapons project—its pace, scope, funding—changed in August 1945 with the bombing of Hiroshima and Nagasaki. Stalin immediately summoned secret police chief Lavrenty Beria and placed him in charge of the Special Committee on the Atomic Bomb, a logical choice since the project would be nourished by espionage and shielded by security, Beria's bailiwicks. (Beria, in turn created "Department S" to handle intelligence materials and appointed the same two men who had been in charge of Leon Trotsky's assassination to head it.) In addition, much of the actual physical work—from the mining of uranium to the construction of "installations," as the secret bomb-lab cities were called—would be performed by Gulag slave labor, again Beria's domain.

Even in his school days Beria was nicknamed "The Detective" for his uncanny ability to solve puzzling thefts, sometimes due to his innate skills or more often to the fact that he himself was the culprit. A virtuoso with the blackjack in his hands-on days, he rose to become leader of the secret police in Georgia. Beria affected a pince-nez and had a beautiful, cultured wife but still preferred raping high school girls, so deeply were pleasure and terror connected for him. Still, it is difficult to say which is worse: a secret police chief with a predictable twist of vice, or one who has no vices whatsoever.

One evening Sakharov, Tamm, and Tamm's grandson had tossed out the big question: what if Trotsky had come to power, not Stalin? Tamm said there would have been ten times fewer victims. Sakharov said a hundred. The grandson, the most distant in time from the events, said a thousand, every generation a power of ten. Then Tamm compared Beria to Trotsky—maybe one pince-nez had reminded him of another—because of their rationalism, the ability to get right to the heart of any problem and to pose intelligent questions. Khariton, who reported directly to Beria, said that though he was the "personification of evil in modern Russian history," Beria could also be "courteous, tactful and simple when circumstances demanded it." Or, as one of Kurchatov's deputies put it: "For us Beria's administrative abilities were obvious. He was unusually energetic. Meetings did not last for hours, everything was decided very quickly."

He was even capable on occasion of a Solomon-like wisdom. When the Politburo decided to divide the Soviet coal industry into two sectors, eastern and western, Beria proposed that the two officials in charge arrange the division themselves. At a subsequent meeting, Beria asked if there were any complaints. The official in charge of the western sector said that he had none, whereas the official in charge of the east had plenty, complaining that the official in charge of the west had taken all the best people and vacation resorts. Fine, replied Beria, since the official in charge of the west had no problems with the division, he would now become the official in charge of the east and vice versa.

One exception to any such positive views was Peter Kapitsa, who could not abide Beria. Kapitsa, confident by nature, had imbibed the air of scientific and political liberty while working with Ernest Rutherford in Cambridge from 1921 to 1934, when he was tricked back into Soviet Russia. Perhaps that was what gave him the courage in late 1945 to perform an inconceivably dangerous act: denouncing Beria to Stalin. In what was essentially a letter of resignation, Kapitsa said that Beria was unfit to run the atomic weapons program, because his attitude was insufficiently respectful of the power of science and scientists: "it is high time that comrades like Comrade Beria should begin to learn respect for scientists and scholars." That fundamental disrespect made itself felt in two ways: in poor manners (Beria canceled nine appointments in a row with Kapitsa) and, more important, in Beria's malign distrust of scientists. Kapitsa reminded Stalin that the success of the atomic bomb project, the USSR's highest priority, depended on one thing alone: "that greater trust must be established between the scientists and the government."

Stalin showed the letter to Beria. Offended by the insults and alert to any challenge to his own power, Beria wanted to arrest Kapitsa. "I'll dismiss him if you like," conceded Stalin, "but don't *you* touch him." The thirties were over; these were modern, postwar, atomic times; you didn't just kill a Kapitsa. House arrest was sufficient punishment.

A realist, Beria swallowed the insult and the order. Like everyone else in the court of the Kremlin, his fate depended on Stalin's favor. And Beria could fawn with the best of them while continuing to pursue his own purposes. Stalin's daughter, Svetlana, who spent a portion of her childhood on Beria's lap, saw him as her father's Iago, a "magnificent modern specimen of the artful courtier, embodiment of Oriental perfidy, flattery and hypocrisy who succeeding in confounding my father, a man whom it was ordinarily difficult to deceive." Though Svetlana knew that her father and Beria were "in a good many things . . . guilty together," she always had a feeling that Beria was tricking Stalin, then later on "laughed up his sleeve" at him.

Kapitsa was right about Beria. His malign distrust erupted in late December 1946 when watching Kurchatov test-start the first nuclear reactor. As described by two of the participating scientists, Kurchatov "put the winch into motion, lifting the control rod. The clicks increased and gradually transformed into a continuous roar. The light beam of the galvanometer ran off the scale. The people on hand shouted 'It's going!' meaning the chain reaction. Beria had seen massed assemblies of Stalin's Organs fire racks of screaming Katyushas; here were only clicking counters and flashing lights. 'And that's all?' he challenged Kurchatov. 'And nothing else?' He wanted to see the famous machine for himself and asked to approach the reactor. 'No,' Kurchatov told him. 'You can't go down there now. It's dangerous to your health.' Beria began suspecting that Kurchatov was swindling him."

Stalin took an entirely different attitude. He had already inspired and rewarded Kurchatov with an elegant eight-room Italianate house on the grounds of Laboratory No. 2—marble fireplaces, parquet floors—even bringing in a team of Italian craftsmen to do the finishing touches on the interior. And besides, as Stalin said about some other recalcitrant physicists, "We can always shoot them later."

* * * *

Very little impinged on the mind of Andrei Sakharov in early summer 1948—not the Berlin crisis, not even the purge of the geneticists. The expunged blackboards and sealed suitcases quickly ceased to register; he was working with "total absorption." "Once," he says, "as I was standing in line for the cashier at the public baths, mulling over certain questions of gas dynamics (I couldn't stop thinking about them), I realized that an explosion in an ideal, cold gas can be described hydrodynamically by a function with a single variable if certain simplifying assumptions were made." Then Sakharov wandered off, washbowl and towel in hand, into the din and steam of the public baths, still distracted by those "certain simplifying assumptions."

It wasn't long before he submitted his first secret report marked S-1 (Sakharov 1). And two months later he had a breakthrough idea that radically changed the direction of the group's research. Those hours of unbroken absorption had yielded a series of insights that now constellated into a workable idea, soon known as the "First Idea." Sakharov proposed an "alternate design for a thermonuclear charge that differed from the one pursued by Zeldovich's group in both the explosion's physical processes and the basic source of the energy released." Nicknamed the "layer cake," Sakharov's idea proposed using a spherical assembly of concentric shells with an A-bomb as the ignition, tamped by a layer of uranium 238, which retards the expansion so that more reaction can occur. The atomic weapon kicks off a thermonuclear reaction in the surrounding fusion fuel, composed of lithium deuteride and lithium tritide. That in turn causes a fission reaction in another layer of uranium 238. Only about 20 percent of the yield of such a device would be truly thermonuclear, and the reaction is usually described as fission-fusion-fission. The same idea (nicknamed the "Alarm Clock") had occurred to Edward Teller in 1946 and to a British scientist who has remained anonymous. In normal times, in normal science, one of them would have been the first to win credit by publishing his findings, the only serious consequences those to the vanity and careers of the other discoverers. But these were not normal times, and this was not normal science.

Tamm vigorously supported Sakharov's "layer cake" design. Busy as he was preparing for the first Soviet A-bomb test, Zeldovich immediately recognized the value of Sakharov's idea when they met and Sakharov explained it to him. The expansive Zeldovich soon invited Sakharov to his home, where he introduced him to his family, "light-heartedly remarking that the greatest blessing in the world is a good-natured wife," to which his wife responded with a smile that seemed to Sakharov "rather tense." Zeldovich made no great secret of his love affairs, and all that was apparently lacking for his own perfect well-being was that his wife be happy about them as well.

But Zeldovich wasn't just being expansive that day; he was also taking Sakharov's measure, even questioning him about his health and family history. The "higher powers" soon took a renewed interest in him as well. In fall 1948, after the "First Idea" was accepted by the Tamm group, Sakharov received a significant raise in pay and an invitation from another KGB general. He urged Sakharov to join the Communist Party, which he explained was not so much a "privilege" as a "responsibility," the chance to be the part of a "great cause." The general himself even offered to recommend Sakharov for membership, giving him to understand that this was an honor in itself and a guarantee of acceptance.

Sakharov declined. Suddenly surprisingly political, he cited his misgivings about the "arrest of innocent people and the excesses of the collectivization campaign." To Sakharov's critique of the Terror of the late thirties and the destruction of traditional Russian agriculture and the class of better-off peasants known as kulaks, the general responded with the party line: "The Party has severely condemned the mistakes committed during the Terror, and they have all been rectified. As for the kulaks, what were we supposed to do when they came at us with shotguns?" The general did not go into detail about how many of the kulaks had shotguns or how the mistakes of the Terror had been "rectified," and they parted with their views exchanged but unchanged, the general to the end requesting that Sakharov give "very serious consideration to his suggestion."

It was also a time of choice for Klava, who was pregnant with their second child, due in the summer of 1949. The poverty of the postwar years and the frequent illnesses of their daughter Tanya had forced Klava to devote herself to motherhood. But their situation had changed by the time their second child was on the way. There was money, and there would have been even more privilege if Klava and Andrei, unlike each other in so many ways, had not been equally lacking in any savvy about manipulating the system. If she was going to return to work or study, as wives of the intelligentsia usually did and were expected to do, the time would be shortly after the birth. Klava chose not to. She may have decided, as some Soviet women did, that the raising of children was the worthiest of tasks. But she may also have felt deeply outclassed by the elite of the intelligentsia to which her husband belonged by birth and by ability. Sakharov's attitude at the time was a mix of the protective and the passive. "Klava seemed to lack the stamina needed for the demands of life. . . . I am to blame (if any blame attaches) for not insisting that she study or find a job." But in those "happiest years," Klava and his mother even having made peace, there was little impetus to question the consequences of her choice.

In early 1949 Tamm and Sakharov were summoned to the offices of Boris Vannikov, who ran the government agency in charge of the atomic weapons program, the Ministry of Medium Machine Building, a term whose camouflage blandness is rivaled only by the British equivalent, Tube Alloys Research. Vannikov, "cautious, clever, and cynical," had changed his last name from something more obviously Jewish and was fond of making anti-Semitic jokes to prove he was one of the boys. Except for a brief arrest during the war, he had climbed to the top of Stalin's bureaucracy with scarcely a slip. There was also a fourth man present at the meeting. Sakharov took him to be Beria's representative, though he seemed a bit extraneous since Vannikov reported directly to Beria. The conversation quickly turned to the Installation, the nuclear weapons laboratory.

"Sakharov," said Vannikov, "should be permanently transferred to work at the Installation. The project needs him."

So agitated by this prospect that he forgot to use the obligatory phrase "Soviet science," Tamm protested that it was not in the country's best interest to limit someone so adept in so many fields as Sakharov to applied research. As if on cue, the direct Kremlin line rang. Vannikov tensed as soon as he heard the voice on the other end. "Yes, they're with me now. What are they doing? Talking, arguing." He paused to listen, then replied: "Yes, I understand. . . . Yes, I'll tell them." Hanging up the phone, Vannikov addressed Tamm and Sakharov. "I have just been talking with Lavrenty Pavlovich," he said, referring to Beria by his first name and patronymic, no last name required. "He *requests* you accept our offer."

"Things," said Tamm as soon as they were out of the office, "seem to have taken a serious turn."

CHAPTER SEVEN

THE SAVIOR OF RUSSIA

ONE measure of the latitude accorded physicists in Stalin's Russia was that something like six months passed between Beria's "request" and the day when Andrei Sakharov actually first set foot on the Installation. (Some of that time was for a detailed check on Sakharov's background, family, politics.) Beria's request was of course ultimately neither Beria's nor a request. Whether or not he cleared this particular action directly with Stalin is irrelevant since, like the physicists, Beria too had some considerable latitude of his own. Yet Beria's request was not entirely a command either. Physicists could be quite balky creatures, as Kapitsa had already demonstrated when refusing to work with Beria. Physicists had a strong sense of "class consciousness," important for the teamwork required on a project such as building the bomb but which also had its inconvenient side: they stuck together. In 1938 Kapitsa had been instrumental in freeing the eminent physicist Lev Landau after he had been imprisoned for adding his name to a leaflet that accused Stalin of hijacking the revolution. Kapitsa had known just what note to strike with Stalin—the utilitarian. As he wrote in protest of the arrest of another physicist, Professor Fock: "Fock's arrest is crude treatment of a scientist, which just like rough treatment of a machine, is bound to damage performance." Sakharov was a valuable scientific instrument and even at this early stage was being handled with due respect and care.

It would, however, be unrealistic to exaggerate the size of the interstices in which Soviet physicists moved and maneuvered. A good corrective to that impression, though one that goes too far in the other direction, is Nikolai Krementsov's *Stalinist Science:* "Soviet physics is often portrayed as having somehow managed to stand apart from Stalinism and preserve its 'intellectual autonomy' from party interference. . . . In the late 1940s, physicists had no more autonomy than did biologists or any other scientific group: the institutional structure, personnel, and research agendas of Soviet physics were, according to the system's rules, a prerogative of the party apparatus. It was the party apparatus, not the physics community, that decided to embark upon a full-scale atomic project. Despite the numerous petitions by physicists before and during the war, this decision was made by the Politburo, and only after Hiroshima and Nagasaki. The Politburo also determined the line of research, overruling Soviet physicists who wished to create their own improved atomic bomb [ordering them] simply to copy the American design for the atomic bomb. Furthermore, physicists, no less than biologists, were subjected to strict 'thought control' and obliged to perform rituals of 'obedience and devotion' to the party."

When Sakharov and Tamm went back to work at FIAN, their connection was stronger than ever because of the increased pressure both were now under. Tamm's protective attitude toward Sakharov, displayed during the meeting with Vannikov, also bore the seeds of guilt. If there was anything Tamm could have done to ensure that Sakharov's genius was not squandered on applied science, that time had passed long before the phone call from Beria. When Tamm was instructed to begin research on the possibility of constructing a thermonuclear weapon and to check the calculations made by the Zeldovich group, it was only natural that he select his most gifted students for the task. That, however, would have been the only time he could have helped shield Sakharov from the project, whose "serious turn" he had clearly not foreseen. But the higher-ups already had their eye on Sakharov, and any attempt to excuse him from weapons work would have had a whiff of the subjective, if not the subversive, about it. The humdrum may also have played its part: by involving Sakharov in atomic work, Tamm may also have been trying to qualify him for a decent apartment.

If Sakharov was initially "flattered" by the security measures implemented at FIAN, the personal attentions of the security chief had to have a similar effect. It was a common enough reaction. After the war, Zeldovich, in the uniform of a Red Army captain, had been dispatched to Germany to collect information on the V-2 rocket and, Sakharov says, "was once invited to dinner by the KGB boss of the Soviet zone, who in effect exercised power over half of Germany. Zeldovich recalled their meeting with some fear, but also with a tinge of admiration, a sin of which we were all a bit guilty at that time."

Such sins aside, Sakharov's confidence and his authority were on the ascent in the Tamm group for the most solid of reasons: actual achievement. And as a result of "explaining things to generals," his ability to communicate his ideas was improving. Though some of his co-workers still found Sakharov hard to follow, as if he were reasoning toward conclusions already reached by intuition, others had quite the opposite reaction, saying that Sakharov "distinguished himself through the clarity and correctness of his thought, and the conciseness of expression of his ideas." Achievement evoked authority, a style of leadership beginning to emerge in Sakharov. He did not have the fireball energy of Zeldovich, the drive and wiles of Kurchatov, or the vigorous charisma of Tamm, but he won people over by his immense, polite patience that lasted until points of principle, or pride, were touched upon.

In June 1949 Sakharov rented half a dacha outside Moscow near the main railroad line; it made the commute to FIAN easier, and Klava was due in late July. If the summer of 1948 had shocked Sakharov with its contrast between serene family outings beside the "sparkling water" and the "bizarre and fantastic" world of nuclear weapons research, that contrast was even starker in the summer of 1949 as they waited for the birth of their second child.

Late in June an official M-1 car pulled up in front of the dacha, and an officer, all spit-and-polish, emerged to inform Sakharov that Minister of Machine Building Vannikov needed to see him "immediately."

Sakharov was whisked to Moscow, where Vannikov told him: "We have to leave for Khariton's place right away." Vannikov was using a euphemism for a euphemism, the atomic weapons lab usually being referred to as the "Installation," one of the meanings of the Russian noun *obyekt,* which also means "object" and, in military parlance, "objective." But this was not yet a permanent transfer. Sakharov would only be spending something like a week there

—it was more orientation than assignment. At the end of their very brief and brisk conversation, Vannikov informed Sakharov where to go for further instructions.

The address proved that of a "Vegetable and Fruit Warehouse," which it may well have also been. It did seem like a typical Soviet establishment of that sort, with one man dozing in a chair while two others played dominos. A "pale nervous" man provided Sakharov with a pass and told him which train he was to take that evening and precisely which car to board.

Cordoned off by men in uniform and plain clothes, that car was Vannikov's own personal car, meaning of course that the state had made it available for his exclusive use, a perk that could vanish in the wink of an eye. Sakharov did not sleep well during the overnight trip but not because he was suffering from any anxiety about his pregnant wife or about entering the next chamber of the secret world: "what kept me awake was a new and challenging idea, the possibility of a controlled thermonuclear reaction."

At dawn they arrived in Sarov. Like everything in Russia, Sarov had changed a great deal since those July days in 1903 when Tsar Nicholas and Alexandra had attended the canonization of Serafim, simultaneously praying for the birth of a male heir and stirring up the mystic nationalism that led to war and revolution. The Sarov monastery sheltered three hundred monks until it was shut down in the twenties, at which time Serafim's remains vanished, exhumed and confiscated by the regime. The monastery was converted into an orphanage, then a prison camp.

Like everything else in Russia, Sarov, a town of around three thousand people, and its environs, had not changed much at all. One of Sakharov's future colleagues at the Installation, Venyamin Zuckerman, was taken by the "dense woods, the fine hundred-year-old pines, the monastery with its cathedral and white bell tower on the high river bank," while another, Lev Altshuler, who, when making the same trip, thought himself transported back to the Russia of before Peter the Great and recited to himself the lines of the nineteenth-century Slavophil poet Fyodor Tyutchev:

These villages so forlorn,
And so ungiving this earth,

Land where long suffering was born,
You land of Russia's birth.

The foreigner's eye will scorn
To comprehend or even see
The glowing of your mystery
In a nakedness of all pride shorn.

The cross his burden and affliction,
The Tsar of Heaven, guised as a slave,
Walked throughout these lands and gave,
Gave all Russia benediction.

Sakharov's group was moving too fast for any more than blurred first impressions: as soon as they arrived at the station they jumped into the two cars waiting for them and sped away at "breakneck speed." But their pace was soon slowed by the roads (Gogol said fools and bad roads were Russia's abiding elements), which did allow Sakharov to take in some of the landscape and to witness the first casualty of his involvement in weapons work. "The pale light of dawn illuminated tumble-down peasant huts, their roofs of old straw or half-rotted shingles, torn rags hanging on clothesline, and collective farm cattle—dirty and scrawny even in summer. The car ahead of us ran over a hen as we raced on through fields and stunted groves. Suddenly the driver slammed on his brakes: we had reached the 'zone'—two rows of barbed wire strung on tall posts and separated by a strip of plowed land."

The papers of the people in the lead car were checked by two officers who then saluted, waved the car in, and began to approach the car with Sakharov and Vannikov. Jangled by thirty-five miles of ruts and bumps, cursing a blue streak, Vannikov ordered his driver to "step on it!" The driver obeyed, scattering the two officers like hens and bearing Andrei Sakharov onto the territory of Arzamas-16, the Installation.

A vast hexagon (ninety square miles when completed), it had originally been named Arzamas-60 because of the sixty-kilometer distance from Sarov, but that precision was considered a breach of security and the name was changed to 16, which signified absolutely nothing. The whole issue about distance from Sarov was, however, a bit academic, since in 1947 the town of Sarov had disappeared from Soviet maps, spirited away like Serafim's bones.

Vannikov's conversation with an Installation official upon their arrival made no sense whatever to Sakharov.

"Is it here?" asked Vannikov.

"Yes."

"Where?"

"In the storehouse."

In fact, all Sakharov's initial impressions of the Installation were a blend of the cryptic and details as vivid as the nick he gave himself while shaving with a straight razor in the VIP hotel. The stars painted on the walls of the top brass's dining room known as the "generals' mess" for some reason pierced the initial blur, as did of course the sudden appearance of The Beard, Igor Kurchatov, flanked by two bodyguards as he strode toward a waiting limousine and greeted Sakharov volubly: "So, the man from Moscow's here. Welcome!"

In a short while Zeldovich arrived to play Virgil to Sakharov, and many of the details began to fall into place. The "it" Vannikov had referred to was a sample of fissionable material that had recently arrived and would be a component in the first Soviet "device," as all weapons, atomic or thermonuclear, were termed. Kurchatov had paid a personal visit to the Installation because of the urgency of the first and therefore critical test of an atomic bomb. Drawing Sakharov aside, Zeldovich explained: "The 'chiefs' are holding some important conferences. Don't get offended if you're not invited. They don't invite me very often—only when they want my opinion. Try to understand: there are secrets everywhere, and the less you know that doesn't concern you directly, the better off you'll be."

Zeldovich then introduced Sakharov to some of his new colleagues, who eyed him with "open curiosity" and one of whom, Evgeny Zababakhin, he recognized as a former Moscow University classmate he had last seen just after the outbreak of the war. Sakharov was impressed by the working atmosphere—"professional and friendly, yet extraordinarily intense: work went on around the clock when necessary."

The Russian tendency to establish intimacy quickly, heightened by the confined circumstances and the welcome arrival of a new colleague, meant that Sakharov was at once initiated into the history and lore of the Installation, which the scientists had already dubbed "Los Arzamas." He was told that the local peasants had decided that within the mysterious, barbed-wire-enclosed city was a "test model of Communism." Some of the new security people had the same impression when arriving for work, amazed by the free buses and bathhouses. This was hardly communism in its ideal Marxist vision of a society where the state had withered away and each worked according to ability and received according to need. But it was the quintessence of Stalinist communism. As David Holloway observed in *Stalin and the Bomb:* "The building of the atomic bomb was the kind of task for which the Stalinist command economy was ideally suited. . . . It was a heroic undertaking for which the resources of the country could be mobilized, including the best scientists and industrial managers, as well as the slave labor of the Gulag. The project was a curious combination of the best and the worst of Soviet society—of enthusiastic scientists and engineers produced by the expansion of education under Soviet rule, and of prisoners who lived in the inhuman conditions of the labor camp."

Something like four hundred thousand prisoners known as Zeks were involved in the mining and construction projects that created the Soviet atomic weapons empire, but not all of them, as Sakharov was told by one of his new colleagues, Viktor Gavrilov, accepted their fate with fabled Russian "long suffering." Two years before Sakharov's arrival, a group of fifty bold and desperate men led by an ex-colonel had risen up in mutiny, commandeered a truck, and shot their way out of the camp with submachine guns seized from the guards they'd killed. They then fled into the dense woods but were soon ringed by a huge number of Beria's troops, who pounded them with mortar and artillery fire as well as strafing them from aircraft until every last one of them was dead.

Analyzing the problem, the security people realized that prisoners with long sentences ("the first twenty-five years are the hardest," as the Zeks joked) had nothing to lose and were the most likely to choose a few last hours of freedom over slowly being ground into camp dust. But the obvious solution—using prisoners serving short terms for petty crimes (theft, "malingering," gleaning grain from collective farm fields)—posed a problem of its own: how could prisoners who had been exposed to a top secret place like the Installation simply be released back into the general population? But as the saying went, There is no fortress a Bolshevik cannot storm. A solution was found: when their terms were up, the prisoners were simply exiled to the wilderness of the Far East or, even more economically, had new sentences slapped on.

Sakharov ended his visit to the Installation by giving, at Zeldovich's request, a talk on quantum field theory; then the journey was played out in reverse, and he found himself once again in the dacha with Klava. who was now in her ninth month. The birth of their second child in late July could not have differed more from that of their first. Klava did the wash that morning, they took the train into Moscow, and that evening, after only two hours of labor, their second daughter was born, her name, Lyubov (meaning "Love," Lyuba for short), chosen by Tanya, who was four and a half at the time. But now their one room in the communal apartment in Moscow would be impossibly cramped. Sakharov, who had nothing of the operator about him, did still have the wit to turn for advice to someone, Zeldovich, who quite clearly did. Zeldovich advised Sakharov to approach Kurchatov directly. That proved wise counsel, for almost immediately the Sakharovs were allocated a three-room apartment on the outskirts of Moscow, causing Zeldovich to quip that this was the first use of thermonuclear energy for peaceful purposes.

On August 29, 1949, Kurchatov and Zeldovich, among others, had gathered for the test of the first Soviet atomic bomb, an operation code-named "First Lightning." The test was to take place on the vast and desolate steppes of Kazakhstan. Beria of course attended, both his presence and the words of one of his aides having made the terms of the situation abundantly clear: medals for success, bullets for failure.

Kurchatov, who had been given five years to build a bomb, had done the job in four, about the same time it had taken the Americans, though they had had no Klaus Fuchs supplying them with espionage, which, in Fuchs's own estimate, had saved the Soviets a year. Against the wishes of some scientists the American model had been copied almost exactly—the point was not to be original. The only question now was, Would it work? Beria tormented Kurchatov during the countdown: "Nothing will come of it, Igor." Angered yet confident, Kurchatov continued the count until, in the second after zero, First Lightning flashed across the steppes of Kazakhstan. Kurchatov's response could not have been terser: "It worked." Beria jumped for joy and kissed him on the forehead.

But unable to sustain elation and affection for more than a few seconds, Beria reverted to suspicion. He called another outpost where one of his men who had observed the American test at Bikini Atoll had just seen First Lightning. Was it the same as the American test? Beria asked, checking and double-checking to be sure that Kurchatov hadn't "bamboozled" them. "So I may report to Stalin that the experiment was a success? Good! Good!"

It was Beria's great moment. As head of the atomic bomb project, he could now personally bring Stalin an offering of power, thereby elevating his own. But calling Moscow on the direct line, Beria was informed that Stalin was asleep. "It's urgent, wake him up!"

Stalin's voice, always guttural and gruff, was angry when he finally came on the line: "What do you want? Why are you calling?"

"Everything went right," explained Beria, not so pleased with himself as to forget the need for oblique references.

"I already know," said Stalin, who hung up.

Beria flew into a rage, threatening to grind the person who had usurped his privilege into camp dust. But it was a rage fueled by its own impotence because whoever had phoned Stalin had hardly done so on his own initiative and, to add salt to the wound, that call had been made when Beria was on the phone to the observer, checking that Kurchatov hadn't pulled any fast ones. And in late October it was Beria who was assigned the task of drawing up a secret decree, signed by Stalin, awarding the highest Soviet civilian honor, the Hero of Socialist Labor, to Kurchatov, Zeldovich, and Khariton among others, along with cars, cash, dachas, and an array of other privileges. In December, in a letter to Stalin signed by Kurchatov, Khariton, and Zeldovich among others, Beria wrote: "We promise You, dear Comrade Stalin, that we will work with even greater energy and self-sacrifice on further developing the task assigned us and will devote all our forces and knowledge to justify, with honor, Your confidence."

Stalin too experienced some frustrations in connection with the test. He had wished to keep the results secret for a time, but his hand was forced by the president of the United States, Harry Truman. Though initially reluctant to believe that "those asiatics," those "pagan wolves" could have devised an atomic weapon, and so quickly, Truman was moved by the preponderance of scientific evidence to announce the Soviet test, and the end of America's atomic monopoly, on September 23, 1949. TASS, the official news agency of the Soviet Union, followed suit two days later.

Adding to the tension, on October 1, standing by the Gate of Heavenly Peace in Beijing, Mao Tse-tung proclaimed the founding of the Chinese People's Republic—the East was Red. It was also in October that Truman, spurred by the Soviet success, authorized an urgent project to build the Super, the hydrogen bomb—a weapon of whose theoretical existence he had been unaware until then. He made that decision public on January 31, 1950, just as Klaus Fuchs in London was confessing to atomic espionage. Within a week of Fuchs's arraignment Senator Joseph McCarthy of Wisconsin announced that he had a list of 205 communists who had infiltrated the State Department. All the pieces that would define the next fifty years were now in place: the logical insanity of the arms race, the ugly atmosphere of hostility, dread, paranoia.

Stalin celebrated his seventieth birthday on December 21, 1949, in grand style, his image projected onto the clouds above the Kremlin, suggesting both the benevolent father watching over his people and a being that was celestial, even immortal. But Stalin's vitality was on the wane. The Second World War had exhausted his resources. The Yugoslav writer and revolutionary Milovan Djilas, who had met Stalin at the end of the war had been surprised by Stalin's physical appearance, not only his "very small stature" (Stalin was five feet four inches) but his sparse hair, blackened teeth (he feared dentists), and unhealthy "Kremlin complexion," the result of late hours. "Not even his moustache was thick or firm," but still Djilas was struck by Stalin's "yellow eyes," which gleamed with a "mixture of sternness and roguishness."

Turning seventy, Stalin could look back on a mixed year. He had folded his hand in Berlin in January after the successful airlift, and in April, with the founding of NATO, the forces of the West were officially arrayed against him. Marshal Tito of Yugoslavia had broken with him in 1948, but Mao had taken China, an enormous propaganda victory for the cause of communism; yet China, though accepting Stalin's tutelage, would be less controllable than East Germany or Poland. Stalin did now possess the A-bomb but seemed to be of two minds about it: "A powerful thing, pow-erful!" and "You cannot win a war with the atom bomb." In some ways the atomic bomb had made Stalin even more cautious and conservative (it was only Khrushchev, the anti-Stalinist reformer, who brought the world close to the abyss). In early 1950 Stalin was still reluctant to aid North Korea in its mounting desire to attack the South, except to supply it with weapons and other military equipment. But on one subject there was no ambiguity, especially after Truman's announcement of the American H-bomb project: Stalin wanted thermonuclear weapons and wanted them fast. In early March 1950 the grace period of requests, invitations, and visits came to an abrupt end—Sakharov was ordered to move to the Installation "without delay" and without his family, who could join him only after a security check was run on Klava. This time their separation would not be bridged by any correspondence—phone calls, telegrams, or letters could not originate from a place that did not even exist. On March 17, 1950, accompanied by Yuri Romanov, the other student Tamm had summoned into his office that day in June 1948, Sakharov flew on a transport plane to Sarov, from which, according to a poem popular among the scientists, no one ever came back.

* * * *

In a nation of peasant huts and communal apartments, the scientists at Arzamas-16 were allocated single-family houses, such a rarity that an English word, "cottage," was used for them. Like everything else at the Installation, they were built by Zeks who, as a joke, as revenge, would sometimes leave hollow spaces in the walls so that the new inhabitants would learn what cold meant when winter came. Sakharov saw them everywhere—"long lines of men in quilted jackets, guard dogs at their heels." It was a sight that was so familiar as to become invisible after a time, but never quite entirely. Compassion for the "unfortunates," as prisoners have been called since Dostoevsky's time, is a Russian tradition, and the scientists always managed to slip the prisoners some clothes, a bit of food, a few rubles.

But there was no housing yet available for Sakharov and Romanov when they arrived, and for the next eight months they roomed together at the hotel, worked side by side during the day, and relaxed together in the evening. Even though Sakharov was now receiving the enormous salary of twenty thousand rubles a month, there was nothing to spend it on at the Installation, and so his principal diversion was watching Romanov take turns dancing with two of their very few women coworkers. Sakharov was necessarily a spectator, for he could still no more dance than swim.

Everything, both work and play, was enlivened by Tamm's arrival a few weeks later "squinting in the bright April sun" as he emerged from the plane, skis in hand, a knapsack on his back (no doubt containing the Agatha Christie novels he devoured). Tamm immediately organized hikes, tennis and chess matches, and volleyball games, hating to lose even more than he loved to win. He taught the other scientists to play Go and a simple but tricky mathematically based game, Taking the Stones.

But enlivened leisure was hardly Tamm's main contribution to Sakharov's life. It was a time of intense personal closeness, scientific collaboration, and spiritual influence. In Romanov's opinion, Tamm "valued and loved Sakharov more than he loved himself." With a daughter born the same year as Sakharov, Tamm had naturally assumed a fatherly, mentor relationship with the younger physicist but bridled when anyone referred to Sakharov as "Tamm's pupil," saying Sakharov was an "original" and no one's pupil. Sakharov himself, however, was not the least reluctant to acknowledge Tamm's influence: "Perhaps the great fortune of my early years was to have had my character molded by the Sakharov family, whose members embodied the generic virtues of the Russian intelligentsia . . . and to have come under the influence of Igor Tamm."

What Tamm taught by example was the one great strength from

which all others flow: independence of spirit. Tamm, who knew several foreign languages, listened to BBC broadcasts on his shortwave in the mornings, itself a risky enterprise, and then discussed the day's developments with Sakharov and Romanov at breakfast. But Tamm was savvy and Soviet enough to know that the most serious discussions were best reserved for the privacy of his room or, even better, for long walks in the most deserted parts of the vast surrounding woods, one section of which was reserved for other privacies as well and inevitably called "Lovers Lane."

Tamm and Sakharov discussed absolutely everything: the Terror, the Gulag, anti-Semitism, collectivization, communism real and ideal. Tamm activated a part of Sakharov's mind that had been dormant, but not entirely uninterested. "In those early years, Tamm's every word seemed a revelation to me—he already understood so many things I was just beginning to notice." Tamm was still a believer, never losing his faith in socialism, forever the fiery youth who had served on his local soviet. Sakharov was more a socialist by inclination than conviction, and his critique of society was still quite piecemeal, as he demonstrated when refusing the general's invitation to join the party, voicing the standard, isolated objections that would naturally occur to a critically minded but loyal member of the intelligentsia.

Tamm taught not only by example and intellectual dispute but by telling stories. One evening in 1937, at the height of the Terror, Tamm and one of his favorite students argued until dawn about the essential question of the time: were the waves of arrests severe revolutionary justice or a nightmare of criminal madness? The student's position was the watchword of the time: "The NKVD doesn't arrest people for no reason; I've done nothing anti-Soviet so they won't touch me." There could have been no argument that lasted until dawn unless Tamm had taken the opposing point of view, in terms much like those Sakharov would use later when wondering whether this was "Blindness? Hypocrisy? Self-deception as a means of surviving psychologically in an environment of universal terror? The sincere delusions of doomed fanatics?" In any case, Tamm's little teaching tale ended with the student's arrest later that very day; he soon proved one of those who did not survive the critical first years in the Gulag.

But even such important and fundamental questions were only a digression from the immediate, pressing problems of applied physics, though even there they sometimes looked beyond the scope of their own science. Tamm was certain that biology would surpass physics as the great science, making the new discoveries, changing the way people lived. In a gust of enthusiasm, Tamm even said that, if given the choice, he would have become a biologist. Sakharov never "took this literally, for fundamental physics was his true passion: it both tormented him and gave his life meaning."

In 1950 what tormented Tamm and gave his life meaning was the same problem that had robbed Sakharov of sleep when making his first trip to the Installation: a controlled thermonuclear reaction, as opposed to a thermonuclear weapon that was essentially uncontrolled. The idea was to create a fusion reactor in which a magnetic field contains plasma, an electrically charged gas whose natural tendency is to diffuse. The enormous energies of a fusion reactor could be used for peaceful purposes, and the by-products would be cleaner than the highly radioactive wastes of a fission reactor. This vision of clean nuclear energy was a natural outgrowth of their work on thermonuclear weapons, but one they could not give the time they would have wanted.

But it wasn't all work. The main holidays in the Soviet calendar were May Day, Revolution Day, and New Year's, all of which tended to stretch for the better part of a week. Vacation leave was one sign of seniority at the Installation. Tamm went home to Moscow for the May holidays and remained in Moscow until August. Sakharov stayed at the Installation and got arrested. He, Romanov, and another physicist went for a walk in the woods and were so engrossed in conversation that they failed to notice that they had strayed near the forbidden boundary zone. Spotted from a watchtower, they suddenly heard a harsh voice behind them: "Stop or I'll shoot!" They turned to see a squad of soldiers and an officer of the Border Guards with their submachine guns trained directly at them. There was no talk; the explanations could come later. A truck pulled up. They were ordered on and to sit with their legs stretched out in front of them. Any effort to escape, including pulling up their legs, and they would be shot without warning. Keeping their legs straight as the truck bounced back to the army camp was the hardest part, though keeping a straight face probably came a close second. At the camp they were interrogated, remembers Romanov, by "an extraordinary committee of state security. . . . They all probably knew us, but adopting a stern attitude, they asked each of us in turn our first name, patronymic, place of birth, and place of work." After half an hour, says Sakharov, "having satisfied themselves that we were not prisoners on the run, they graciously let us go." For the young scientists it was a holiday lark, and for the security people a snafu that could have cost them dearly. When one of them, Yuri Khabarov, was briefed on his new assignment to Arzamas-16, he was sternly warned: "If anything happens, we'll have your head."

Though Sakharov always found the extreme security measures "ridiculous," the incident had also been a reminder of the limits on their freedom. The barbed wire around Arzamas-16 confined them as well as the Zeks. The physicists were inclined to identify with the Zeks, joking that the posters exhorting the prisoners to WORK HARD FOR EARLY RELEASE applied to them as well. Psychological adjustment to the Installation was a function of how well a person could blank out the prisoners, the barbed wire, the armed guards, the informers, the visits from Beria and other security chiefs. Sakharov was rather adept at focusing his mind on the task at hand, whereas his colleague Lev Altshuler found that the security procedures were perniciously seeping into his unconscious: "The regime of secrecy had an oppressive effect. . . . It was not merely a regime, but a way of life, which defined people's behavior, thoughts, and spiritual condition. I was haunted by one and the same dream from which I would awaken in a cold sweat. I'd dream that I was walking down the street in Moscow with top secret documents and top secret special files in my briefcase. I'd be killed because I couldn't explain how I happened to have them."

Sakharov was also unable to leave the Installation and join his family to celebrate his twenty-ninth birthday on May 21, 1950. The near teetotaler Sakharov ended up celebrating his birthday by being initiated into the Russian tradition of drinking pure alcohol, flavored with honey, at the home of Kolya Dmitriev. Novitiates are inebriated within a matter of seconds. "We listened to music and had a wonderful conversation about the meaning of life and the future of mankind," was about as much as Sakharov could recall. The atmosphere was enlivened by Tamara, whom Kolya had only recently married, spending so much time in bed with her that Zeldovich complained that she was ruining Kolya's brilliant career. Kolya had passed all the university math exams when in high school and entered the university at fifteen, going right to work on the theory of probability. But like many prodigies, he was soon to know tragedy and dissipation. Four years later Tamara would leap to her death from a fifth-story window, and the child prodigy was suddenly a widower with children. His great talent came to very little, blunted by the Installation; he was, says Sakharov, a "master jeweler, not an assembly line worker."

Sakharov was still apart from Klava during the sweet days of the Russian spring, brief as youth, that end in the violent thunderstorms of June. Zeldovich had found the perfect formulation: "The only victims are the wives. Ours are essentially the wives of sailors." Not only did Zeldovich find the perfect formulation of the problem, he found the solution as well. One spring evening Sakharov was strolling the streets of Sarov, the moon causing the bell tower to cast a long shadow on the square in front of the hotel where he was living. Suddenly, Zeldovich appeared, balding, compact, lost in thought, his face "somehow radiant." Seeing Sakharov, he exclaimed: "Who would believe how much love lies hidden in this heart?"

Every last scrap of paper may have been top secret at the Installation, but there were no secrets there. Everyone knew that Zeldovich was involved in a love affair with a woman prisoner, Shiryaeva, an architect and artist by profession, the person who had painted the stars on the walls of the generals' mess that had caught Sakharov's attention on the day of his first visit to the Installation. Shiryaeva, outspoken in the wrong company and sentenced for anti-Soviet slander, had been promptly renounced by her husband. Now she was carrying Zeldovich's child. One of the short-term offenders, she was about to be "permanently resettled" to Magadan, a wilderness of slag in far eastern Russia. Zeldovich, who had been deluged with rubles when awarded the Hero of

Socialist Labor was, for some reason, short of cash in midsummer 1950. He woke Sakharov up in the middle of the night, asking to borrow money to give to Shiryaeva. Having just been paid, Sakharov gave him all he had. A few months later Zeldovich told him that Shiryaeva had given birth to their daughter in a building whose floor was "covered in ice an inch thick."

Even in the camps, nothing halts Eros. As one former Zek recalled with a certain nostalgia for the authenticity of existence in the Gulag: "Life didn't stop in the camps. On the contrary, life was more intense in the camps than anywhere else. And sex was too. . . . A girl of fourteen or fifteen would sleep with someone for his bread ration. . . . But it had to be done fairly quick—around the corner, behind the barracks, inside the barracks at night. . . . The toolmaker had a huge chest where he kept shovels and crowbars. He had made a false bottom for the chest, which had just enough room to crawl into during the lunch break. There were some grimy prison jackets for them to lie on. The toolmaker charged half a bread ration for that. But there were also romances on a very high level, that sometimes even led to suicide. But it would be disgusting to speak of them in the same breath."

* * * *

On June 26, 1950, North Korea attacked the South. Stalin was operating on two assumptions: victory would be swift, and Soviet involvement would be purely technical. Both were wrong. And he needed a victory. He had won the Second World War, consolidated the new Soviet Empire, and produced an atomic weapon, but he had also had a string of losses—the failed attempt to blockade Berlin, losing control of Yugoslavia, the Americans stealing his thunder by making the first announcement of the Soviet bomb. He was aware that his power as a leader flowed from his demonstrated superiority, not only at the late-night Kremlin banquets but on the world stage. The men around him, especially the ambitious Beria, could turn against him if they sensed he was losing his vitality, charisma, cunning.

In preparation for a new purge, new sins were found: kowtowing to the West, refusal to accept Lysenko's theory of heredity. These were almost identical since certain sciences such as Mendelian genetics and cybernetics, as computer science was known in its infancy, were considered to be merely superstructures that grew directly out of a capitalist base and were not therefore applicable to the Soviet experiment in social engineering. Some scientific positions were treasonous.

In mid-1950 a special security commission came to the Installation to check out the senior personnel. Sakharov was called in and given the Lysenko litmus test, which he failed handily by saying that he rather thought the chromosome theory of heredity to be "scientifically correct."

But the security people labeled as cranks or crackpots those scientists who, while appearing perfectly loyal, openly professed unorthodox opinions, the same words that American security used for their own scientists at Los Alamos. The security evaluations also took scientific importance and ethnicity into account. Lev Altshuler, already haunted by nightmares of being caught with top secret documents, was more volatile than Sakharov, calling Lysenko's theories "ravings." He was also less important and a Jew, of which the Installation had more than enough already.

Once again drawing Sakharov aside and speaking in a confidential hush, Zeldovich told Sakharov that he and his colleague Zababakhin should attempt to save Altshuler from dismissal. And now was the perfect time because KGB General Zavenyagin was at the Installation and had the power to overturn the decision. It seemed like such an obviously good idea that it took Sakharov quite some time to realize that Zeldovich had manipulated him into an action that, given Zeldovich's higher standing, he should have performed himself. At the least he could have offered to accompany them. Zeldovich may have been worried that he'd already cut things a little close with his love affair, or that since both he and Altshuler were Jews they would be accused of "sticking together"; or he might have just been afraid to face the general, who had a formidable reputation from his days running a construction complex in Norilsk in the permafrost of Siberia.

But Sakharov found Zavenyagin interesting and, in time, even likable. "Zavenyagin was a tough, decisive, exceptionally enterprising chief. He heeded the opinions of scientists and understood their role in the project. He made some attempt to study the concepts involved in our work, and from time to time would come forth with technical solutions, usually quite sensible. He was a man of great intelligence—and an uncompromising Stalinist. He had large, black, melancholy Asiatic eyes, a reminder of his Tartar ancestry. After Norilsk, he always felt cold, and he wore a fur coat draped over his shoulders even in a warm room. Surprisingly, given his background, there was a gentleness apparent in his relationships with people—I was later to find myself included in this select group." Sakharov was successful, the intelligence and sincerity of his argument causing Zavenyagin to overturn the decision concerning Altshuler: "You say he's done a lot for the Installation and he'll be useful in the future. Fine. We won't take official action now, but we'll watch how he behaves."

Altshuler would be free to remain at the Installation doing math, having nightmares.

The machinery of Soviet officialdom could work very swiftly if the right person was approached at the right time, as when Kurchatov had quickly secured the three-room apartment for Sakharov, or in the Altshuler incident, from which Sakharov had learned the exact weight of his own worth, the political physics of applied pressure. Otherwise, that machinery tended to move with agonizing slowness. Immediately upon his arrival at the Installation in mid-March Sakharov had submitted the paperwork for his family to join him, and five months later in August he had still received no definite word. Sakharov apparently made no further efforts to speed up the process, though he could have turned to Tamm or to Zeldovich who, however, tended to believe that young physicists got more work done when their young wives weren't around.

There probably was some limited contact between Sakharov and Klava during those months of separation. Tamm could have phoned her when he was in Moscow and brought back her news to him. Later on, when he was freer to travel, Sakharov offered to take letters from people at the Installation and mail them from Moscow; in the best Russian tradition, ways around the most stringent rules were quickly found.

Tamm returned in early August and resumed work with Sakharov on their parallel projects—the creation of a thermonuclear reaction and of a magnetically controlled thermonuclear reactor—either one of which would have been sufficient to absorb them utterly. But as Tamm was fond of saying, "Physics is what physicists do when the working day is over."

Though this was the "heroic" period of their labors, Sakharov could have easily reckoned the cost to him personally. The math was simple. His second daughter, Lyuba, had been born in late July 1949. He had left for the Installation in mid-March, when she was seven and a half months. It had now been something like six months since he'd seen her; he had missed the first steps, the first words.

Finally, in late October, Sakharov was allowed to return to Moscow, all the paperwork done including the extensive background check on Klava's family, which must have amused and disgusted her father, Alexei. On November 9, carrying babies, bedding, and baggage, Andrei and Klava boarded a plane with folding metal chairs. "We took off and after some time (even children were trained not to tell anyone in Moscow how long we were in the air) began our descent. The Installation's two barbed wire fences and the guard towers flashed past under the wings. We were home."

* * * *

"You're blind like newborn kittens," said Stalin to his entourage one late Kremlin night. "What will happen without me?" The question of a successor was no longer unthinkable. Stalin was thinking of it himself. But the real question preying on his mind is what his cohorts might do to hasten the inevitable. He began spending more time at his dacha, surrounded by new walls, more guards, locks multiplying everywhere like iron mushrooms. And even at the Kremlin banquets he watched to see if everyone had tasted their food before he ate sparingly himself. The threat would most likely come from Beria, the most likely weapon poison from his pharmacology. But Beria would not act alone; he would first make alliances. Common sense dictated that it was inefficient to try to guess who might side with whom—better simply to rid himself of them all. His long-standing principle was "No one is indispensable."

Stalin decided to play the anti-Semitism card in his last hand. The fact that he himself was no great friend of the Jews had little if anything to do with it. Khrushchev, who was a member of Stalin's inner circle, observed: "When he happened to talk about a Jew, Stalin often imitated in a well-known, exaggerated accent the way Jews talk. This is the same way that thick-headed backward people who despise Jews talk when they mock the negative Jewish traits. Stalin also liked to put on this accent, and he was pretty good at it." But unlike Hitler, who was fixated on Jews, Stalin also loved to hear Muslims mocked, especially if there was one present, and during the war had exiled entire Muslim nations—the Ingush and the Chechens. Except for a respite during the war the Russian Orthodox Church had suffered greatly. The bloodletting in Stalin's native Georgia had been particularly intense, and of course no one killed more communists than he. In his hatreds Stalin played no favorites.

Initially, the anti-Semitic campaign had simply been one element of the larger campaign against "kowtowing to the West." The Soviet Union had been among the first countries to recognize the state of Israel in 1948—to drive a wedge into British imperialism—and one of the first to fear that Jewish citizens could have a divided loyalty. This "danger" became particularly apparent in October 1948 when Golda Meir was mobbed by thirty thousand Jews at a Moscow synagogue. But the anti-Semitic campaign had already been launched by that time, beginning with the assassination of the famous Yiddish actor Solomon Mikhoels, who had been the head of the Jewish Antifascist Committee during the war. Stalin tested his foreign minister, Molotov, by ordering the arrest of his Jewish wife, who had also served on the Antifascist Committee, and was furious when Molotov displayed unexpected Politburo gallantry by abstaining in the vote to sanction the arrest. It was not only Jews who had divided loyalties.

Officially, anti-Semitism was a punishable crime in the USSR, and so, officially, the campaign had to be against "rootless cosmopolitans." No one was in the least deceived. Minister of Medium Machine Building Vannikov, himself a Jew, liked to recite the popular ditty:

So no one will think you're an anti-Semite,
Use the word cosmopolitan instead of kike.

Stalin had loosed anti-Semitism in Russia knowing that it would be used by some to advance themselves and by others out of pure animosity untainted by any self-interest. At some later point he would turn that force against all his closest associates, as he already had against Molotov.

It was not long before his lightning struck the Installation. Mattes Agrest, the head of a mathematics group, was a deeply religious Jew. The philo-Semitic Sakharov liked visiting Agrest and his family of eight, from a six-month-old son to his seventy-two-year-old father, who reminded Sakharov of the "Jews in Rembrandt's paintings."

Agrest was useful, and until 1951 his Judaism had been overlooked. But the security organs, very active in the campaign against the "rootless cosmopolitans," were now combing through the files of everyone connected with the nuclear weapons program, no place for treasonous Zionists. It was no secret that there were already too many Jews working there—Khariton, who ran the Installation, Hero of Socialist labor Zeldovich, the troublesome Altshuler, even Vannikov, head of the ministry, but he at least had the political sense to be publicly anti-Semitic.

Agrest had already caused a scandal at the Installation by having his newborn son circumcised, the bris performed by his own father, the Rembrandtesque Jew. A circumcision at the Installation! More than enough, it wasn't all. Running a deep background check on Agrest, a security official discovered that before beginning his scientific career Agrest had studied Talmud and Torah at a Yeshiva and, though he had never made use of his diploma, had in fact been ordained as a rabbi.

A rabbi at the Installation having his son circumcised during Stalin's anti-Semitic campaign! A security nightmare. Heads would roll if this monstrous anomaly were not dealt with at once. On January 13, 1951, Agrest was summoned and informed he had twenty-four hours to vacate the Installation.

All the scientists learned immediately of Agrest's dismissal, which was officially for having relatives in Israel, a reason that

Sakharov and the others accepted as "valid grounds." The incident was another gauge of loyalty.

In his conversations with Sakharov, Tamm had made the point that there was "one foolproof way of telling if someone belongs to the Russian intelligentsia. A true member of the intelligentsia is never an anti-Semite. If he's infected with that virus, then he's something else, something terrible and dangerous." Tamm acted on his own principles, protesting the summariness of Agrest's dismissal and winning him a week's time. When the day came, Tamm made a point of announcing that he was leaving work early to help Agrest pack.

Sakharov made two moves of his own. In a situation where words were actions, he walked over to Agrest, shook his hand sympathetically, and said: "How will we manage without you?" More substantial and daring, Sakharov offered Agrest the use of his Moscow apartment, quoting the Russian proverb: "We don't know what we've got till it's gone."

It was two of the Jews, Zeldovich, the head of Agrest's group, and Khariton, the head of the Installation, who struck Agrest as the most demonstratively indifferent to his predicament. When Agrest brought Zeldovich two chairs as a parting gift, Zeldovich did not respond with a single word except to say, when Agrest was already at the door: "Don't bring up your children religious."

* * * *

Shortly after the Agrest incident, Sakharov was summoned by Beria. He had been to Kremlin Office No. 13 before as part of a large group, but this was the first time he had been called in alone. Sakharov did not have the slightest misgivings about the meeting, which he knew would be purely technical, and was at his ease from start to finish.

His pince-nez glinting, a light-colored raincloak draped over his shoulders, Beria was seated at the head of a conference table. He welcomed Sakharov in a manner that "verged on the ingratiating." Still, as if in reminder of who he was, Beria was flanked by his aide, a former Gulag executive. A fourth man was also present, Lavrentiev, a sailor who had sent Stalin a letter with an idea concerning magnetic thermonuclear reactions, and had been plucked from obscurity. It would have pleased Lavrenty Beria's vanity to partake in the discovery of a major scientist, especially since the sailor's name, Lavrentiev, meant "Lavrenty's." Sakharov was there to deliver his opinion, which was that the idea, though respectable, was unworkable. The sailor was dismissed, whisked off the stage of history, though he did go on to more of a midlevel scientific career than might otherwise have occurred. That left Sakharov alone with Beria. Whatever small talk or shoptalk followed did not last long, because Beria, a man with a rather full schedule himself, showed scientists respect by respecting their time. But, after rising to signal the meeting was over, Beria suddenly added: "Is there anything you want to ask me?" He, Beria, had availed himself of the young man's knowledge; perhaps the young man would want to avail himself of his.

Unhesitatingly Sakharov asked: "Why are our new projects moving so slowly? Why do we always lag behind the USA and other countries, why are we losing the technology race?" While the questions were definitely pointed, even more pointed than Beria had expected, they were entirely loyal in their phraseology: why are "our" projects lagging, why are "we" losing?

"Because," said Beria, "we lack R and D and a manufacturing base. Everything relies on a single supplier. . . . The Americans have hundreds of companies with large manufacturing facilities." At the time this nuts-and-bolts answer struck Sakharov as insufficient, though he himself couldn't quite say why. But Beria, pleased with himself and the meeting, offered Sakharov his hand, causing what was hardly among Beria's specialties—the illumination of consciousness. The feel of Beria's hand—"plump, slightly moist and deathly cold"—focused into an insight only that evening, when Sakharov told his parents the story: "their fear made me conscious" of having been "face to face with a terrifying human being."

Though that handshake left a vivid sensory imprint on Sakharov, it was soon forgotten in the pressures of the Installation, where he was the first at work and the last to leave. The heat was greatest on the theoretical groups in early 1951 because the main theoretical problems had yet to be resolved. The workload was enormous: "The theoretical groups . . . established the main directions of research, studied the explosion process, investigated possible variants of the devices, and supervised detailed calculations of the explosion parameters for selected models. The actual numerical calculations were performed by secret mathematical teams in several scientific research institutes in Moscow. The theoretical groups were also responsible in most instances for assigning tasks to other departments at the Installation and affiliated organizations, coordinating their work, and analyzing their results."

Sakharov took a particularly active role in coordinating the work of the theorists and the experimentalists, one of whom bestowed a nickname on him that he found rather flattering—"the Martian." Sakharov enjoyed escaping from thought, diagrams, calculations, equations chalked on blackboards to their "world apart. . . . high voltage equipment, flickering lights on the decoders, and metal gleaming mysteriously with a play of violet color." He visited the different experimental groups about once a month, and it was one of his pleasures to sit and discuss the results of the experiments with them in a "calm, friendly, unhurried manner until nine or so in the evening."

Days were long, hours late. Sakharov's elder daughter, Tanya, remembers her father coming home too tired to notice how nicely Klava had cleaned the house or to be much interested in the little things of the day—Tanya's school, Lyuba's colds. He never scolded, never praised, and his daughters wanted both as proof of love. After being supercharged all day, his mind did not come to a halt just because he had crossed the doorstep of his home. Sometimes, when telling Tanya a bedtime story, Sakharov would pause, then vanish into thought for several minutes as his daughter stared at him with anxiety and annoyance. He could be an absent father without leaving the room.

But sometimes he could shake off the weariness and excitement and fully engage with Klava and his family. There were picnics, hikes, skiing trips, Klava emerging from the woods, laughing as she packed a snowball. Klava soon lost interest in skiing, but Sakharov would take Tanya instead, who loved it when he was at home and even more when he spent time with her. On May 21, 1951, Klava gave a party for Andrei's thirtieth birthday, serving a cake with candles, an innovation in Russia, their cottage packed with guests who remembered the evening as one of "wit, spirit, and wine."

Those, however, tended to be the exceptional moments. Neither Andrei nor Klava was outgoing; neither had the knack or taste for making a home that was often visited or for being sought-after guests. Some of the other physicists found Sakharov sweet but dull, though those who connected with him discovered he could be sparklingly witty. Others, like Agrest, while liking Klava, found her a "simple countrywoman." It was a milder version of that first encounter with her mother-in-law, not good enough, not true intelligentsia.

Klava began to feel quite isolated at the Installation. And since she had no outlet for her mental energies, her chess games with Andrei assumed special significance. She'd win as often as she'd lose. Even though she was now in Soviet terms a rather wealthy woman, she did not feel comfortable having servants doing the housework; that was her job. And even having money was not enough in that society, where goods and services had to be hunted down, strings pulled. Part of the problem was the Soviet distribution system through which goods moved intermittently, jam-ups causing months of shortages abruptly broken by outpourings that had all the providential suddenness of grace. Another part of the problem was Andrei and Klava's ineptitude at manipulating that system. Tanya was fourteen by the time they were finally able to obtain a refrigerator from Moscow; until then they had kept their food on the windowsills, and Klava had to shop every day. She did catch on fairly quickly to the fact that she could simply bring prepared food home from the "generals' mess," a ten-minute walk from their cottage. But they were innocents when it came to securing good medical care for their frequently ill children and for themselves, Andrei still suffering from stomach problems dating back to his bout of dysentery in Ashkhabad and Klava from the poisoning at the munitions factory, which still caused her to collapse in pain. Klava was even more reluctant than Andrei to avail herself of the privileges that came with their wealth and position. When he gave her money to buy a fur coat, even he was disappointed that she could not help but choose the plainest and cheapest model. "Klava, you had a lot of money, you could have bought a beautiful fur coat!" he exclaimed in amused indignation. "This kind suits me fine," she replied. But he was no different. When out of embarrassment for him his colleagues arranged to have his cheap, shabby overcoat disappear, he went and bought another one just like it.

Unlike Andrei, Klava was highly susceptible to the intrusions of security into her life and her home. He was either oblivious to them or amused by all their furtive scurryings. But Klava, whose mind was occupied with little besides domestic tasks and chess matches, easily fixated on a cigarette butt left by security men in their house, accident or insolence, hard to say. Such incidents only made her feel more isolated at the Installation, itself isolated by barbed wire, secrecy, and woods so dense that coming home from visiting neighbors they would sometimes hear wolves howling in the night.

* * * *

"I'm finished. I trust no one, not even myself," mumbled Stalin on the porch of his summer house, failing to notice that Khrushchev and long-time Politburo member Mikoyan were standing close enough to overhear him. Or maybe he had noticed them and was just testing them, as he sometimes did at the late-night banquets when he would complain that he was growing old and watch to see how vociferously his cohorts would object. He tested them in other ways, cajoling Khrushchev to do Ukrainian folk dances to the tune of a scratchy Victrola or having pastries slipped onto Mikoyan's chair while he was standing for a toast, always good for a laugh. But things weren't going well. After a good initial start, the Korean War had become dangerously complicated, the Chinese entering the war in November 1950. Truman had threatened to use the atomic bomb, Red hatred running especially high in the United States when the Rosenbergs were found guilty of atomic espionage in March 1951. In their progress toward a nuclear weapon, the Americans continued to test new weapons, a large fission device having been exploded at Eniwetok in May 1951. And the Soviet nuclear bomb was still a year or more away. By the summer of 1952 Stalin was losing patience. Beria was dispatched to the Installation and warned a group of scientists that included Sakharov of the price of failure: "we have plenty of room in our prisons."

Winter came early to Russia that year; heavy snow fell in the first days of November. On November 1, 1952, a world away in an atoll of red-orange coral and blue water where sharks meandered, the United States exploded the first hydrogen-based thermonuclear device, Mike-1, whose immensity (twenty feet high, weighing 164,000 pounds) made it undeliverable as a weapon. The Russians were caught off guard by the explosion, larger, earlier, and more advanced than anticipated, which only ratcheted up the tension at the Installation.

Contrary to long-held American opinion, the Soviets had no technology able to analyze the fallout and on that basis reconstruct the firing mechanism. Not that Sakharov didn't try. Learning of the American test, he went out to collect some of the fresh snow, having calculated that by then radioactive traces would have been borne by high-altitude winds from the Pacific to the woods of Arzamas-16. He brought the snow back to the lab, where a chemist began the process of concentrating the radioactive fallout, in the process causing a little tragicomedy of Cold War science. Distraught over personal problems, utterly distracted, the chemist poured Sakharov's radioactive snow concentrate down the drain. Now all that mattered was ensuring that security not learn of this treasonous lapse. Beria's recent threats of prison, the surprise American test, the hard fact that it would be at least a year before the Soviets could test a nuclear weapon of their own, meant that it was no time to be caught pouring top-secret state property down the drain.

At the same time that Sakharov was gathering snow, Stalin was performing the first of the two highly enigmatic acts he would commit that winter. Stalin dismissed his longtime right-hand man Poskryobyshev, so trusted that Stalin even gave him blanks with his signature and who, according to Churchill's interpreter, was about "five feet tall, with broad shoulders, a bent back, large head, heavy jowls, long hooked nose and eyes like those of a bird of prey." Stalin was weakening, suffering dizzy spells and memory lapses—once even forgetting his defense minister's name—and so concerned with his mortality that he even gave up smoking. His judgment came into question—later that winter some of his close associates were shocked that Stalin would seriously consider even for a moment a plan to have a Costa Rican assassinate Tito with a jewel box that, when opened, released a lethal poison gas.

The second enigmatic act of that winter was the dismissal and arrest of General Vlasik, long the leader of the guards responsible for Stalin's personal safety. Vlasik quickly succumbed to the pressures of torture and signed the required confession.

But in both cases the signal was the same: Stalin no longer trusted the people closest to him. The signal was meant for those in his immediate entourage whom he suspected of plotting against him. Beria and Khrushchev were the most ambitious, and the dapper Armenian Mikoyan, of whom it was said that he needed no umbrella because he could dodge the raindrops, would line up with whatever side he thought had the better chance of winning. But Stalin still had control of the political machine, and on January 13, 1953, he pulled one of its levers—TASS announced that a cabal of Jewish doctors had been murdering Soviet leaders by medical means. That the Kremlin hospital could have been so successfully infiltrated constituted a grievous lapse of security—at best, Beria had failed woefully.

Panic radiated from Moscow across the country's eleven time zones—if today Jews were murdering Russians in the Kremlin, then absolutely anything could happen tomorrow—and quickly reached the Installation. Even though its actual effects were "muted" there, the atmosphere was much the same: "Passions grew more frenzied with each passing day, and people began to fear that pogroms were in the offing."

The first two months of 1953 are the murkiest time in all Soviet Russia's never overly transparent history. In his *Enigma of Stalin's Death,* the Chechen author Abdurakhman Avtorkhanov makes the case that Beria and Khrushchev were engaged in a conspiracy to kill Stalin before he began purging the ranks of his entourage. The operation was supposedly code-named "Mozart," an allusion to Pushkin's play *Mozart and Salieri;* as Salieri poisoned Mozart out of envy for his natural genius, Beria would poison the Mozart of dictatorship. Though Beria did have means and motive, all that can be really shown is that Beria, like Sakharov, could quote Pushkin for his purposes.

Still, when Stalin was found unconscious in his dacha, soaked in his own urine, Beria was unable to conceal his jubilation except when Stalin's eyes opened and Beria fell to his knees, kissing the leader's hand. For the first ten to twelve hours Beria prevented any medical attention being given to Stalin in the hope that nature or poison would take its course. Then every effort, from inter-muscular injections to leeches, was made to save Stalin. But, as bad luck would have it, the Kremlin's best doctors were under arrest.

Not that they could have probably saved him either. By hook or by crook, death had come for Joseph Stalin. And no sooner did it arrive than Beria departed at a run, calling out for his car and racing to the Kremlin safe that contained Stalin's personal papers, which may have included a will that named his successor, but Beria quickly reduced all those documents to atoms.

The official day of Stalin's death, March 5, was followed by four days of mourning; on the last of them, at noon, the entire Soviet Union came to a standstill, and factory sirens wailed for three minutes in tribute to the leader who had industrialized the wooden Russia of the peasants. The only break in the solemnity came when hundreds were trampled to death in the stampede to see Stalin lying in state.

All Russians remembered where they were when they heard of Stalin's death, and what they felt. Most reacted with sorrow, some with fear for the future, and a few with a private, vindictive glee. Tamm, one of those worried about the chaos Stalin's death might plunge the country into, brought his wife out to the Installation, which he thought a safer place than Moscow. Sakharov did not bring or summon Klava. What he did do was write her a letter, saying: "I am under the influence of a great man's death. I am thinking of his humanity." Containing genuine awe at "death's universal dominion," Sakharov's sentiments are a mix of Russian solemnity and Soviet emotional kitsch. Later, recalling these words with shame, Sakharov was able to analyze his reaction dispassionately. He found it to be an amalgam of several elements: the lack of a "big picture"; the belief that historical upheavals are impossible without suffering, which was the message of Soviet propaganda but also that of Russian history itself; the power of collective emotion on a thirty-one-year-old who was still more "impressionable" than he cared to admit.

If Sakharov's emotions were distorted by the "hypnotic power of mass ideology," his intellectual reaction to Stalin's death as a political event was clear and logical. To the question on everyone's lips of what would happen now, Sakharov replied: "Nothing, of course. Everything will go the same way. A complex system is governed by its own internal laws and supports itself." Sakharov's abstract analysis concealed an emotional justification. "I needed, as anyone might in my circumstances, to create an illusory world, to justify myself." In laboring to achieve nuclear parity, Sakharov believed he was ensuring the safety of his country and the world, doing the duty of the intelligentsia, actively pursuing the good as he saw it. "I still believed that the Soviet state represented a breakthrough into the future."

The Zeks at the Installation were cheering. Zeks might rise up and slaughter their guards, but they never cheered. Only three weeks after Stalin's death, in late March 1953, an amnesty had been proclaimed, Beria's doing, though it was limited to criminals, not political prisoners. A short while later, on April 4, Tamm came running to Sakharov's office panting and shouting from the doorway: "They've freed the doctors! Have we really lived to see it? Have we really lived to see this moment?"

Tamm, as usual, was the first to know, having picked up the news on his shortwave. But it was all there in black and white in *Pravda* and *Izvestiya* a few hours later in the form of a communiqué from Beria's Ministry of Internal Affairs: "The accused were all arrested without any legal basis. They have been completely exonerated and have been freed from imprisonment. The persons accused of improper methods of investigation have been arrested and charged with criminal responsibility." Zeldovich too was "jubilant," though he put his own spin on events, saying: "And it was our own Lavrenty Pavlovich who brought all this to light." It never hurt to speak well of your superior, but the last sentence of the communiqué indicated both a purge of the police and a political power struggle of which that purge was only a reflection.

Georgy Malenkov was officially the new leader. Assessments of him range greatly, from utterly dismissive—"a man without a biography. . . . He had no image of his own, nor even his own style. He was an instrument of Stalin pure and simple"—to laudatory: "a man of formidable intelligence, ability, toughness, and ambition." The former evaluation seems the more likely, since Malenkov was quickly chewed up in the power struggle—which was between Beria and Khrushchev. Beria had no aspiration to replace Stalin as leader, certain that the Russian people would not accept two Georgians in a row, but he was moving to shore up his own power. The writer Konstantin Simonov observed Beria and Khrushchev together: "Beria underestimated Khrushchev, his qualities—his deeply natural, pure masculinity, his tenacious cunning, his common sense and his strength of character. Beria, on the contrary, considered Khrushchev a round-headed fool, whom Beria, the master of intrigue, could wrap around his finger."

Khrushchev was already moving against him: "We're heading for disaster. Beria is sharpening his knives." But, ironically, it was Beria's liberal policies toward East Germany, which led to a workers' revolt in June 1953 and had to be crushed by Soviet tanks, that won Khrushchev the support of the military and the hardliners. Suspecting nothing, casually dressed, Beria was arrested at a Politburo meeting on June 26 by Marshal Zhukov, whose soldiers had entered Moscow in case KGB troops moved to defend their leader.

On July 10 at the Installation, Sakharov noticed that the signs on Beria Street had been replaced with cardboard ones reading "Kruglov Street." Official word of Beria's arrest and his replacement by Kruglov came a few hours later on the radio. The municipal department at the Installation, in charge of street signs, had been informed of Beria's fall even before the physicists, who had to wait like the rest of the nation.

Near the end of July Sakharov was invited to a meeting of the City Party Committee; although he was not a party member, his standing at the Installation warranted his inclusion. Bound in a "blood-red" cover, a long letter detailing Beria's crimes was distributed and read. In keeping with tradition, a few of those crimes may have been political falsifications, but the preponderance of them were all too real—rape, torture, murder. Worst of all, from the party's point of view, was the execution of party members and heroes of the Russian Revolution, the lists signed by Beria and Stalin, this first linking of their names in criminal complicity causing, as Sakharov was told, "a great moan" to arise from a party meeting at one factory, and prefiguring denunciations to come.

Beria was held in an underground military bunker beneath an apple orchard by the Moscow River until December when, begging for his life, he was executed, his body supposedly burned on the spot. The deaths of Stalin and Beria were doubly fortunate for Sakharov. In the long run it would spare him the burden of knowing that he had bestowed nuclear weapons on such men, and in the short run the new atmosphere proved that the system he still considered a breakthrough to the future was not only self-perpetuating but capable of self-correction. Jubilation was in the air in the spring and summer of 1953. Tamm, ever the optimistic socialist, was particularly sanguine; even Sakharov felt that a "new era seemed to be dawning."

But there was barely a second's time to savor hope—the test of the first thermonuclear weapon was only weeks away, the scheduled date August 12, 1953. Sakharov, now considered too valuable to fly, made the long journey to the test site in Kazakhstan in Khariton's private railway car. Beria's replacement was issuing no threats of prison. If the physicists failed, the bitterness would be pure.

But, arriving at the test site, Sakharov discovered that a failure had already occurred, one of foresight. Victor Gavrilov, who had a taste for the dramatic—it was he who had confided the story of the Zek revolt to Sakharov on his first visit to the Installation—now alerted everyone to the most obvious thing of all: "We had all been so busy preparing the device, organizing the test, and performing calculations that we had simply lost sight of the fallout problem."

Not blinded by monomaniacal devotion and ambition, Gavrilov was, for his troubles, promptly nicknamed the "Evil Genius" by the new head of security. Several teams went immediately to work on the "fallout problem," aided by the "Black Book," an American manual on the effects of nuclear explosions. The numbers were run and pointed clearly to only two choices: cancel the test and reduce the fallout by switching from a tower-dropped bomb to one dropped by plane, a process that would require a delay of at least six months; or immediately evacuate the tens of thousands of Kazakhs who were downwind from the test.

Further calculations indicated that in a mass evacuation of that magnitude some twenty to thirty people could be expected to lose their lives. Glimpsing himself in a mirror, Sakharov was "struck by the change—I looked old and gray."

The test would take place; the Kazakhs would be moved.

On August 5, in the midst of the evacuation, before the device had even been rigged on the tower, Malenkov, in his brief stint as Soviet leader, announced welcome news—a whole series of liberal reforms—and some exceedingly unwelcome news for Sakharov and his colleagues: he declared that the Soviet Union had already successfully tested a thermonuclear weapon. "Malenkov's remarks would have raised the level of tension if we had not already been keyed up to the maximum."

The Kazakhs finally evacuated, the device was installed on the tower in the midst of the sandy, stony steppe where only wormwood and feather grass grew, and in whose distances vast lakes and mountains shimmered, mirages.

Alarm bells rang at 4:00 A.M. on August 12. It was still dark, but Sakharov could see the "headlights of trucks sweeping across the horizon." The day had come.

* * * *

Though it was his idea that was being tested, that idea had not proved successful yet, and so Sakharov was placed in one of the more "junior" observation points located some twenty miles from ground zero. Tamm, the senior man, was invited to join the VIPs. But Sakharov felt no hesitation in walking over to their post to confer with Tamm and noticed that the "chiefs were just as nervous as we were," cracking bawdy jokes to break the tension. Sakharov was back at his own post when the loudspeakers began the countdown at ten minutes to go. "Two minutes to go" meant it was time to put on their dark goggles.

"One minute." In sixty seconds he would know how good he was.

"We saw a flash, and then a swiftly expanding white ball lit up the whole horizon. I tore off my goggles, and though I was partially blinded by the glare, I could see a stupendous cloud trailing streamers of purple dust. The cloud turned gray, quickly separated from the ground and swirled upward, shimmering with gleams of orange. The customary mushroom cloud gradually formed, but the stem connecting it to the ground was much thicker than those shown in photographs of fission explosions. . . . The shock wave blasted my ears and struck a sharp blow to my entire body, then there was a prolonged, ominous rumble that died slowly away after thirty seconds or so. Within minutes, the cloud, which now filled half the sky, turned a sinister blue-back color."

O Pushkin, you son of a bitch you!

That release of energy transformed Andrei Sakharov's life. Now, when he returned to the VIP observation post, it was as the hero of the day. Malenkov called to congratulate everyone, Sakharov in particular. Bowing to Sakharov, Kurchatov said: "It's thanks to you, the savior of Russia!"

Then a crew of them went out in open cars to see exactly what they had wrought. An eagle with badly singed wings was put out of its misery by an officer with one "well-aimed kick." At two hundred feet from ground zero the convoy came to a halt, and only two men, Sakharov being one, got out to walk across a "fused black crust that crunched underfoot like glass toward the concrete supports with a broken steel girder protruding from one of them—all that was left of the tower."

Then the honors came cascading. On October 22, 1953, Andrei Sakharov was nominated for full membership in the Academy of Sciences, bypassing the usual period of candidacy (Tamm's had lasted twenty years), and elected unanimously. At thirty-two he was the youngest Academician ever inducted. Unlike Western scientific academies, mostly honorific bodies, the Soviet Academy of Sciences was both an important part of the country's administrative structure and did the best science in the country, Tamm's group being part of FIAN, the Physics Institute of the Academy of

Sciences. Sakharov was also awarded the highest Soviet civilian distinction, the Hero of Socialist Labor, and the Stalin Prize, which came with a staggering five hundred thousand rubles. The medals were pinned on Sakharov's chest at Kremlin receptions.

But in November Sakharov fell ill with a fever of 106, delirium, severe nosebleeds. The official diagnosis was acute tonsillitis, but Sakharov had to wonder if he hadn't received an overdose of radiation when strolling "nonchalantly" around ground zero. Before the onset of the fever, Sakharov had been summoned to Moscow and asked to write a report on his conception of the next generation of nuclear weapons and delivery systems. Sakharov knew that he should have simply replied that he had to wait until his colleagues were back from vacation; matters of such importance should not be based on one person's views. But in fact he wrote up the report on the spot. Two weeks later Sakharov attended a meeting of the Politburo, which approved the idea for the device he had so "incautiously" proposed, setting a vast enterprise into motion.

Why had he taken such an uncharacteristic act? The success of the test had put his "self-confidence at a peak"; having the energy that drove the suns within his grasp had infused him with "euphoria." Still "outwardly modest," inwardly he was "actually quite the opposite." His was a pride that singed the wings of eagles.

CHAPTER EIGHT

COMPLICITIES

AT a picnic in Kazakhstan after the test of the hydrogen bomb, Tamm, taking advantage of the flush of success and the looser times, asked the new Minister of Medium Machine Building to be transferred back to Moscow. The idea for the bomb was Sakharov's; young, energetic, devoted, he was now the better man to run the show. The minister agreed; Sakharov accepted.

Heading the theoretical department would entail a new welter of functions and responsibilities, but it would also mean that Sakharov was replacing his mentor, a man old enough to be his father and with whom the original connection had come from his own father. It was the end of a long sonhood.

Tamm was quickly proven right. Sakharov became an outstanding leader, even "ideal" as one subordinate put it. Called half affectionately, half formally A.D.S. by his colleagues, Sakharov could not invigorate work and leisure like Tamm but was able to create a unique atmosphere of exacting rigor and relaxed liberty. He could pass from the utterly abstracted to the immediately available in the blink of an cye. His benevolent smile, his mild manner, and his faultless manners put people at their ease and allowed them to be as objective as he was about the problem at hand. He was so willing to entertain any idea that his sudden implacable certainty could come as a surprise, though its impact was softened by a smile—but a different one, amused, dismissive—and the words "Yes, perhaps you are right."

His easy, immense self-confidence was both his principal strength and failing, as Khariton, the director of the Installation, noticed: Sakharov's weaknesses as a scientist "if there were any—sprang from his strength. He felt his own strength and could not imagine anyone understanding better than he. One time one of our colleagues found the solution to a gas dynamics problem which Andrei Dmitrievich was unable to. This was so unexpected and so unusual that he set about with exceptional energy to search for flaws in the proposed solution. And it was only some time later that, not finding them, he was forced to admit the solution was correct."

Sakharov had great pride but no vanity. And so he was particularly adept at reconciling squabbles in which each person considered himself the injured party. His moral authority was, says Romanov, "indisputable." It derived from his humane generosity—he was liberal in granting passes to workers with a good reason to leave the Installation—from his innate sense of justice, and from his odd "Martian" alienation from pettiness.

A force for harmony with his colleagues, Sakharov soon became a source of trouble for officialdom. The first discord arose from what one colleague called the "magnificent duet" of Sakharov and Zeldovich. Sakharov could lose himself as utterly in talking science with Zeldovich as he had in fantasy in his courtyard days; Zeldovich was the Grisha Umansky of the Installation. The younger scientists delighted in the differences between the two men. Zeldovich might come to work with his Hero of Socialist Labor medal pinned to the jacket of a good suit; Sakharov might arrive wearing two different shoes and not be particularly interested when the disparity was pointed out to him. Zeldovich drove to work himself in his sporty white Volga, swooping to a stop and jauntily backing into his space. His entrance was equally boisterous, Zeldovich firing off orders, jokes, and greetings as he strode to his office. Sakharov would dismiss his driver at a distance from the building and cover the rest of the way himself, lost in thought by the time he was walking quietly down the corridor, pausing every few steps to touch the wall, as if marking off lengths.

Zeldovich and Sakharov's offices were adjacent and identical: each was divided into areas for work and relaxation, each had the two attributes of highest importance—an electric calculator and a massive safe—though only Sakharov had a high-frequency phone, a direct line to all leaders. Zeldovich attended his younger colleagues' parties, downing the hot new drink, the Bloody Mary. Everyone knew Zeldovich; very few knew Sakharov by sight. He was semilegendary, being known only for his achievements and a few eccentricities—the mismatched shoes or his habit of wearing galoshes in any weather. But he had a perfectly logical reason for that: he was susceptible to colds, especially if his feet got wet—why take a chance when slipping on a pair of galoshes was all that was entailed? That habit also came to the attention of S. A. Akhtyamov, the head of the guards in his sector: "Andrei Dmitrievich could show up wearing galoshes even in good weather. . . . He walks up to the sentry booth, stops, looks. . . . I can see he's already thinking about something. He's not even in his office yet and he's already working! We made efforts so that the soldiers would recognize people like Sakharov by sight. . . . Sometimes it happened that instead of his pass Andrei Dmitrievich would pull out some piece of paper or a notepad, but the guard would immediately say: 'Please come in!' "

Zeldovich was always quick to grant the greater talent to Sakharov, the natural, the Mozart. But difference made their duet dynamic, their hours together the best of the day. Sakharov had devised the "First Idea," which had been significantly supplemented by Vitaly Ginzburg's "Second Idea"—that lithium deuteride in the form of a solid chemical compound be used as the thermonuclear fuel. Now, Sakharov and Zeldovich had co-sired a "Third"—"radiation implosion," using the x rays generated by the fission of an atomic bomb to compress the thermonuclear fuel and yield more fusion. A better idea, it spelled immediate trouble: now they would have to convince the ministry that the design Sakharov had so "incautiously" proposed at the height of his hubris had been greatly superseded by this new insight, that Soviet science had its own logic of development that did not always operate in sync with the needs of the Soviet state.

Well aware that he was "technically . . . guilty of flagrant insubordination," Sakharov still pressed his case to Malyshev, the new Minister of Medium Machine Building, a short, ruddy-faced man who blew his top and began shouting, calling Sakharov and company "reckless gamblers toying with the country's fate." But Malyshev was up against Academician and Hero of Socialist Labor Sakharov, who was backed by Kurchatov, head of atomic science, Khariton, head of the Installation, and scientists like Zeldovich, all of them Heroes of Socialist Labor. The Heroes would ultimately triumph, but not immediately and not without travail. Malyshev was not able to make a decision on the spot—too much money was involved. The budget simply did not allow them to build a "classical device" as planned and simultaneously go ahead with Sakharov's new breakthrough. And not only did Malyshev lack sufficient authority, he soon lacked even his job. In 1955 Khrushchev made a first move against Malenkov, toppling him from his post as chairman of the Council of Ministers and toppling all his people in turn, Malyshev included. Within a year Malyshev was dead from leukemia—he had been the other person who had walked with Sakharov from the convoy across the black crunching glass toward ground zero.

The Politburo met to set a date for the testing of the new device. Malenkov was still part of the inner circle but had the pallor of those who are being drained of power. It was Khrushchev, pacing restlessly, his hands thrust in the pockets of his blue jeans, who dominated the meeting, though he still lacked the reflexive assurance of those fully in charge. The date set was fall 1955.

That wasn't much time. Months burned like paper at the Installation. The work immediately became so "feverish" that Sakharov could not wait "for the resolution of all theoretical questions or the final calculations" and began increasingly to rely on his intuition. But it was not only a matter of time constraints. Sakharov had now demonstrated his scientific genius to himself as well as to others, and believed in its lightning insights. That gift was evident to those around him. One spoke of Sakharov's ability to "foresee a result before doing any calculations" and "to represent with extreme clarity the details of the behavior of an electron or a nucleon . . . he could mentally transform himself into these particles, as if his very skin could feel what it was to be them."

The state now not only considered Sakharov too valuable to fly; he was in the summer of 1954 assigned two "secretaries," the official euphemism for bodyguards, sometimes called "guardian angels" by their charges. One was a KGB colonel who had served in Stalin's personal guard before being transferred to the Baltics, where he was, as he put it, "in arrests." Sakharov found him "tactful, never intrusive, always obliging." The other "secretary," a young lieutenant taking a correspondence course in law to better himself, also tried to better Sakharov politically, but to zero avail. Both bodyguards were trained to fire their Makarov pistols from their pockets without wasting the precious seconds needed to draw them. Still, tactful and obliging as both men were, Sakharov had irksomely to alert them every time he went from one place to another, whether at the Installation or when visiting his wife and children in Moscow. They hovered at the margins of family outings, lounging by the waiting cars.

Relations between the guarded and the guards varied greatly. The expansive Kurchatov enjoyed their company and would converse with them nonstop. Zeldovich, on the other hand, took every opportunity to torment his keepers in order to demonstrate their absurdity and uselessness, once even jumping into a river, knowing that the older man assigned to him could barely swim. Khariton's bodyguards felt "almost like part of the family" and wept unashamedly when departing for a new assignment.

Klava could not abide them. She, who would not allow servants in her home, was now encumbered with those presences, their discretion itself an irritant. But they were affixed to her husband like his titles and medals. Klava's solution was to spend more time in Moscow in their roomy apartment in a wooded, peaceful part of the city that did not oblige her to pass through rings of barbed wire. Their marriage was being defined by separations.

* * * *

When the physicists weren't working, they were disputing that favorite subject of the intelligentsia—the fate of Russia. The writer Ilya Ehrenburg, always attuned to the times, quickly found a name for the new era, "The Thaw," the title of his novel that appeared in April 1954. By August the Hermitage Museum had for the first time displayed its Picassos, Matisses, Van Goghs. Among those delighting in the liberating daring of those artists were many of the country's future dissidents, still young, just beginning to struggle their way free of the Soviet mind-set.

Prisoners had begun returning from the Gulag, first a trickle, then a stream, though not yet a flood of millions. In kitchens, over tea or vodka, the Zeks told their stories, the shape of a fate chiseled in a few strong strokes, the real history of Soviet Russia.

Trained by Tamm in both theoretical physics and independent social thought, Sakharov had now replaced him in both. Not that anyone, Sakharov included, had the time or the inclination to subject the system, its assumptions and acts, to fundamental scrutiny. Besides, things did indeed seem to be changing for the better, the idea of "thaw" having the exact right resonance in a country where every year is mostly winter. Sakharov and Zeldovich introduced a thaw of their own, doing away with the prohibition on scientists discussing their work except on a need-to-know basis and never outside the lab or department. Everyone knew that the best ideas often came when people were just sitting around and talking. Within limits of their own choosing, the authorities had already acknowledged that reality—*The Bulletin of American Scientists* was made accessible on the principle that Soviet scientists should be aware of what their American opponents were up to. However, another principle was operative here, the law of unintended consequences; the *Bulletin* also contained reports on developments in molecular biology, which provided Sakharov with hard evidence that Lysenko and company were not only wrong but dangerously wrong.

It was at this time that Sakharov wrote his first letter of defense, a genre in which he would in time become adept. A play, *The Guests* by Leonid Zorin, which assailed the "new class" of bureaucrats whose arrogance was only outdone by their greed, had itself come under particularly ferocious attack. Nudged into action by Zeldovich, who "himself eschewed any public role in this affair," Academician Sakharov added the weight of his voice to the discussion. The old Zeldovich two-step had worked again.

But all Sakharov's authority would evaporate if the Third Idea proved a dud, in which case he would be more than "technically guilty" of insubordination.

In the fall of 1955, accompanied by his two "secretaries," Sakharov crossed a platform thick with KGB at Moscow's Yaroslavl Station. The theoretical physicist and the men ready to kill and die to protect him moved through the crowd of uniformed and plainclothes agents to their railroad car, which had been attached to the Moscow-Peking Express. A song about Soviet-Chinese friendship came through the public-address system; its refrain was "Stalin and Mao." Stalin was still in the air.

The vast domed sky of the Kazakh steppes was the ideal amphitheater for explosion. But that was all that was ideal. The "Evil Genius," Victor Gavrilov, who had pointed out the downwind dangers of radiation at the first test, now struck again. He saw the one thing they had all had been blind to: the heat radiation from the thermonuclear explosion might well incinerate the plane that dropped the bomb. True, some precautions had been taken—the plane had been coated with white reflective paint and even the red stars had been left off, so as not to create any weak spots on the surface. But no one had done the actual calculations. "One night, unable to sleep, I calculated the bomb's trajectory, the airplane's exposure to heat (in calories per square centimeter), and the likely effect of the heat on the plane's surface. Of course, there were others employed to do that work, but it's a pleasure to figure out such things for oneself."

That sleepless night Sakharov succeeded in transmuting anxiety into pleasure through the exercise of reason, but the respites were brief. Too much was riding on this test. He sought solace in nature, watching the ice break up on the Irtysh River, which flows south to north through Siberia: "The dark, turbulent waters of the Irtysh, dotted with a thousand whirlpools, bore the milky-blue ice floes northward, twisting them around and crashing them together. I could have watched for hours on end until my eyes ached and my head spun. Nature was displaying its might: compared to it, all man's handiwork seems paltry imitation."

From the very start the fall 1955 tests had a malign Dostoevskian doubleness about them. There were to be two tests, that of the classical device, whose successful explosion in early November interested Sakharov so little it barely registered with him, and the test based on the Third Idea, in which he had a great interest in every sense. A dry run on November 18 proved that the parachute attached to the device would slow it sufficiently for the plane to fly out of blast range. The actual test was set for two days later, on November 20. But no sooner had the plane taken off with the device in its bomb bay than low clouds darkened the sky, effectively preventing optical sighting of the bomb and optical measurement of the explosion. In favor of aborting and rescheduling, Sakharov and Zeldovich reported to the command post, where they had to state their conviction in writing that it was safe for a bomber to land, even to crash-land, with a nuclear bomb still on board. It was the perfect question for a theoretician, since there were no experimental data on the matter.

Sakharov and Zeldovich took their stand, but it was Kurchatov who had to make the decision. "One more test like this and I'm retiring," said Kurchatov, who agreed with Sakharov and Zeldovich that the bomber could even crash-land with no danger of its bomb exploding. The theory was almost put to the test when the runway suddenly iced over, but an army unit plowed a path for the plane, and a "very long day" was over.

The test was moved two days back, to November 22. The chancy landing, the delay, had keyed people up past a point they thought possible.

The weather was clear on the twenty-second; the storm had passed, leaving a light snow on the feather grass of the steppe. The white bomber rose into the sky, wings swept back, reminding Sakharov of a "sinister predator poised to strike." He, Zeldovich, and a few others who might be needed for last-minute consultation spent the last hour before the test on a low observation platform near the command post. Five minutes before the drop, the loudspeaker clicked on, and the controller began the countdown in an appropriately solemn voice that mounted in excitement until all three hundred seconds had finally been counted away: "The bomb has dropped! The parachute has opened!"

Sakharov turned his back to the explosion, but only the better to see it. He had chosen not to wear his dark safety goggles—fine and significant detail was lost with them on, and by the time you removed them and your sight adjusted, the critical moments had passed. This test he had to see with his own eyes.

"I stood with my back to ground zero and turned around quickly when the building and horizon were illuminated by the flash. I saw a blinding yellow-white sphere swiftly expand, turn orange in a fraction of a second, then turn bright red and touch the horizon, flattening out at its base. Soon everything was obscured by rising dust which formed an enormous, swirling, gray-blue cloud, its surface streaked with fiery crimson flashes. Between the cloud and the swirling dust grew a mushroom stem, even thicker than the one that had formed during the first thermonuclear test. Shock waves crisscrossed the sky, emitting sporadic milky-white cones and adding to the mushroom image. I felt heat like that from an open furnace on my face—and this was in freezing weather, tens of miles from ground zero. The whole magical spectacle unfolded in complete silence. Several minutes passed, and then all of a sudden the shock wave was coming at us, approaching swiftly, flattening the feather grass.

"'Jump!' I shouted as I leaped from the platform. Everyone followed my example except for my bodyguard. . . . he evidently felt that he would be abandoning his post if he jumped. The shock wave blasted our ears and battered our bodies, and we heard the crash of broken glass. Zeldovich raced over to me, shouting: 'It worked! It worked! Everything worked!' Then he threw his arms around me."

But Sakharov's own joy had a very short half-life. A few hours after the test, he learned that the explosion had caused two deaths. A trench had collapsed on a young soldier. Elsewhere, the residents of a settlement that the scientists had calculated was out of the danger zone had taken refuge in a primitive bomb shelter. Thinking themselves safe after the flash on the horizon, they too wanted a look with their own eyes and had gone outside. Acting out of impulsive curiosity or maternal caution, one mother decided to leave her two-year-old daughter playing with her blocks in the shelter. And so the little girl was alone when the shock wave flattened the concrete like feather grass.

Windows shattered for a hundred miles around, spraying splinters of glass into the meat at a packing plant and into the eyes of children who had been drawn to their windows by the flash. Ironically, the change toward better weather had caused a temperature inversion, which spread the shock waves farther than had been calculated, as would not have been the case two days before when they decided to abort.

Examining his own conscience about the deaths of the soldier and the girl, Sakharov concluded: "I did not hold myself personally responsible for their deaths, but I could not escape a feeling of complicity."

That evening was the banquet to celebrate the successful explosion, the table lined with bottles, the walls by bodyguards. Marshal Nedelin, who had directed the operation and was acting as master of ceremonies, nodded to Andrei Sakharov, signaling that he would have the honor of the first toast. Standing, glass in hand, Sakharov said: "May all our devices explode as successfully as today's, but always over test sites and never over cities."

Silence fell over the table as if Sakharov had made an "indecent" remark. Grimacing, the marshal himself rose glass in hand to tell a story that would put things, and Sakharov, back in their place: "An old man wearing only a shirt was praying before an icon: 'Guide me, harden me. Guide me, harden me.' His wife, who was lying on the stove, said: 'Just pray to be hard, old man, I'll take care of the guiding.' And so," said Nedelin, "Let's drink to getting hard."

Isolated in the silence that followed the story and in the too loud buzz of conversation that ensued a few seconds later, Sakharov felt "lashed by a whip." He drained his glass of brandy and never said another word for the rest of the evening. Not only because the mood had been spoiled but because he had been shocked into a realization: "We, the inventors, scientists, engineers, and craftsmen had created a terrible weapon, the most terrible weapon in human history, but its use would lie entirely outside our control." And that could not sit well with either Sakharov's conscience or his pride.

* * * *

"You won't be able to keep it a secret!" shouted one Politburo member after another, attacking Khrushchev for proposing to denounce Stalin's crimes at the Twentieth Party Congress scheduled for February 1956. "Word will get out about what happened under Stalin, and then the finger will be pointed straight at us."

The worse their crimes, the more vehement their opposition. Their fear was based on slabs of fact. They were, after all, the same people who had feasted with Stain in the Kremlin, laughed when he laughed; their hands had done his will and applauded the killing. Disclosure was suicide.

Political to his bones, Khrushchev was also sincere in his struggle against Stalin: "It was as though we were enchained by our own activities under Stalin's leadership and couldn't free ourselves from his control even after he was dead." For the sake of the country and of his own career, which nicely coincided, Khrushchev knew Russia had to be purged of Stalinism, not Stalinists. Who wasn't a Stalinist?

Unlike many of the other members of the Politburo, who seemed to have formaldehyde in their veins, Khrushchev had retained his humanity, his gusto. And that gave him an edge in the power struggles. His passion was more than just ambition; he did want to see some justice done. Khrushchev was a family man, unlike Stalin who, after the suicide of his wife in 1932, was a bachelor tyrant for the twenty-one years until his death. A good wife may have kept Khrushchev human—an invisible contribution, but an invaluable one.

Politically, Khrushchev's background could not have been better: mine worker, communist by conviction, labor organizer, first-rate war record against the Whites and against the Nazis, took part in the building of the Moscow Metro, party boss of Ukraine (which always needed bossing and was so nationalistic and anti-Stalin that many of them had even welcomed Hitler's armies).

A man of the people, Khrushchev was also in some sense a populist. There may have been no public opinion in Russia, but that did not mean the public had no opinion. And the masses had feelings that could be roused to great effect, as had happened during the revolution and the war. And there'd be no tricking them either by blaming it all on Beria. Beria had not been head of the secret police in 1937. Khrushchev rejected any falling back onto the "Beria version . . . Beria didn't create Stalin, Stalin created Beria."

In some ways the surviving Politburo was as much haunted by Beria as by Stalin. Beria too had tried liberal experiments in East Germany, and that had led to a workers' revolt. A bloc had formed against Beria, who was then arrested and executed—who was to say he was the last? And then by the perverse logic of Soviet power politics, some favored Khrushchev delivering the speech, so that the party's full wrath would be drawn down on him.

But 1956 was, in the words of one Russian writer, "the year of passion," and Khrushchev's passion prevailed.

On February 25, 1956, in a speech to the Twentieth Party Congress in the Great Hall of the Kremlin, Khrushchev delivered his denunciation. The speech was, however, supposed to remain secret, copies to be circulated exclusively to members of the party and the Soviet elite, which of course included Academician A. D. Sakharov. Described by others as "utterly loyal" and "apolitical," and by himself as "politically passive," Sakharov found no fault with Khrushchev's general line: Stalin had deviated greatly from Lenin's revolution; the only proper course was to return to Leninist principles and practices. A monster of egotism, Stalin had replaced collective leadership with the "cult of the personality." He had remained aloof from the people, had not been prepared for the German invasion. But his greatest crimes were against the party—of the 139 people elected to the Central Committee, Stalin had executed 98, something like 70 percent. There had been terror under Lenin too, but Leninist terror was fundamentally different: it was a response to very real enemies, the Whites, who used terror against him, forcing him to reply in kind. Otherwise, said Lenin as quoted by Khrushchev, "We would not have lasted two days." It was time to return to the Leninist straight and narrow. The Great Hall erupted, according to the official memorandum, in a "tumultuous" standing ovation.

Though like all Russians Sakharov was shocked by the revelations, other feelings moved him more. He was still under the spell of the tongue-lashing at the banquet: "The ideas and emotions kindled at that moment . . . completely altered my thinking." The deaths of the soldier and the little girl had stung him. Stalin had said: "One death is a tragedy, a million is a statistic." But now Sakharov would prove him wrong with tragic statistics—he began calculating the probable number of deaths and genetic mutations caused by the radioactive fallout from an atmospheric test. Coincident with the beginning of his research, *Pravda* launched an attack on Mendelian genetics that "shocked and puzzled" Sakharov.

How could such stupid science continue to exist? Obviously, Lysenko had powerful backers, and his quack genetics appealed to the quick-fix mentality of leaders. But Lysenko's ideas were also difficult to dispute—a harvest wasn't a bomb, which either exploded or didn't; harvests had always been unpredictable in Russia, influenced by everything from insects to sun. At this same time, accompanied by Zeldovich, Sakharov visited a biologist who had to breed fruit flies at home because genetics was simply prohibited at his institute. In fact this was a scouting mission initiated by Kurchatov, who was about to throw his weight behind a plan to build a new microbiology lab.

But all that was a digression, though one that foreshadowed battles to come. Sakharov now confronted a "moral dilemma," as his initial calculations of death by fallout over the generations made it plain that the "total number of 'anonymous' victims will be staggering." The moral dilemma only sharpened as the figures became more precise: ten thousand deaths for each megaton tested. According to his calculations, around fifty megatons had already been tested worldwide, meaning some five hundred thousand deaths. In 1957 he put his findings into an article, "Radioactive Carbon from Nuclear Explosions and Non-threshold Biological Effects," which was published first in the Soviet scientific journal *Atomic Energy* in 1958 and later reprinted with Khrushchev's personal authorization in a more popularized form as "The Radioactive Danger of Nuclear Tests."

The more accessible version was published by the propaganda agencies and distributed by Soviet embassies to serve a political purpose: denouncing the American "clean bomb," a thermonuclear weapon that supposedly produced no radioactive fallout. Sakharov pointed out that the "clean bomb" was a fiction, since it produced the radioactive isotope carbon 14, which had a half-life of five thousand years. And he gladly echoed the general line and its high-minded rhetoric: "The Soviet state was compelled to develop nuclear weapons and conduct tests for its security in the face of American and British nuclear weapons. But the USSR's goal is not universal destruction, but peaceful coexistence, disarmament, and the banning of nuclear weapons." But what engaged him most deeply was the question "What sort of moral and political conclusions should be drawn from the figures cited?" By his willingness to include the moral dimension Sakharov was "just beginning to stray from the official position," though at this stage the angle of deviation was barely perceptible to others or to himself.

Sakharov had been awarded his second Hero of Socialist Labor at the end of 1956 and invited to the Kremlin for a New Year's Eve of champagne and fireworks. To the higher authorities, he was the eccentric scientist whose high-minded toasts needed to be slapped down to earth. And Khrushchev, who had personally signed off on Sakharov's article after a few changes were made to reflect new shifts in policy, was much too busy to detect any untoward drifts in his unfailing bomb maker. In fact it was Khrushchev who now irritated Sakharov, not to mention the Politburo, with his erratic zigzag behavior. Some of that was his own nature, of course, but much of it had to do with his self-appointed task of loosening the screws and springs of the Stalinist system. How much tension could be released short of disaster?

A few months after the "secret speech"—so unsecret that it was quickly published by the U.S. State Department—the workers revolted in Poznan, Poland, under a familiar revolutionary banner—Bread—and the quite counterrevolutionary—Soviet Troops Out. And in late October 1956 the passions were Hungarian: not only did the Hungarians want Soviet troops out; they wanted out of the Warsaw Pact, founded in 1952 to counterbalance NATO. As Mikhail Heller and Aleksandr Nekrich put it in their book *Utopia in Power,* the Hungarian Uprising was "crushed by the treads of Soviet tanks and the indifference of Western nations."

One of the people for whom the crushing of the Hungarian Uprising was a significant career move was Yuri Andropov, who played a particularly treacherous role in the proceedings. Anyone with any investment in the Soviet system had to be blind to the real significance of the events in Poland and Hungary. Even though the streets were flooded with workers, the imperialists and the bourgeoisie had to be behind it. Hadn't the Hungarians fought with the Nazis in World War II? Weren't the Poles always Russophobic aristocrats? These events had to seen as isolated incidents, though that did grow difficult as the uprisings occurred with clockwork twelve-year regularity: Hungary 1956, Czechoslovakia 1968, the Poland of Solidarity 1980. Perhaps twelve years is the amount of time the collective memory needs to forget the reverberations of tank engines off Baroque walls.

For some young Russians, such as future dissident Vladimir Bukovsky, the invasion of Hungary was their Kronstadt, the end of a faith: "After the tanks with the red star, dream and pride of our childhood, had crushed our peers in the streets of Budapest, everything we saw was stained with blood. The entire world had betrayed us, and we no longer believed anyone. Our parents had turned out to be agents and informers, our military leaders were butchers, and even the games and fantasies of our childhood seemed to be tainted with fraud."

But some of the intelligentsia still looked with favor on Khrushchev—he had allowed Matisse to be shown at the Hermitage and more important was releasing hundreds of thousands of prisoners from the Gulag. The contrast couldn't have been sharper: Stalin put them in, Khrushchev let them out. In April 1957 the great poet Anna Akhmatova felt free enough to begin committing to paper her long poem *Requiem,* which opens with a description of her standing in line every day for seventeen months in front of the Leningrad prison, where her son was being held during the Great Terror. A woman whispers to her with lips "light blue from the cold . . .":

> "Can you describe this?"
> And I said: "I can."
> Then something like a smile passed fleetingly over what had once been her face.

Akhmatova was also to say: "I am a Khrushchevite because Khrushchev did for me the noblest thing one human being can do for another; he gave me back my son."

And, just as in her poem, she was speaking for more than herself.

* * * *

In July of 1957 Khrushchev had disposed of the last of his rivals—Malenkov, Molotov, and the noxious Stalinist Kaganovich—by branding them an "antiparty" group. But to demonstrate that he was genuinely reverting to Leninist ways, he expelled none of them from the party, merely demoting them to humiliatingly lower positions, though ones still in keeping with their experience—Molotov, the former Soviet foreign minister, being made ambassador to Mongolia. But Khrushchev's clemency also notified future rivals that they did not have to fear arrest and death from him.

On October 4, 1957, the USSR launched *Sputnik,* the first Earth-orbiting satellite. Small—184 pounds, twenty-two inches in diameter—*Sputnik* made an immediate and enormous impression. It could zoom around Earth in an hour and a half, beeping signals back continuously on two frequencies. And less than a month later the Soviets sent a dog, "Laika," into orbit. Space was the future, and the Soviets had entered the future first. Sakharov's belief in the "breakthrough" nature of his society seemed increasingly justified.

The new Soviet leadership was also attuned to another frequency of the future: the liberation of women. Valentina Tereshkova would soon be the first woman in space. A progressive position on the "women's question" had always been part of Soviet propaganda and practice. Stalin had liberated the women of Central Asia from the chador, but only as part of smashing a patriarchal culture to reforge it to his liking. Still, from the lineup on Lenin's Tomb it was plain that political power remained a male preserve. And so it came as something of a shock when a woman, Ekaterina Furtseva, joined Khrushchev's inner circle as Minister of Culture.

Inevitably, there were off-color jokes about the two of them. Inevitably, the joke-loving intelligentsia would, choosing their company with care, repeat them. Occasionally, that company was not chosen with sufficient care. Someone informed on Isaac Barenblatt.

Barenblatt, a prominent endocrinologist who had treated Khrushchev himself as well as Sakharov's wife, Klava, had not only repeated the jokes with suspect gusto but had even added that Khrushchev had no moral right to judge Stalin—he too was up to his elbows in blood. Barenblatt's son Grigory, a mathematician, worked with Zeldovich and turned to him for help. Zeldovich first took independent action before enlisting Sakharov, who invited the son to come to discuss his father's case. Arriving at the appointed time on a warm summer evening, Grigory Barenblatt saw Sakharov crossing the courtyard with one colleague, issuing instructions. A younger colleague waited nearby and struck up a conversation with Barenblatt. When Sakharov was done, he walked over and found Barenblatt being peppered with questions by the young scientist. Sakharov, who took security seriously, rebuked the young man: "Your curiosity is quite out of place. . . . You may go; I have no further need of you today."

Sakharov wrote a letter defending Barenblatt senior to Khrushchev, who assigned the matter to Mikhail Suslov, the cardinal of ideology. Tall, thin, ascetic—even his eyes were "pale, almost white,"—Suslov, invariably correct in all dealings, was a hardliner, not only insisting on the violent suppression of the Hungarian Uprising, but flying to Budapest to oversee it. His vast domain was all that concerned thought and word, including all schools, propaganda, television, the press, the political administration of the Soviet army—to name but a few of his provinces. And so he was the proper person for Khrushchev to refer this rather odd matter to—the bomb maker wasting his time defending a Jew for telling off-color anecdotes.

In his meeting with Suslov, Sakharov took a commonsense approach: "During the war, Barenblatt had proved himself a faithful defender of our system—what were words compared to deeds?"

Suslov, who made no distinction between words and deeds where politics was concerned, listened with a "slightly condescending air" and replied that what Barenblatt had said was "inadmissible." They went back and forth on it a few more times than Suslov was prepared to tolerate, and the atmosphere "began to assume a certain ominous quality." Anxious not to alienate such an important scientist, Suslov agreed to take another look into the case, then handed Sakharov a typed document with a red No Copies stamp on the margin. It was the Politburo's decision to announce a unilateral test ban in March. Sakharov was elated and irked. Elated because no matter what the political motivations, every megaton untested meant ten thousand lives saved. Irked because he felt that the Installation should already have been notified so that "loose ends" could have been tied up. Suslov then switched to the third item on his agenda—the genetic consequences of testing—which gave Sakharov the chance to vent his outrage against Lysenko and his followers, calling them "reckless opportunists and schemers" and to go on record in support of Kurchatov's idea for a new molecular biology lab.

Even though the encounter with Suslov caused Sakharov to say later "What monsters rule us!" it had to be counted a successful meeting. Sakharov had struck a blow against Lysenko, learned the good news of the test ban, and had even apparently done Barenblatt some good, for he was released after serving one year of a sentence that could have run to two and a half, which Sakharov called with both accuracy and irony "relatively mild treatment."

Sporadically, Sakharov was learning the science of political influence. Accompanied by Khariton he went to see the man in charge of military research and development, Leonid Brezhnev, a lover of good suits and fast cars, who greeted them expansively: "So, the bomb squad's here!" Before they could even make their case to challenge pending changes in resource allocations, Brezhnev regaled them with a story about his working-class father, who thought that people who invented new weapons should be hung on a high hill as an example to others. "And now," said Brezhnev, "I'm involved in that dirty business myself, just as you are, and with the same good intentions. Right, let's hear what you have to say."

Brezhnev listened, adjudged their budgetary objections sound; the appropriate measures were taken. Responsive leadership. But the leadership of the United States and Britain were not responsive to the Soviet test-ban initiative and exploded several devices in the summer of 1958, claiming that the unilateral Soviet ban was a political gimmick because the Soviets had just completed their latest series of tests. That was not true. In fact, Sakharov and his colleagues at the Installation were now thrown into multiple technical quandaries: "Could devices so nearly ready for our arsenal be renounced? Could a portion of them, at least, be accepted without testing? Was it possible to design new devices, perhaps with inferior characteristics, that could become part of the arsenal without prior testing? Or was it altogether inconceivable to accept untested devices under any circumstances?"

All the technical theorizing was rendered moot a short while later when Khrushchev reversed the moratorium and ordered testing resumed, an abrupt switch that Sakharov found destabilizing, "completely unacceptable, both politically and morally." Sakharov wrote up a list of five proposals: a year's wait before resuming testing; redesign existing devices so they could be deployed without testing; renounce the doctrine of adopting no weapons without testing; invest in computers to perform calculations in place of tests; create new, experimental methods for modeling various functions to eliminate the need for full-scale testing.

Khrushchev would not like the idea. Sakharov needed allies. He turned to Kurchatov, knowing The Beard was the "only person who had any chance of influencing Khrushchev, and the one official in our ministry who might be sympathetic" to the five points. Ailing and without long to live, Kurchatov had lately been expiating his atomic guilt by defending genetics and cybernetics while also investigating the peaceful application of atomic energy. It was a sunny September day, and they sat outside in the park around Kurchatov's cottage. Kurchatov's dog crawled around Sakharov's feet as he made his case. Agreeing with everything, Kurchatov said he would fly to Yalta, where Khrushchev was on vacation, and make the case to him in person.

No sooner had Sakharov and Kurchatov worked out their plan of action than Kurchatov's secretary appeared with a camera. The brilliant young physicist and the great old man—that would make a terrific shot for the photohistory of Kurchatov he was compiling. In the photographs, Kurchatov stares directly into the camera yet remains unreadable. Sakharov is cocksure.

Khrushchev found something displeasingly cheeky about the whole business, not to mention having his vacation disrupted. Kurchatov at once fell out of favor, Sakharov's proposals were rejected, the tests went ahead. Though the tens of thousands would die invisibly, never knowing why, the tests were, as Sakharov noted with pained objectivity, a "great success and important from the technical point of view."

* * * *

In late 1958 the thunderings in Russia were not exclusively nuclear. On October 23 Boris Pasternak was awarded the Nobel prize for his novel *Doctor Zhivago.* No sooner had he cabled his acceptance—"Immensely thankful, touched, proud, astonished, abashed"—than he came under furious attack for placing a "weapon in the hands of the enemy," one definition of treason. A matter of state importance, the affair was debated in the Central Committee, where Pasternak was described as an "internal émigré" who should "actually become an émigré and leave for his capitalist paradise."

Mistakes are a prime ingredient of history. It was something of a mistake to award Pasternak the prize for *Zhivago;* his talents as a poet far exceeded his as a writer of prose—it was as if Yeats had inadvertently written a best-seller. It may also have been premature for the Swedish Academy to probe the exact extent of change in Soviet Russia, and premature for Pasternak to exalt the role of the individual and question the very value of the revolution. The novel was clearly Pasternak's bid for the greater renown he knew he deserved.

The charisma of Pasternak's genius fascinated others and, in a certain disinterested way, himself as well. The only time Stalin picked up the phone and called a poet, it was Pasternak he chose. It was 1937, the peak of the Terror. Pasternak was not fazed in the least and began that conversation as he began all others—by saying he couldn't hear well because of all the noise in his communal apartment—children were hooting and hollering in the corridor the day Stalin called. Stalin immediately took the moral high ground, chiding Pasternak for not coming to the defense of his fellow poet Osip Mandelstam, who had been recently arrested: "If I were a poet," said Stalin, "and a poet friend of mine were in trouble, I would do anything to help him."

Pasternak's response was a mix of sensible defense and a tendency to skitter into nuance. Sensing the conversation coming to its end, Pasternak asked if they could meet and talk. "About what?" asked Stalin. "About life and death," replied Pasternak. Stalin hung up. And thus ended the first of the two most significant phone calls in Soviet history, the second coming a half century later, to Sakharov.

Pasternak renounced the Nobel prize: "Considering the meaning this award has been given in the society to which I belong. . . ."

But that renunciation was considered insufficient penance by the Moscow Writers' Organization, which called for Pasternak to be deprived of his citizenship. Pasternak wrote to Khrushchev: "I am linked with Russia—by birth, life and work. I cannot imagine my fate separate from and outside Russia. . . . A departure beyond the borders of my country would for me be equivalent to death, and for that reason I request you not to take that extreme."

Sakharov paid especially close attention to the Nobel prizes that year because Tamm had just become the first Soviet physicist to win the award. Here too, to Tamm's chagrin, a mistake was involved; he had been awarded the prize for the "discovery and explanation of the Vavilov-Cherenkov Effect," work he had done at the end of the thirties. The Nobel committee had overlooked what Tamm thought his best work—on beta forces.

Though the Pasternak affair may have consisted largely of mistakes, it was not in the least accidental. Rather, it was the inevitable result of a society attempting to redefine the shifting boundaries of the permissible. It created a new template in Soviet culture: a manuscript is sent abroad, creates a sensation, the author is attacked, exile is the threat and, sometimes, the punishment. In time, with certain variations, both Khrushchev and Sakharov would find themselves in that position. Pasternak was neither arrested nor exiled. But the hostility, ostracism, and humiliation had broken his heart. He kept a public silence, confining his feelings to the privacy of the page:

> But what wicked thing have I done,
> I, the murderer and villain?
> I made the whole world weep
> Over my beautiful land.

* * * *

Though Khrushchev had abruptly switched his position in 1958, a de facto moratorium on nuclear testing lasted from 1959 through mid-1961, sullied only once by vainglorious France. Those two and a half years were thus also a time of moratorium between Sakharov and Khrushchev, the principal bone of their contention having been removed. But Sakharov continued to gnaw. Or be gnawed at. On a workday at the Installation in early 1961, Sakharov dropped in on a colleague with a copy of *The Voice of the Dolphin,* political fantasies by the Hungarian-born physicist Leo Szilard, who had worked on the American bomb. Exactly how that volume, in English, had made its way through the barbed wire of Arzamas-16 will probably never be known, but it does illustrate the degree to which Soviet society had become permeable. In the most powerful story, and the one Sakharov urged his colleague to read, "My Trial as a War Criminal," a victorious Soviet Union indicts the scientists who worked on the bombs that destroyed Hiroshima and Nagasaki. But Sakharov was all too aware that Soviet bombs had already sent enough lethal radiation into the atmosphere to kill tens of thousands, a less dramatic homicide but no less criminal for that.

The year 1961 was erratic, even by Khrushchevian standards. It saw both the first man in space, Yuri Gagarin, and the building of the Berlin Wall. In midyear Khrushchev suddenly decided to resume testing, a decision that "as usual, came as a surprise to those most directly affected"—such as Sakharov, who was convinced that these tests were not necessary. Kurchatov had interceded for Sakharov in the last clash, but Kurchatov had died in 1960 at the age of fifty-seven. There was no one else Sakharov could turn to. This time he was on his own.

And this time it was Sakharov's vacation that was cut short by Khrushchev. Sakharov had already observed Khrushchev at an earlier Kremlin meeting, noticing his intelligence, his appealing willingness to change his mind, his brash assertiveness, his need for the last word. This time Khrushchev was abrupt and commanding: tests would be resumed because the international situation had worsened and the USSR was behind the United States in testing. Khrushchev had a third reason, political and therefore most compelling: he wanted the test to coincide with the Twenty-second Party Congress, where he would renew his attack against Stalin and Stalinism.

Then the scientists were allowed to speak. When Sakharov's turn came, he "volunteered the opinion" that there was "little to gain from a resumption of testing at this time." Certain that his remark had "registered" and noting that it had met with a total lack of response, Sakharov moved on to speak of some of the Installation's futuristic projects, such as nuclear-powered spacecraft. But Sakharov did not digress long. No sooner did he return to his seat than he borrowed a piece of paper from his friend Zababakhin and wrote Khrushchev a note, which opened with adroit politics: "I am convinced that a resumption of testing at this time would only favor the USA. Prompted by the success of our Sputniks, they could use tests to improve their devices. . . . Don't you think that new tests will seriously jeopardize the test ban negotiations, the cause of disarmament, and world peace?"

Khrushchev glared at Sakharov, folded the note in quarters and put it in his jacket pocket, waiting until the last speaker had finished to invite everyone to a banquet in an hour's time.

As a sign of respect the scientists were seated first at a table for sixty, lined with wine and mineral water, bright salads, and black caviar against fine linen. When everyone was seated, all eyes went to Khrushchev, who alone was still standing, wineglass in hand, about to make the opening toast. But he set the glass aside and took out Sakharov's note, which he waved in the air: "Here's a note I've received from Academician Sakharov," said Khrushchev, without actually reading it aloud. "Sakharov says that we don't need tests. But I've got a briefing paper which shows how many tests we've conducted and how many more the Americans have conducted. Can Sakharov really prove that with fewer tests we've gained more valuable information than the Americans? Are they dumber than we are?" Having made his argument that quantity is the stuff of quality, Khrushchev grew increasingly incensed: "But Sakharov goes further. He's moved beyond science into politics. He's poking his nose where it doesn't belong." And he was red in the face by the time he concluded: "Sakharov, don't try to tell us what to do or how to behave. We understand politics. I'd be a jellyfish and not Chairman of the Council of Ministers if I listened to people like Sakharov!"

The room went silent. No one looked at Sakharov. Only Mikoyan, the man who could dodge raindrops and had outlasted Stalin, grinned down at his plate in some private amusement. Sensing the mood losing its celebratory tone, Khrushchev proposed a rousing toast to their future successes, to which everyone drank but Sakharov. His isolation was complete for the rest of the evening until one person, Yuri Zysin, his fellow physicist and sometime skiing companion, came over to express his support.

Not only would bombs be tested but one of them would be a "Big Bomb," fifty megatons, several thousand Hiroshimas condensed into a single device. Sakharov could do nothing about that decision, but there were decisions that he could make, such as that to "test a 'clean' version. . . . By reducing the fission component, we would minimize the casualties from fallout, but radioactive carbon would still cause an enormous number of victims over the next five thousand years." It clearly wasn't over for either of them, because Khrushchev was soon demanding public repentance. "Does Sakharov realize he was wrong?" Khrushchev asked Khariton at a meeting, but Sakharov, who was also present, answered for himself: "My opinion hasn't changed, but I do my work and carry out orders."

Khrushchev muttered in disgruntlement, then went on to speak about world tensions, not helped by the building of the Berlin Wall, a necessary measure to keep educated East Berliners in and profiteering West Berliners out.

To Sakharov's surprise, Khariton not only shared his opinion but was willing to go to Brezhnev to make the case against testing. But Khrushchev was immovable. He needed the grandeur and authority of explosion behind him at the party congress, when he would propose that Stalin's body be removed from its place beside Lenin, a sanctioning on the highest symbolic level of Khrushchev's distinction between the two. Now there was nothing left for Sakharov but to do his work and carry out orders.

In October, in Moscow to discuss some of the final calculations, Sakharov showed up without advance notice at his parents' dacha outside Moscow just as his mother was making apple jam from the apples trees his father, Dmitri, had planted and tended. As they drank tea with apple jam that fall day at the dacha, Dmitri told his son he was composing music again, writing what he wanted, playing as he felt. Taking Andrei out for a little walk, he showed him one of his latest experiments, one that combined his love of gardening and physics, an elegantly simple means of showing how water rises from roots to leaves. Almost in passing, he mentioned that recently he'd had a sharp pain in his chest but hadn't wanted to worry Andrei's mother with it.

Sakharov was back at the Installation the next day. The Big Bomb was scheduled to be tested in the Far North on the last day of the party congress. The device was already being assembled in a special workshop on a platform car that would be attached to a locomotive, then travel on a straight-line track to the waiting bomber. In this time of final preparations, Sakharov himself made a sudden, paradoxical move. He proposed testing not one bomb but two. This "Extra" bomb would, along with the Big Bomb, provide so much useful data that no further tests would be needed for a long time to come. It was a case where more is less.

The Minister of Medium Machine Building was outraged by Sakharov's request. No additional nuclear charge had been requisitioned by the Installation, none had been allocated, and none would be this late in the game. But Sakharov was on something of a tear: "for the first and only time in my life I performed miracles of string-pulling." He scrounged parts and glued them together with epoxy, mickey-mousing apocalypse. But then all that receded in importance when he received a letter from his mother saying that his father had suffered a severe heart attack. Even though telephoning from the Installation on the eve of a test was extremely difficult, Sakharov managed to get through to his mother, who assured him that his father was in no imminent danger. He would stay at his post until the Big Bomb was exploded, but not wait for the test of the Extra.

Sakharov had witnessed the other tests with his own eyes, but this one would be abstract; it would take place near the Arctic Circle, the only sign of success the radio silence that ensued because the "ionized particles released by a powerful explosion interfere with radio transmission; the more powerful the explosion, the longer the communications blackout."

To distract himself from anxiety about his father and the test, Sakharov sat at his desk doing busywork, but waiting is waiting and wears through any disguise. Finally, the phone rang and he heard a voice excited by a magnitude of silence: "There's been no communication with the test site or the plane for over an hour. Congratulations on your great victory!"

As soon as everything was set for the test of the Extra, he left for Moscow, calling his mother from the airport when he arrived to say he was going straight to the hospital. By then, as Sakharov knew, the Extra had already been tested, and he could not resist making a call to learn the result. He was informed that his record was still perfect. No extraneous questions would be nagging his mind when he was at his father's bedside, but those few minutes on the phone cost him guilt as a son.

Dmitri was now seventy-two, old in that unsparing Soviet Russia. But his spirits were good. He described his life in the hospital with humor and insight, observation still affording him pleasure. But he struck more somber notes as the conversation progressed. Dmitri was worried about how his wife would fare without him, and about his other son, the more fragile Yura, who needed care and attention. But the doctors gave Dmitri a reasonable chance of surviving the heart attack well enough to resume a normal life. This could not possibly be the last real conversation.

Still, taking no chances, Dmitri gave his son the bittersweet blessing of honesty, a last bonding in disappointment: "When you were at the university, you said that uncovering the secrets of nature could make you happy. We don't choose our fate, but I'm sorry that yours took a different turn. I imagine you could have been happier."

In the Sakharov family, to be the inventor of the thermonuclear bomb and twice awarded the Hero of Socialist Labor was both too much and not enough.

Not knowing quite what to say, Andrei agreed with his father that, yes, we don't get to choose our fate.

But as Dmitri demonstrated when his son visited him again in December, you do get to choose how you meet that fate, what you draw upon, what you sacrifice. Dmitri had suffered another heart attack but had not informed his doctors. After more than two months in the hospital, all he wanted was to go home. Learning of this from his father, Andrei Sakharov had to take yet another oath of secrecy.

Dmitri was supposed to have been carried up the four flights to his apartment in a chair, but, arriving, he absolutely refused. If this was to be the last time he went in his front door, it would be under his own steam.

He had two days of happiness at home. Before going to sleep on the night of December 15, and referring to Andrei by his childhood name, Dmitri said to his wife: "You don't have to call Adya yet."

Andrei was in fact already in Moscow, intending to visit the next morning, but his father died suddenly in the night. Neither got the fate they would have chosen.

* * * *

Khrushchev kissed Sakharov. All the rancor and ire had been incinerated by the Big Bomb, which Sakharov, doing his work and carrying out orders, had delivered on time for the finale of the party congress, a single explosion exceeding all the bombs dropped in World War II, atomic included. In the splendors of the Kremlin, Khrushchev himself pinned a third Hero of Socialist Labor to Sakharov's suit jacket front. Sakharov was given the seat of honor at the banquet table between Khrushchev and Brezhnev, who took Sakharov's hand in both of his and would not release him until his profusions of gratitude were fully expressed. But Khrushchev hadn't quite forgotten the note and now referred to it in his opening address, but only to say that in the meantime Sakharov's work had been excellent.

But Khrushchev's kiss, forgiveness, approbation meant little to Sakharov with his father only two months dead, the traditionally marked fortieth day after the death having been only weeks ago. He went right from the banquet to see his mother, who gasped at the sight of her son in all his "regalia."

In some ways his parents had misunderstood him. It was the "regalia" that symbolized his power to influence the course of events in ways that, though sometimes unintended, were often to the good. Khrushchev had used the authority of Sakharov's Big Bomb to oust Stalin from Lenin's tomb, the deed done in the dead of night, a dump truck covering the fresh grave with a load of fresh cement. And the scientist in him was still avid. Sometimes too avid. Concerned that the military had no effective means of delivering a Big Bomb, Sakharov dreamed up a "giant torpedo, launched from a submarine and fitted with an atomic-powered jet engine that would convert water to steam. The targets would be enemy ports several hundred miles away. . . . When they reached their targets, the 100-megaton charges would explode underwater and in the air, causing heavy casualties."

Sakharov, who had calculated that, from testing alone, each megaton caused ten thousand deaths, had now been carried away by the grandiose idea of a hundred-megaton giant torpedo. Since torpedoes were naval business, Sakharov consulted with an admiral, who did not at all give Sakharov the hearing he had expected. As a true military man—a type Sakharov had known in his grandfather—the admiral believed in "open battle" and was disgusted, outraged by the "idea of merciless mass slaughter."

Sakharov was humbled, brought back to his senses. He who had preached morality to the military was now preached it by them.

His balance restored, Sakharov in the summer of 1962 was moved to take action against injustice, his sense of symmetry violated by a press account of an old man sentenced to death for counterfeiting a few coins. Appending his titles, he wrote a letter to the Attorney General and received a respectfully prompt reply to the effect that crimes against the socialist economy were a form of treason and in any case the old man had already been executed. In science he knew only the sweetness of success; this incident left him with a "bitter taste."

That summer after his father's death was a time for taking stock. He was now forty-two, fleshier, his hairline receding. He was the father of three, a son having been born in 1957 and named Dmitri, the name of Andrei's father. Sakharov was Andrei Dmitrievich and his son was Dmitri Andreevich, the generations interlocked by patronymics. He had been married to Klava for almost twenty years. The girl who swam the Volga and was trailed by suitors had now become hypersensitive to slights, susceptible to jealousy. Fearing that her husband may have fallen under the influence of the womanizing Zeldovich and that all his honors may have turned his head or made him attractive to scheming women, Klava made jealous scenes—a chance remark from a waitress to Sakharov at the "generals' mess" caused an eruption that could only sadden and baffle Sakharov. Except by Klava, Sakharov could hardly be imagined slinking about the Installation at night. Though misplaced, the emotion was real enough—those who are not happy with themselves are always jealous of those who are. The best of Klava went into caring for her children, who remember her with love as the better parent.

That summer the United States and Britain resumed testing. It was a political and technical necessity for the Soviet Union to enter on another round as well. Sakharov accepted the reality but not the fact that the USSR intended to perform duplicate tests, for which there was no scientific necessity whatsoever.

But the testing of nuclear weapons is never purely a scientific matter; it is entangled in politics on every level from back corridor to world stage. In 1955 a second Installation, Chelyabinsk-70, had been created east of the Ural Mountains. The principle was "socialist competition," rivalry in pursuit of a common goal. There were noticeably fewer Jews at that Second Installation, which, among themselves, the authorities referred to as "Egypt," which made Arzamas-16 "Israel" and its dining hall the "Synagogue." Each Installation would be allowed to test its own bomb.

During that summer overshadowed by his father's death and the looming tests, Sakharov had again to confront the reality that such tests were "a crime against humanity, no different from pouring disease-producing microbes into a city's water supply."

As the Installation director's car, driver, and "secretary" waited, Sakharov strode back and forth with Khariton, making his case against duplicate tests. Khariton declined support on grounds of conflict of interest: "My intervention would give people the wrong idea. Their design differs from ours, and from their point of view, and the Ministry's point of view, that justifies testing both devices." But he let Sakharov know that he was free to act as he saw fit.

Sakharov went next to Efim Slavsky, the new head of the Ministry of Medium Machine Building, whom Sakharov respected as an engineer, considering him an intelligent man. Slavsky, a tough Soviet boss who had put in ten years as a commissar in the Red Army cavalry, was not easy to convince, but in the end he accepted the merits of Sakharov's position, agreeing that two tests were not necessary provided, of course, that the first was successful. Then

Slavsky asked which Installation had the better bomb. Sakharov said that his Installation's was preferable, being "simpler and more reliable." So, Slavsky pointed out, it was only a question of which Installation's bomb would be tested—maybe Sakharov himself could sort it out with the Second Installation.

Sakharov flew out to the Urals to meet with the leader of the Second Installation, his old university colleague, Zababakhin, who had gone with him to defend Altshuler and had given him the paper to write the note to Khrushchev. Sakharov made a presentation, using colored charts as visual aids. But Zababakhin, who had been joined by a half dozen colleagues, would not be moved and would not look Sakharov in the eye as he said: "You can do whatever you want so long as our device is tested first. But if yours is first, we'll insist that our device be tested too."

Sakharov lost his temper, shouting that Zababakhin's position was "tantamount to murder!"

Sakharov returned to Moscow and informed Slavsky that the Second Installation's device should be tested first. What mattered most was that there be no duplicate test. "I've already agreed to that," said Slavsky.

The tests would, however, prove not only duplicate but duplicitous. The scientists at the Second installation modified their design enough so that Slavsky no longer had to view the tests as redundant; besides, they were convinced that their bomb would prove of especially enormous power. The actual explosion was not a failure but a disappointment, which only made Sakharov's own argument for the First Installation's "simpler and more reliable" device now seem all the more relevant.

Arriving back at Arzamas-16, Sakharov was shocked to learn that their device was to be tested the next day. Sakharov again went to Khariton, who again declined to intervene but did not object to Sakharov using his high-frequency phone to call Slavsky to save time. Sakharov accused Slavsky of breaking their agreement. Slavsky argued that the difference in weight between the two devices justified the test; the lightness of Sakharov's would also increase the delivery range of the missile bearing it. Sakharov countered that more than a hundred thousand people would "die for no reason. . . . It's pointless and . . . criminal."

"The decision is final," said Slavsky.

"If you don't call it off, I can't work with you anymore," said Sakharov. "You've double-crossed me."

"You can go to hell if you want," shouted Slavsky. "I don't have you on a leash."

That left only Khrushchev, who was in Ashkhabad. It took several calls to get through.

"I'm listening, Comrade Sakharov."

But then, as in a nightmare, the words wouldn't come, and the connection itself was poor, staticky, voices doubling, fading.

"I don't understand," said Khrushchev. "What do you want from me?"

Then finally Sakharov found the words to request that the duplicate test be postponed until a government commission had heard both sides. Khrushchev promised to turn the matter over immediately to one of his deputies, Kozlov. Sakharov thanked him.

At half past eight the next morning, still in his cottage, Sakharov received a panicked call from his secretary: Kozlov had just telephoned Sakharov at his office. Either Kozlov was giving the matter the immediate attention Khrushchev had promised, or he knew that Sakharov almost always arrived at nine and wanted to get on record as having tried to communicate without actually having to do so. It took almost an hour to track Kozlov down, and all Sakharov heard from him was the general line: the more tests, the sooner the imperialists would agree to a ban, and so the more lives saved in the end. His own argument was turned back on him.

In defiance and desperation, Sakharov called the KGB general on site in charge of the test in the hope of winning a day's time from him, especially if weather conditions were at all unfavorable. But the general informed Sakharov that, on orders from Minister of Medium Machine Building Slavsky, the test had been moved up four hours; the plane had already crossed the Barents Sea and was now approaching target.

"It was the ultimate defeat for me. A terrible crime was about to be committed, and I could do nothing to prevent it. I was overcome by my impotence, unbearable bitterness, shame, and humiliation. I put my face down on my desk and I wept."

CHAPTER NINE

CRITICAL MASS

BARKING edicts, Khrushchev was deaf to the whispers of conspiracy around him, one in which Sakharov's name would ultimately figure. Increasingly, Khrushchev struck many in the inner circle as high-handed and foolhardy, less a reformer than an undoer. He had alienated the Chinese with his constant attacks on Stalin, causing Mao to reply: "The criticism of Stalin's mistakes is correct. We only disagree with the lack of a precise limit to the criticism. We consider that out of Stalin's ten fingers, only three were bad." And it was one thing if the Hungarians or the Poles revolted, quite another when Russian workers took to the streets.

Due to a mix-up that sounds like the premise of a play by a Soviet Gogol, prices were raised and salaries cut on the same day in June 1962 in the southern city of Novocherkassk. The traditional capital of the Don Cossacks, Novocherkassk was an odd place for a disturbance to break out, but the place always did have an uncommon history. After Russia defeated Napoleon in 1814, the Cossacks had ridden into Paris and were so taken with the city they didn't want to waste time eating, constantly yelling at the waiters "Make it quick!," "bistro" in Russian, lending a name to what now seems the most French of institutions. The Cossacks were so smitten by Paris that when they returned to their capital they constructed both an Arc de Triomphe and a Place de l'Etoile in the midst of the steppe. Later French architectural influences resulted in what might be called Cossack Art Nouveau. But all those fine buildings had suffered a precipitous Soviet dilapidation by the time army troops were firing on the workers of the city in June 1962, killing them by the dozens. Mostly non-Russian troops were used to avoid unwarranted sympathies and the dangers of fraternization. Asiatics suppressed Russians and vice versa, one of the advantages of a multiethnic empire.

The trouble in Novocherkassk was bad enough, but it wasn't isolated—there were demonstrations and disturbances in other cities and at the Moskvich Auto Plant in Moscow itself, though none of them was suppressed so bloodily. That was little consolation to those members of the party and bureaucracy whose ideal was Stalinist stability without Stalinist terror. But Khrushchev wasn't about to let them or the rest of the country forget where Stalinism had already once led—to the brick execution cellars and the Gulag. And he soon found a convenient reminder. A manuscript, "One Day in the Life of Ivan Denisovich" by Aleksandr Solzhenitsyn, had come across the desk of Aleksandr Tvardovsky, a poet and editor of the liberal journal *Novy Mir* (New World). He decided to read it at home, in bed. But after ten lines, Tvardovsky stopped: "Suddenly I felt I couldn't read it like this. I had to do something appropriate to the occasion. So I got up. I put on my best black suit, a white shirt with a starched collar, a tie, and my good shoes. Then I sat down at my desk and read a new classic."

This was not a short poem about the harvest that any editor could approve but a matter for the Central Committee, where it was hotly debated—Suslov, the cardinal of ideology, and Kozlov, Khrushchev's deputy and heir apparent, being vehemently opposed. Khrushchev retorted: "There's a Stalinist in each of you, there's even some Stalinist in me. We must root out this evil." Again Khrushchev's passion prevailed, and the novel was published in the magazine and as a separate book that sold a hundred thousand copies within days.

But that was small potatoes compared to the Cuban missile crisis in the fall of 1962. Khrushchev's political logic was fundamentally sound: why shouldn't the USSR have missiles in Cuba when the United States had them in Turkey, which bordered directly on the Soviet Union without even the famous "ninety miles" between them? President John F. Kennedy did not, however, share that logic, and the resulting confrontation cost Khrushchev face in the Politburo, where yielding was equated with defeat. It did not matter to those men of the purest power politics that in some sense Khrushchev had actually succeeded: the American missiles were subsequently quietly removed from Turkey, and a pledge of further nonintervention in Cuba was secured. Objectively, Khrushchev had maneuvered his country into a more secure position, but he had misread his opponent and brought humiliation to the Soviet Union in the international arena. A better player would not have gotten himself into that position in the first place—Stalin had had the bomb for five years and never come close to having to use it.

Sakharov, who thought Khrushchev had rather proved his "mettle" during the crisis, now also found himself involved in nuclear politics again.

The previous summer one of Sakharov's colleagues had reminded him of a proposal, originally made by President Dwight Eisenhower, to ban nuclear tests on land, in the sea, and in space, leaving aside the issue of underground explosions, always more difficult to monitor. Sakharov had floated the idea to Minister of Medium Machine Building Slavsky well before their falling-out. Now, in early 1963, Sakharov found himself necessarily allied with Slavsky, who called him to say: "No matter what happened between us in the past, life goes on and somehow or other we've got to get back on good terms. I'm calling to let you know that there's a great deal of interest at the top in your proposal." Khrushchev proposed the idea in a speech delivered on July 2, 1963, in East Berlin (he too was a "Berliner," replying to Kennedy's famous speech made only the week before). The two leaders who had brought the world close to nuclear conflagration agreed to the ban, which was signed in Moscow on August 5 and became effective as of October 10, a hot line having been installed in the meantime to further reduce the dangers of which the whole world had gotten a heady whiff the previous fall.

Back in harmony with his country's policy, Sakharov was proud of the role he had played in what was officially known as The Treaty Banning Nuclear Weapon Tests in the Atmosphere, in Outer Space and Under Water. As he was aware, the lives saved were all too calculable.

By then his mother was in her final days. After Dmitri's death she both wilted and mellowed, speaking about people and the past with what seemed to Sakharov a "new tolerance." She died in mid-April 1963 just after Easter and was buried with the rites of the

Russian Orthodox Church. With both parents now gone, the world became an even lonelier place for Andrei Sakharov.

It may have been his mother's death, reminding him of the limitations on time, or it may have been his father's remark that Sakharov's real happiness lay in the pure pursuit of science, or it may have been the slackening of work at the Installation, but in that same year Sakharov began work on his first cosmological paper, "The Initial Stage of an Expanding Universe and the Appearance of a Nonuniform Distribution of Matter." The questions he asked were "why galaxies, stars, and planets are precisely as we observe them and not otherwise, and exactly how they formed," how the "small initial inhomogeneities grow."

Small inhomogeneities were also growing in the personal universe of Andrei Sakharov. He was moving both toward purer science and toward the highly impure world of social activism, though once again the ban on testing had removed the source of direct clashes with the establishment. But in June 1964 Sakharov's "indignation toward Lysenko" had "boiled up once again."

One of the candidates for full membership in the Academy of Sciences in the spring 1964 elections was Nikolai Nuzhdin, whom Sakharov saw as "one of Lysenko's closest associates, an accomplice in his pseudo-scientific schemes and in his persecution of genuine scientists." Sakharov made no secret of his opinion, prompting one Academician to say: "Yes, I know he ought to be blackballed. But you wouldn't really dare speak out at the General Assembly, would you?"

"Why not?" replied Sakharov.

In fact there was a good reason for Sakharov not to act alone: Tamm and a few other leading scientists were working out their own concerted plan of attack. For some reason Tamm had not informed or included Sakharov, even though the sins of Lysenko and the future of biology had been recurrent themes in their long conversations at the Installation. For his part Sakharov had grown used to acting alone, both because that was his nature and because he rarely could gather support. He was well aware of the weight his titles and achievements would carry in that setting—he was the youngest member ever elected to that august body and had been elected unanimously. There was also something to be said for just speaking for yourself, as an individual. In any case the war against Lysenko had not yet been won by any means, and Sakharov was glad to be able to duel for the honor of science.

During the opening remarks and reports, Sakharov sketched out his main points on the cover of his program. Lysenko was seated nearby. Always dressed like a good communist with a medal on the lapel of his blue suit jacket, Lysenko had the foxy face of a peasant who has come to the city and learned to part his hair on the side; his light gray eyes were bright with the dementia of his wrong science.

Nuzhdin was nominated and described as an "eminent scientist," a phrase so offensively untrue that Sakharov at once requested the floor. Addressing his fellow Academicians for the first time, Sakharov said: "The Academy's Charter sets very high standards for its members with respect to both scientific merit and civic responsibility. Corresponding member Nikolai Nuzhdin . . . does not satisfy the criteria. Together with Academician Lysenko, he is responsible for the shameful backwardness of Soviet biology and of genetics in particular, for the dissemination of pseudoscientific views, for adventurism, for the degradation of learning, and for the defamation, firing, arrest, even death of many genuine scientists.

"I urge you to vote against Nuzhdin."

The hard-liners in the audience were the first to react, shouting imprecations, but applause was only seconds behind, cascading down from the rear, where guests and corresponding members were seated. Lysenko glared at Sakharov, who was descending from the stage's center stairs to applause that now obliterated the shouts and catcalls.

Sakharov returned to his seat passing Lysenko who, now in a fury, shouted: "People like Sakharov should be locked up and put on trial!"

When the uproar died down, Tamm and his group launched their own attack, calling Nuzhdin a "forger of true science." Glad to see allies enter the battle, Sakharov was also embarrassed to think that he might have inadvertently stolen some of their thunder. After an hour of stormy debate, the Academicians left the auditorium for the ballot boxes in the lobby. Sakharov found himself rather enjoying the pleasures of the hero—the smiles, the handshakes, the excited words of praise.

Nuzhdin's candidacy was roundly rejected—114 to 23.

Sakharov had succeeded in reminding his fellow Academicians that "science remains a keystone of civilization and unwarranted encroachment on its domain is impermissible." The Politburo cannot rule on genetic transmission. And, in doing so, harms it own interests. All this so infuriated Khrushchev that he threatened to shut the Academy down: "First Sakharov tried to stop the hydrogen bomb test, and now he's poking his nose in again where it doesn't belong."

It was to be the last of their clashes. At around the same time that Sakharov was speaking openly at the Academy elections, Leonid Brezhnev was speaking quite conspiratorially to Semichastny, the head of the KGB, about the aging, erratic, and dangerous Khrushchev, finally coming to the point: "Can we poison him?" According to his own self-serving account of the meeting, Semichastny protested indignantly: "I'm not a plotter, and I'm not a murder." These weren't Stalin's times anymore; those methods "just won't fly."

Brezhnev, a hail-fellow-well-met party functionary, preferred assassination because it would spare him the necessity of assembling a coalition against Khrushchev and coming out against him in a showdown at the Politburo. Irresolute by nature, he dreaded confrontation. But now he had no choice but to begin the slow work of building a constituency. Some allies were won more easily than others—to Suslov, Khrushchev must have always seemed the bull in the china shop of ideology. By early fall Brezhnev had his bloc.

In October Khrushchev was on vacation on the Black Sea. Several times Brezhnev picked up the phone to summon Khrushchev back to Moscow for an emergency Politburo meeting, and every time he replaced the receiver without having made the call. Suslov finally put the call through himself and, shakily backed up by Brezhnev, summoned Khrushchev to Moscow.

Enraged when not officially greeted at the airport, Khrushchev raced to the Kremlin and burst in on the Politburo meeting: "What's going on here?"

When Suslov informed Khrushchev that the subject of the meeting was his dismissal, Khrushchev left the room and phoned the defense minister, ordering him as commander in chief to arrest the conspirators at once. But as a true communist, the minister of defense was against any cult of personality and would act only on orders from the Central Committee. KGB chief Semichastny proved equally principled.

Khrushchev rushed back into the meeting to argue with Suslov, who read out the fifteen counts against him, listing in detail all the "grave mistakes." His worst sin was weakening the party, by the "wanton proliferation of administrative bodies," while at the same time concentrating a "great deal of power in his own hands" and abusing it. The humiliation over Cuba, the break with China, which the previous day had exploded its first atomic weapon, corruption, a series of harebrained schemes—the list was long. The fourteenth count was failing to heed Academician Sakharov's protest against "Lysenko's nonsense."

Khrushchev refused to go quietly. The meeting was adjourned. But reality set in overnight, and the following day Khrushchev resigned for "reasons of health" and was replaced by Brezhnev. Arriving home that evening, Khrushchev threw his briefcase in a corner and passed judgment on his own career: "Well, that's it, I'm retired now. Perhaps the most important thing I did was just this—that they were able to get rid of me simply by voting, Stalin would have had them all arrested."

Not that he hadn't tried.

* * * *

Among the many problems Brezhnev inherited was the invaluable yet vexing Andrei Sakharov. Brezhnev's father had thought that inventors of new weapons should be hung on a high hill, but he himself took a more therapeutic approach: "Sakharov has some doubts and inner conflicts. We ought to try to understand and do all we can to help him." Brezhnev's understanding reflected party reports on Sakharov, which characterized him as ideologically immature and apolitical and thus easily influenced. Brezhnev's solution: Sakharov should join the party. The invitation was proffered in private by a regional party secretary. Making a distinction that could not have been to the liking of any communist, Sakharov declined, saying he "could be of greater use to the country . . . outside Party ranks."

The invitation was a signal and a test. Sakharov had been put on notice by the new regime: he was being monitored as a potential problem. And though Sakharov would continue to write "non-Party" in the appropriate slot when updating his questionnaire, that status had now been redefined by yet another active refusal.

Still, to both the party and to Sakharov, this incident was only a minor perturbation. Sakharov forged ahead with bomb work, though his motives had changed: "I still believed that my presence at the Installation might prove decisive at some critical moment, and this was one of the reasons I didn't leave to 'do science' like Zeldovich." He would rather make his second return to pure science, but again he felt his duty lay elsewhere. But 1965, Brezhnev's first full year in office, was not a year of any particular crisis. The First and Second Installations were still competing, though now some of the focus had shifted to underground explosions, nuclear-powered space flight, and nonmilitary applications: explosions to open mines, release oil reserves, cap blowouts. Military applications were not slighted. At around this time Sakharov began researching the problems of antiballistic missile systems (ABMs), which he found cheaply circumventible and highly destabilizing. Once again his research made him aware of the "horror, the real danger, and the utter insanity of thermonuclear warfare."

But his main worry now was Klava. She had suffered severe gastric bleeding in September 1964 and again in April 1965. He thought this may have been a delayed consequence of the hydrogen sulfide poisoning she had experienced at the munitions factory during the war. The dreary round of doctors began. Ulcers were diagnosed.

Klava was under great stress. After forty-five years of marriage, Klava's parents had now divorced, her mother going to live with her sister in Leningrad, her father staying on in Ulyanovsk in the one-room wooden house with its tiled stove. It could only mean that in the end her mother still loved the soldier who had died in the First World War more than she loved her husband. And she had never much loved Klava either, always preferring the younger sister. According to Klava's daughter Tanya, Klava's own mother had said something unforgivable: "It would have been better if Andrei married your sister; they would have made a better pair." Klava now barely spoke to either her mother or her sister.

* * * *

Sakharov was hardly the only problem that Brezhnev had inherited from Khrushchev, now melancholically tending his tomatoes in retirement. Khrushchev had been the first to buy capitalist wheat with Soviet gold, and that in a country that before the revolution had consistently exported grain to the West. Brezhnev found himself doing the same. To win the peasants' favor, he allowed an increase in the size of their private plots, all that kept people from going hungry. Food held back from market by supporters of the Brezhnev coup in order to lay a groundwork of dissatisfaction with Khrushchev was now released—sausage to assuage discontent. To gain the scientists' support, Brezhnev accepted their verdict on Lysenko, whose theories certainly hadn't helped agriculture much. Industry was not in a growth phase, but that was the result of Khrushchev's "harebrained" schemes and should straighten out in time.

But what needed immediate attention was ideology. People had lost the habit of belief and obedience. Khrushchev had subjected Stalin's twenty-four-year reign to withering criticism, and his own ten years had now to be presented in a less than positive light. Next year would be the fiftieth anniversary of the revolution, but if thirty-four of those fifty years were bad, that meant that the Soviet experiment was precisely 68 percent a failure. Any peasant knows that a harvest that is 68 percent bad is no cause for celebration.

From the very beginning Lenin's party was a minority that imposed its will on society because of its obligation to lead the country into the first just society on earth. Party rule was legitimized by the success of the revolution and sanctioned by Marxist doctrine. The revolution and party rule were in theory the inevitable results of a near mystical trinity of historical law—the dialectic—in which thesis gives rise to antithesis, the clash of opposites resolved in synthesis: capitalism gives rise to the working class, which destroys it and creates socialism. The validity of the doctrine and the value of the revolution were beyond dispute. Discussion was one thing, heresy another. On this point Brezhnev was clear: "Our Party has always warned that in the field of ideology there can be no peaceful coexistence."

In September 1965 two writers were quietly arrested in Moscow. Andrei Sinyavsky and Yuli Daniel had probably first come to KGB attention when they served as pallbearers at Pasternak's funeral in 1960. (After the death of a Soviet leader, the emerging leaders could be guessed by their roles at the funeral, and in Pasternak's case something of the same held true for the literary opposition.) Sinyavsky and Daniel had been sending manuscripts abroad and publishing under pen names—Abram Tertz and Nikolai Arzhak—but the vigilant intelligence agencies had tracked them down. And if this trial was to set an example, they had chosen well. Sinyavsky especially fit the profile—he questioned everything, merrily turning the dialectic against itself:

> So that prisons should vanish forever, we built new prisons. So that all frontiers should fall, we surrounded ourselves with a Chinese Wall. So that work should become relaxing and a pleasure, we introduced forced labor. So that not one drop of blood be shed any more, we killed and killed and killed. . . . Sometimes we felt that only one final sacrifice was needed for the triumph of Communism—the renunciation of communism.

But the country had changed in the Khrushchev years; there were now a small number of people who had lost their fear. Among them was a brilliant, madcap mathematician, Alexander Esenin-Volpin, the son of Sergei Esenin, the popular poet who at thirty—disillusioned with Russia, communism, and love—had slit his wrists and written a poem in blood: "In this life to die is nothing new, / but living's nothing newer either." Esenin-Volpin had sized up the new leadership, which, though headed by Brezhnev, was in fact a rare instance of collective leadership: "Stalin's successors . . . lacked the imagination and courage to follow in the footsteps of their leader." He also hit upon an ingenious, obvious strategy—to always act in accordance with the Soviet Constitution and call on the authorities to do the same. Participating in a demonstration in Pushkin Square on December 5, 1965, Soviet Constitution Day, he held up his banner: "Respect the Soviet Constitution!" That of course did not prevent him from being arrested along with the rest, who were mostly there to protest the arrest of Sinyavsky and Daniel. The attempt to hold authors legally and politically responsible for sentiments of fictional characters struck the intelligentsia as Stalinist. Stalin's daughter, Svetlana Alliluyeva, was particularly incensed: "Now you can be tried for a metaphor, sent to a camp for a figure of speech!"

Official announcement of the arrests came only in January 1966 in an *Izvestiya* article whose title, "The Turncoats," could easily have prejudiced a trial had it, of course, been prejudiciable.

None of this made any impression on Sakharov at the time, though he too had caught the scent of Stalinism back in the air. A party congress was scheduled for the end of March. Stalin had been denounced in two previous party congresses; if the Stalinists wanted to reassert themselves, now was the time. A collective letter against any such rehabilitation began circulating among the intelligentsia, both the scientific and the cultural—the physicist Kapitsa would sign, the ballerina Plisetskaya. Now, after the acclaim Sakharov had won for his speech at the Academy elections, the letter came to him, and he added his name.

In February, Sinyavsky, who had brought the entire socialist experiment into question, was given the maximum sentence, seven years; Daniel, perhaps because of a good war record, five. Two young men made a transcript of the trial, which would soon be available in samizdat, the underground press—carbon copies made by selfless typists. Samizdat meant self-publishing; as one dissident, Vladimir Bukovsky, said: "I write it myself, censor it myself, print and disseminate it myself, and then I do time in prison for it myself."

Excellent books were circulating in samizdat; like the collective letter, they came to Sakharov. The dissident Marxist historian Roy Medvedev brought him portions of his *Let History Judge,* the first comprehensive history of Stalin's reign that Sakharov had encountered. He did not necessarily agree with all the book's conclusions, but it gave him material for his own "big picture." And for harrowing detail he had *Journey into the Whirlwind* by Eugenia Ginzburg, a memoir by an ardent communist who, when she was arrested, became suddenly clear in mind: "For the first time in my life I was faced by the problem of having to think things out for myself—of analyzing circumstances independently and deciding my own line of conduct."

By the fall of 1965 Sakharov was doing just that, signing another appeal, this one against the pending enactment of Article 190-1 of the Russian Criminal Code: Circulation of Fabrications Known to Be False Which Defame the Soviet State or Social System. That along with Article 70 (Anti-Soviet Propaganda) were clearly going to be judicial weapons for suppressing anything perceived as dissent, a fetish for legalism one of the Stalinist perversions. The letter was signed by twenty people, leading figures in the arts such as the novelist Vladimir Voinovich and the composer Dmitri Shostakovich as well as such scientists as Tamm and Zeldovich.

This time Sakharov not only signed the letter but a few days later on his own initiative telegrammed his concerns directly to the chairman of the Supreme Soviet. He did not receive a reply, but that did not in the least mean his telegram had gone unnoticed. At best he would have received a formal response, and he was about to get more interesting mail.

Home in Moscow in early December 1966 Sakharov found two unsigned sheets of onionskin paper in his mailbox. The first of them recounted the arrest and forcible psychiatric hospitalization of a man who had worked on Constitution II, a draft of a new constitution that was meant to stimulate discussion on political issues. Of course, the Soviet Union already had one of the world's most liberal constitutions, even allowing any of its constituent republics to secede, though as Stalin had remarked: "Recognition of the right to secede does not mean the recommendation to secede." Similarly, anyone who thought a new constitution was needed was clearly insane and should be placed in a psychiatric institution, a punishment now used with increasing frequency.

The second sheet contained an invitation to attend what had been slyly designed as the mildest of all possible demonstrations: a little before six in the evening on December 5 people would gather around the base of Pushkin's statue on Pushkin Square, the last great square on Moscow's main thoroughfare, Gorky Street, before Red Square itself. At the stroke of six everyone would remove their hats and observe a minute of silence out of respect for the Constitution and solidarity with political prisoners. Anyone watching would only see people removing their hats. Not that the appropriate agencies weren't fully aware of the impending event, its true significance, and its organizer, Esenin-Volpin.

Sakharov decided to attend. Klava didn't object, though she did say it was "an odd thing to do."

He took a taxi to Pushkin Square from their Moscow apartment that Klava so much preferred to the Installation. But in both places Sakharov was an object of constant security and surveillance, a fact of life that he both accepted and disregarded. Dusk fell early that December day in Moscow, and the KGB had to use infrared film. What the cameras saw was a few dozen people gathered around the base of Pushkin's statue, some of whom removed their hats at the stroke of six while the others, being KGB, did not.

Sakharov recognized none of his fellow doffers, though it was their presence, not their identity, that mattered. After a minute of silence, people put their hats back on but did not disperse at once. Sakharov too wanted to prolong the moment and read aloud to himself the lines of Pushkin's poem inscribed on the base of the statue:

> And long by the people will I be loved
> For I have struck the chords of kindness
> And sung freedom's praise in this cruel age,
> Calling for mercy to be shown the fallen.

* * * *

Nineteen sixty-seven was the fiftieth anniversary of the revolution, an occasion, in the party's view, best marked with enthusiastic unanimity. But, in an irritating twist, the trial of Sinyavsky and Daniel had not silenced the intelligentsia but emboldened them.

Alexander Ginzburg and Yuri Galanskov, the two young men who had compiled the transcript of the trial, sent copies of their white book to high officials in the government, the KGB, and the

Supreme Soviet. If one is acting constitutionally, there can be nothing to hide. The government saw it otherwise, and in mid-January Galanskov was arrested. A small group protested the arrest on Pushkin Square and were themselves arrested. Ginzburg was picked up the next day.

Finally, the Sinyavsky-Daniel case, the most important of the time, both in content and consequence, came to Sakharov's still "hermetic" attention and was quickly consequential for him as well. He signed a letter to Brezhnev in defense of Ginzburg and Galanskov, his "first intervention on behalf of specific dissidents."

Sakharov had refused to join the party, spoken out against re-Stalinization, and now had signed this letter. "We'll be taking measures," said Minister of Medium Machine Building Slavsky, speaking of Sakharov to a party group at the Second Installation. Measures were taken: Sakharov was soon removed from his post as head of the Department of Theoretical Physics but remained deputy science director of the Installation, and his salary was reduced by 45 percent. Strong gestures, but ones that also meant he was not yet considered a lost man. Yet there had been a qualitative change in his life; accolades had turned to retribution.

In April Stalin's daughter, Svetlana Alliluyeva, defected. Russia brings the most improbable people together; Svetlana had been friends with Sinyavsky when they both were researchers at the Institute of World Literature. Svetlana had a taste for literature, and a talent, as she demonstrated in two books, one of which, *Twenty Letters to a Friend,* offers a unique daughter's-eye view of the tyrant. The arrest and trial of her friend for a "figure of speech" meant that Russia was persecuting its writers again, and she'd had enough of that to last a lifetime, even if defecting meant leaving her children in Moscow. Her dramatic break was both big news and symbol. The USSR countered with a propaganda barrage, alleging corruption and CIA complicity.

A fatal breach of security had occurred. In the inner circle, Suslov demanded that Semichastny, the head of the KGB, be replaced. Semichastny had demonstrated his loyalty to Brezhnev when refusing to arrest the Politburo on Khrushchev's orders, but losing Svetlana during the celebration of the fiftieth anniversary of the revolution was unpardonable. Semichastny was replaced by a dynamic, intelligent comrade, Yuri Andropov, who would become Sakharov's nemesis.

But there would be no immediate encounter, Sakharov having turned his attention from the present to the future. A futurological essay had been commissioned from above—the task was to project the development of science and technology in the upcoming decades and to touch on "more general questions if . . . so desired." Sakharov found that he did so desire. Though the essay was published in a collection titled *The Future of Science* on a limited-distribution basis, it was in fact only the first draft of a much greater work.

The essay had stimulated Sakharov's desire to finally elaborate a "big picture" of his own, but time for speculation was limited. The Installation was still demanding and daily life full of interruption and crisis. Klava's health was not improving, as Sakharov could see for himself when back in Moscow in late June and early July. And pressing, specific injustices required reaction. While in Moscow he was given a letter by Daniel's wife, Larisa Bogoraz, who had visited her husband in the labor camp and found his situation "desperate." Sakharov took the letter with him back to the Installation and called Andropov on the high-frequency phone in his office. He may have been in a state of semidisgrace, but his prestige was still great enough to get him through to Andropov, who assured him that the matter was already under advisement.

It was a happy coincidence of Sakharov's nature and the basic stance of the nascent human rights movement to act openly and legally. Yet Sakharov was not without a trace of guile. When Andropov asked Sakharov to send him the original letter written by Daniel's wife for his "collection," Sakharov pretended he "had misunderstood, and mailed him a retyped copy."

Six weeks later Sakharov received a call from the deputy attorney general, acting on Andropov's request, informing him that Daniel and Sinyavsky would be released as part of an amnesty celebrating the fiftieth anniversary of the revolution.

Sakharov would later be told that political prisoners were excluded from the amnesty at the last moment, meaning that he had not initially been lied to, but one thing was clear: the regime would pardon thieves and rapists but not heretics like Sinyavsky and Daniel.

* * * *

Sakharov's string of fortuitous writing assignments continued. The subject he now was given was the present-day role and responsibility of the intelligentsia. Reality, he concluded, was far too complex to be dealt with by anything less than complete freedom of discussion. But even that attempt at radical common sense immediately encountered obstacles. The woman who did his typing begged off doing any further work for him. She claimed family matters; he suspected a visit from the KGB. And the newspaper that had originally commissioned the article now found it too hot to handle without permission from the top. Accordingly, Sakharov sent the manuscript to Suslov along with a letter that argued against ABMs on purely economic grounds. ABMs were three to ten times more expensive than equivalent offensive systems. The United States could afford them; the USSR could not. He cited statistics, indicating he had begun doing research on the American economy.

The article itself, "World Science and World Politics," was an amalgam of the daring and the retrograde, an artifact of a mind struggling to liberate itself. He could write: "The credo of progressive scientists and the progressive intelligentsia the world over is the open and unconstrained discussion of all problems, including the most pointed." But he was also capable of concluding the article: "Shoulder to shoulder with the working class, opposing imperialistic reaction, nationalism, adventurism, and dogmatism, scientists and the intelligentsia should be aware of their power as one of the main supports of the idea of peaceful co-existence."

Whatever the exact mix of sincerity, tactical considerations, and rituals of linguistic fealty in Sakharov's article, Suslov found it "interesting, but unsuitable for immediate publication since its ideas might be interpreted incorrectly."

The urge to write something with scope was stronger than ever as Sakharov was now ingesting information from a great variety of sources. He learned a great deal both from what he read in samizdat and from conversations with the people who brought him the typescripts or the collective letters to sign. Writing on the role and responsibility of the intelligentsia, he was writing about himself.

Again immediate causes kept him from work on his own "big picture," but in time nearly every such interruption would prove of value. As elsewhere in the world, in the fifties and sixties people in the Soviet Union began to become aware of the destruction wrought by pollution. The emerging Soviet environmental consciousness focused on a holy of holies, the Siberian Lake Baikal, more than a mile deep, holding one-fifth of the earth's freshwater and so capacious that it would take all the world's rivers a year to fill it. Over one thousand species of the lake's flora and fauna were found nowhere else.

Siberia was an ambivalent symbol for Russians: the place of exile and hard labor but also of pristine nature, the wide open spaces where a person could breathe freer with every mile of woods he put behind him. Now Siberia was being despoiled by the shortsighted state. For reasons of bureaucracy and prestige, a hazardous rayon-cord factory for airplane tires was still scheduled for completion on Baikal even though the aviation industry had already changed over to metallic cord. There were also dangers from deforestation and timber rafting, industrial effluents, chemical wastes.

The despoliation of Baikal had been an issue for ten years by the time Sakharov was visited by a student from the Moscow Institute of Energy who represented the Young Communist League Committee to Save Baikal. It was precisely the sort of cause that would magnetize ardent and intelligent Young Communists and cut across the usual political lines. Sakharov began attending meetings at the Young Communist League's building, where he learned that, even after it was determined that the location was seismically active, the rayon-cord factory project had been assigned to the Ministry of Medium Machine Building. "Do you know who's in charge of the murder of Baikal? Your own Slavsky!"

Sakharov not only joined in the collective effort but took two independent actions of his own, as was lately becoming his habit. After first conducting independent research, he declared himself fully ready to support the committee's well-documented position:

"We proposed that the lake shores be closed to new industry and that existing enterprises be moved. We calculated the expense of such relocation and showed that it was not excessive—far less than had already been spent on the Baikal project."

His second independent action, done for "good measure," was to call Brezhnev in what would be their last conversation. Affable and waffling as always and pleading overwork, Brezhnev promised to hand the matter over to Kosygin, the official head of state, who proved all too willing to accept assurances that the water purification system and other safeguards were completely reliable.

The outcome was utterly predictable: construction continued apace, and toxic wastes wreaked havoc.

Sakharov wasn't the only one marring the fiftieth anniversary of the revolution with his protests and appeals. Aleksandr Solzhenitsyn was publicly calling for the abolition of Glavlit, the office of censorship, which "is not provided for in the Constitution and is therefore illegal." The reaction against him, which had been mounting since the publication of *One Day in the Life of Ivan Denisovich* in 1964, now became vicious. He was said to have surrendered to the Germans during World War II, in which he had in fact fought bravely until he was arrested for making an unflattering remark about Stalin in a letter. A lecturer in Moscow delivered the ultimate anathema: "The person known to you as Solzhenitsyn is really Solzhenitser, and he's a Jew."

Solzhenitsyn's call for an end to censorship elicited a passionate response in Czechoslovakia not only from the intelligentsia but from the new leadership. The forty-six-year-old Alexander Dubcek, who became first secretary of the Czech Communist Party in January 1968, had exactly the kind of background to recommend him to the Soviet Politburo: the son of a communist who had gone to the USSR in the thirties to build socialism, Dubcek had graduated from the Higher Party School in Moscow and fought as a partisan in World War II. As a member of the Czech Communist Party, he was in liaison with the Soviet KGB. But he was in fact that most dangerous sort of committed communist, one who believed in the system—that it could and should be reformed. It was not long before Brezhnev was disillusioned: "Our reliance on Dubcek has not been justified," he said to KGB head Andropov, who replied: "The situation is serious. It is reminiscent of the Hungarian events."

Sakharov was also closely monitoring those developments on the BBC and the Voice of America, having bought a shortwave radio like Tamm. "What so many of us in the socialist countries had dreamed of seemed to be finally coming to pass in Czechoslovakia. . . . Even from afar, we were caught up in all the excitement and hopes and enthusiasm of the catchwords: 'Prague Spring' and 'socialism with a human face.' "

In early 1968, which had been designated the Year of Human Rights by the United Nations, Sakharov was invited to write another far-ranging article. Three forces—the events in Prague with their euphoria of imminence, Sakharov's own need to order disparate insights into a coherent whole, and the smarting bitterness of past insults and failures—fused to generate a work that far exceeded anyone's expectations, the author's included. He wrote like his father, slowly, painstakingly. Not only because exact formulation was always difficult, but because the writing could proceed no faster than another process, what Chekhov had called squeezing the slave out of oneself drop by drop. Sakharov would quote him on that. His passion for accuracy and symmetry far outweighed any for grace of exposition, which caused him to wince in his self-critical modes. But the work was exhilarating, and to one person who worked closely with him at this time on another project Sakharov seemed "a perfectly happy person." Style was a minor matter compared to what mattered most: to speak freely. Like the breaking events in Czechoslovakia, Sakharov's thinking increased exponentially in scope and daring. The work he was writing not only coincided with the Prague Spring; it was his own Prague Spring.

After several drafts, by late April Sakharov felt that his long essay, "Reflections on Progress, Co-Existence, and Intellectual Freedom," was ready to be typed and shown. Certain that his activities were of more interest than ever to the KGB and with a sharp distaste for the surreptitious, Sakharov gave his manuscript to a typist at the Installation. She had secret clearance, which obliged her to report any such questionable manuscript to the KGB's ideology department. And, as in many of Sakharov's choices, here too there was a strong dose of enlightened common sense: why make any effort at concealment when the whole idea is to make the thing public?

Sakharov flew back to Moscow from the Installation in late April with a typed copy in his briefcase that he swapped that very evening to Roy Medvedev in exchange for the last chapters of his monumental indictment of Stalin. No longer just a reader of underground literature, Sakharov had become a samizdat writer. A few days later Medvedev returned with comments from readers to whom he had shown the essay and with the offer to have carbon copies made and circulated. To foil any KGB bugs in Sakharov's apartment, their conversation took place in writing. Medvedev warned that a copy might end up abroad. Sakharov replied that he had already "taken that into account."

Sakharov's first effort at comprehensive independent thought went out into the world in May 1968 in a very small edition but one that quickly found its way to a great variety of readers, Andropov among them. Summoning Khariton to his office, Andropov opened his safe only long enough for Khariton to glimpse Sakharov's incendiary typescript. Andropov wanted the head of the Installation to deliver a message of his displeasure to Sakharov. Khariton fulfilled the request in the first week of June when Sakharov was traveling aboard his private railroad car back to the Installation. Khariton told Sakharov that Andropov's agents were finding the essay everywhere; it turned up in every raid. And Andropov wanted it withdrawn. Amused by the realization that Andropov had not trusted Khariton enough to let him actually read the essay, Sakharov offered him the chance to make up his own mind, handing him a copy.

They met the next morning.

"Well, what did you think?"

"It's awful."

"The style?"

"No, not the style. It's the *content* that's awful!"

"The contents reflect my beliefs. I accept full responsibility for circulating my essay. It's too late to withdraw it."

Andropov and Khariton were not the only people concerned about the effects of Sakharov's manuscript. Klava, in worsening health, was of two minds. Of course he was free to do as he saw fit, but the consequences for the family also had to be taken into account. Sakharov was still protected to some degree by his rank, but his letter to Brezhnev had already cost him his post and nearly half his salary. This time it could be even worse.

To some extent Sakharov was blinded by his own sincerity. He had written his essay as an act of loyal opposition, out of the conviction that the "government I criticized was *my* government." But the key factors were still imponderables—the Brezhnev regime had not yet showed its hand, and Sakharov could not predict the explosive power of his first literary "device."

On the evening of July 10, at the Installation, through short-wave static, he heard that the Dutch newspaper *Het Parool* had published an essay by A. D. Sakharov. Now that there was no longer the slightest possibility of turning back, he experienced "the most profound feeling of satisfaction."

An effort to apply scientific rigor to human affairs, "Reflections" was intended to invite discussion. Its central premise, contained in the "General Statement," is that the danger of nuclear war has created a change in human affairs so great as to constitute a discontinuity with the past. War, which for Clausewitz had been a continuation of politics by other means, had now been transformed into "universal suicide." From that fact all sane policy must flow. Since the "division of mankind threatens it with destruction . . . any action increasing the division of mankind, any preaching of the incompatibility of world ideologies and nations is madness and a crime."

Critical and complex, these problems require complete intellectual freedom for their solution. But that freedom is under attack by stupefying mass culture, ideologies such as Stalinism, fascism, and Maoism (the Cultural Revolution then running rampant in China), and bureaucratic regimes relying on censorship. Resurgent Stalinism is the enemy at home. Sakharov called for Stalin's expulsion from the party, rehabilitation for all victims, and a full disclosure of his crimes. Sakharov wasn't waiting for that full disclosure and estimated that 10–15 million were put to death under Stalin; by 1939 half of the party—1.2 million members—was under arrest. And of those, 1,150,000 never returned. "Is it not highly disgraceful and dangerous to make increasingly frequent attempts . . . to publicly rehabilitate Stalin, his associates and his policy."

Though sensitive to the differences, Sakharov saw Hitlerism and Stalinism as, in essence, one, a point of view that went very much against the Soviet grain. They both led to the same end: "We shall never forget the kilometer-long trenches filled with bodies, the gas chambers, the SS dogs, the fanatical doctors, the piles of women's hair, suitcases with gold teeth, and fertilizer from the factories of death."

Sakharov divided his essay into three parts, the "General Statement" and the two main sections: "Dangers" and "The Basis for Hope." The dangers, as usual outnumbering the hopes, include police dictatorships, trials like those of Sinyavsky and Daniel, Ginzburg and Galanskov, which he cites by name, and the "crippling censorship. . . . Dozens of brilliant writings cannot see the light of day. They include some of the best of Solzhenitsyn's works, executed with great artistic and moral force." Other dangers demonstrate that the real problems of the mid–twentieth century are by their very nature international. Industrial pollution and fallout ride winds and rivers without respect to boundaries and are no better for being localized, for example, the "sad fate of Lake Baikal." Flash points like the Middle East and actual wars like that in Vietnam, where the Tet Offensive was raging, are painful, glaring instances of division. He was unequivocally against the war in Vietnam, his position coinciding with the Soviet government's: the "forces of reaction . . . are violating all legal and moral norms and are carrying out crimes against humanity." His attitude toward Israel was more ambivalent: the War of Independence in 1948 and the Six Day War were justified; the 1956 campaign and the treatment of refugees were not.

It is when dealing with another danger—hunger, overpopulation, and the psychology of racism—that Sakharov first loses perspective. He sees the Malthusian dilemma reaching a crisis within some twelve years and says that a "fifteen-year tax equal to 20 percent of national incomes must be imposed on developed nations." Knowing that the "white citizens of the United States are unwilling to accept even minimum sacrifices to eliminate the unequal economic and cultural position of the country's black citizens," he nevertheless calls on Americans to accept a "serious decline in the . . . rate of economic growth . . . solely for the sake of lofty and distant goals, for the sake of preserving civilization and mankind on our planet."

It is clear from the letter Sakharov wrote to Suslov and from citations in "Progress" that Sakharov did some research on the United States; he refers to Norbert Wiener's *Cybernetics,* a *Scientific American* article of March 1968, economic statistics. But his image of America is incomplete and necessarily confused, if only because the America of 1968 was itself confused, a country at war with Vietnam and with itself. Martin Luther King was assassinated just as Sakharov was finishing his manuscript, the riots in American cities delighting the KGB—"American blacks were setting the cities on fire all on their own," as one general put it. And Bobby Kennedy was murdered two days before Sakharov heard on his shortwave radio that his essay had been published.

Still, Sakharov held the United States in both too high and too low regard. He was a member of the Russian intelligentsia who did not care what sort of coat he wore and who for relaxation read Goethe—the epigraph in "Reflections" is from *Faust:* "He alone is worthy of life and freedom / Who each day does battle for them anew!" Sakharov was naturally repelled by America's wealth, its unjust distribution, and its stupefying popular culture. Though Sakharov in his essay has some harsh words for the United States and a good many for Mao's China, his prime concern is for his own country, on whose fate hinges the fate of socialism. As a man of "profoundly socialist" views, he worries that "socialism and the glorification of labor" might lose out to the "egotistical ideas of private ownership and the glorification of capital" if the Soviet Union fails to deal with humanity's crises in a free and open manner, as the Czechs are now doing. He is speaking from "within," for the good of the cause. For that reason, and as a true member of the intelligentsia, he is morally and intellectually obliged to do more than rebuke and warn; he must also demonstrate positive alternatives and the means of implementing them.

Divisiveness is the disease, says Sakharov, and convergence the cure. Capitalism and socialism, roughly equal economically and militarily, each with its glorious achievements and hideous failings, must adopt the best features of each other's system, a process begun by downplaying differences and expanding common ground. In the West the agents of change will be the "leftist reformist wing of the bourgeoisie," and in the East the democratic socialists, the people of the Prague Spring. The resulting hybrid will be more socialist than capitalist, Sakharov stressing that although "convergence" is a "Western" term, he employs it with a "socialist and democratic meaning."

Socialism's real enemy is not capitalism but dogmatic Marxism. Sakharov in fact accepts the initial principle of the dialectic: "The capitalist world could not help giving birth to the socialist." But what Marx could not foresee and what the dogmatic Marxists refuse to see is that the danger of nuclear war has rendered dialectical clash into suicide.

And the dialectic itself is demonstrably faulty—capitalism, having given birth to socialism, shows no sign of being sapped of its powers and becoming decadent. On the contrary, capitalism is proving surprisingly healthy and flexible enough to adopt socialist welfare practices, thereby also obviating any possibility of revolution. As "typical representatives of the reformist bourgeoisie," Sakharov singles out Cyrus Eaton, Franklin Roosevelt, and "especially, President John F. Kennedy."

Well aware of the "primitiveness of his attempts at prognostication" and inviting "positive criticism," Sakharov presents a Four-Stage Plan for Cooperation. Since he accepts the premise that socialism is the driving force of history, it is only natural that the first stage, 1960–1980, revolves around events in the socialist camp. Ideological conflict between "Stalinist and Maoist forces" and the "realistic forces of leftist Leninist Communists" will lead to a worldwide schism in the world of the left. Ultimately the "realists" will prove victorious—this is, after all, an avowedly "most optimistic" scenario. Their foreign policy will be one of coexistence, while domestically they will concentrate on economic reform, democratization, and, if necessary, a multiparty system, though Sakharov says he is "not one of those who consider the multiparty system to be an essential stage in development or, even less, a panacea for all ills." Between 1972 and 1985—dates overlap in this scheme—the leftists in capitalist societies, aided by the workers and intellectuals, follow the example of the socialist world and launch their own reforms, peaceful coexistence their foreign policy as well. He calls not only for battle against racism and militarism but for "collaboration with socialism on a world scale and changes in the structure of ownership."

When both countries have reformed their systems, they are ready to cooperate in "saving the poorer half of the world" by means of the 20 percent tax. A worldwide socialist utopia will begin to emerge: "Gigantic fertilizer factories and irrigation systems using atomic power will be built. . . . gigantic factories will produce synthetic amino acids."

Since the reasons for distrust have been removed and the two systems are actively cooperating, this third stage, 1972–1990, is the obvious time for disarmament to become a reality.

The ultimate convergence is world government, which comes into being during the fourth stage, somewhere between 1980 and 2000. Humanity will then be free to concentrate on its two great tasks: the exploration of space, requiring "thousands of people to work and live continuously on other planets and on the moon, on artificial satellites and on asteroids whose orbits will have been changed by nuclear explosion"; and the breakthroughs in biology, which "will make possible effective control and direction of all life processes at the levels of the cell, organism, ecology, and society, from fertility and aging to mental processes and heredity."

With his usual blend of self-deprecation and self-confidence, Sakharov had hammered out a vision of a "better alternative" that would appeal to every "honorable and thinking person who has not been poisoned by narrow-minded indifference." He had few doubts as to what charges would be leveled: "I can just see the smirks about political naivete and immaturity." And he had no doubts that any of it would come to pass without freedom of speech and deep structural reforms; the words that he used in Russian, glasnost and perestroika, were themselves to have a great future.

Naivete is the risk that a noble mind runs, though perverse history often reveals the cynics as fools of trust. Sakharov's naivete was largely in the fair hearing he seems to have expected from the authorities. But Minister of Medium Machine Building Slavsky was livid: "You've got to disown this anti-Soviet publication. . . . You write about the mistakes of the personality cult as though the Party had never condemned them. You criticize the leaders' privileges—you've enjoyed the same privileges yourself. . . . You pit the intelligentsia against the leadership, but aren't we, who manage the country, the real intelligentsia of the nation? . . . You have no moral right to judge our generation—Stalin's generation—for its mistakes, its brutality; you're now enjoying the fruits of our labor and our sacrifices. . . . Convergence is a dream. We've got to be strong, stronger than the capitalists—then there'll be peace. If war breaks out and the imperialists use nuclear weapons, we'll retaliate at once with everything we've got."

Sakharov replied that he had no intention of withdrawing his essay, which faithfully reflected his opinions and which in fact "warned against exactly the kind of approach" Slavsky was taking "in which life-and-death decisions are made behind the scenes by people who have usurped power (and privilege) without accepting the checks of free opinion and open debate."

On a single day, July 22, 1968, Andrei Sakharov went from the secret world to world fame when the *New York Times* printed "Reflections" in full. The high-clearance obscurity in which he had dwelled for twenty years was so great that when Sakharov was attacking Lysenko's candidate at the Academy elections, he was almost challenged by the head of the Central Committee's propaganda department, who had no idea that Sakharov was the "father of the H-bomb" and sat back down upon acquiring that information. Now that information was front-page news, although Harrison Salisbury, assistant managing editor of the *New York Times* and an expert on Russia, was wrong about some of the other particulars, thinking that Sakharov "would appear to be a member of the Communist party" and that it was "barely possible that Sakharov is Jewish."

Andrei Amalrik, who was instrumental in passing Sakharov's manuscript to foreign correspondents, said of his own futurological essay, *Will the Soviet Union Survive until 1984?* that for Westerners interested in his country it "should have the same interest that a fish would have for an ichthyologist if it suddenly began to talk." So too should Sakharov's essay—and the creator of the H-bomb was no minnow. He was the subject of a front-page article in the *New York Times* on July 11, the day after "Progress" was published in *Het Parool,* and by the end of the month the *New York Times* had published eleven pieces on him, including the full text of the essay on the twenty-second. In the following year more than eighteen million copies of his essay were printed in book form worldwide, which put him, as Sakharov later noted with a mixture of surprise and authorial pride, "in third place after Mao and Lenin, and ahead of Georges Simenon and Agatha Christie." Sakharov received response both at home and from abroad; he was particularly pleased by the letter from the German theoretical physicist Max Born, who praised his courage and shared most of his ideas except for Sakharov's high opinion of socialism, which Born had always "considered a creed for idiots."

Sakharov's strongest appeal was to the dread that every thinking person suffered, the necessarily intermittent realization that life and civilization could be incinerated in an hour's time. E. B. White ended his paean to his beloved city, "Here Is New York," with a vision of nuclear annihilation ("A single flight of planes . . . can quickly . . . burn the towers, crumble the bridges, turn the underground passages into lethal chambers, cremate the millions") while hoping that the United Nations, just founded as he wrote in 1948, would house the "deliberations by which the planes are to be stayed and their errand forestalled." Apart from any of his misapprehensions about America—and America's about him—Sakharov had quite accurately calculated the mind-set of the Western intelligentsia.

There were, however, some interesting misinterpretations. Since Sakharov was not immediately punished, his essay was viewed in some quarters (as a KGB report to the Central Committee noted) as a "deliberate leak by the Soviet side to prepare public opinion for far-reaching initiatives in Soviet-American cooperation now being planned."

The KGB was aware of Sakharov's essay soon after it was typed at the Installation in mid-April and knew that it had begun circulating in samizdat in May. On May 22, 1968, KGB chief Andropov made an official report on the subject to the Central Committee in a top secret "special dossier." It is purely descriptive, informative, although some of its information is comically incomplete—Andropov was able to append Sakharov's essay only beginning with page 6, the other five having somehow gotten lost in the KGB shuffle. His report of June 13 was, by contrast, both analytical and prescriptive. He characterized Sakharov as "apolitical" and "susceptible to outside influence" and concluded that "in order to eliminate the opportunities for anti-Soviet and anti-social elements to exploit Academician *Sakharov's* name for their hostile purposes, and in order to prevent him from committing politically harmful acts, we consider it would make sense for one of the secretaries of the Central Committee to receive *Sakharov* and to conduct an appropriate conversation with him." A basic stance and strategy is implied: Sakharov has been duped but is worthy and capable of correction. Minister of Medium Machine Building Slavsky's tirade of late July may have been that "appropriate conversation," but Suslov, whose bailiwick was ideology, had been "nauseated" by the essay and urged action.

In early August, Sakharov was about to leave Moscow for the Installation with Klava and the children to spend the summer together as they did each year when he received word that Khariton wished to speak with him, privately, at his home. Khariton came right to the point: Slavsky was against Sakharov returning to the Installation, fearing a provocation against him.

"That's absurd—by who?"

"Those are Slavsky's orders. You're to remain in Moscow for the time being."

Sakharov understood this was "tantamount to being fired," and in fact he would never set foot on the Installation again in any official capacity. The vacation was aborted, and he was effectively unemployed. Klava's worst foreboding had come true and did nothing to improve her health. She was growing thinner, paler, and suffering abdominal pain with increasing regularity and severity.

Stranded in sweltering Moscow, on August 21 Sakharov went out to buy a newspaper and received a shock of his own. At the "urgent request" of Czech party and government officials, twenty-nine divisions of Warsaw Pact troops including 7,500 tanks and

1,000 planes had invaded Czechoslovakia in the early morning hours.

"This is the tragedy of my life!" exclaimed Dubcek, who as a sincere communist had believed that proclaiming the "leading role of the party" and adhering to the Soviet foreign policy line was enough to assure Moscow of his fealty. But, as usual, Moscow preferred caution and violence. In the words of an eyewitness: "At some time after 4 A.M. on 21 August a black Volga from the Soviet embassy drew up at the Central Committee building which was soon surrounded by armored vehicles and tanks. Troops jumped out in their claret-colored berets and striped vests, carrying machine guns. . . . The doors of Dubcek's office were thrown open and about a dozen machine-gunners rushed in and surrounded us, pointing their guns at the back of our necks." Dubcek was thrown onto a plane, struck in the back with a rifle butt, and flown to Moscow.

Referring to Dubcek by his nickname at a Kremlin confrontation, Brezhnev's voice trembled with offended self-pity: "I believed in you, I defended you, I said that our Sasha's a good comrade. And you let us all down!"

From a sense of either duty or reality, the Czechoslovak leaders agreed to renounce their democratic reforms. Though occupied and instructed by the government not to resist, the country was still free in spirit. Activists such as the writer Václav Havel went on the air, making for "seven days of government by microphone. . . . The mass media outwitted the occupiers and moved and mobilized the entire society against the powerful will of the aggressors. . . . The country simply refused to pay attention to the occupation."

As Sakharov stood on the sidewalk in Moscow that hot August morning, he knew he was not alone in experiencing the invasion as a wound to his "faith in the Soviet system and its potential for reform." But it was only seven people—five men, two women—one of whom, Daniel's wife, Larisa Bogoraz, was three months pregnant, who gathered by the Place of the Forehead, where beheadings were done in Old Russia and which is situated directly across from the Kremlin. Red Square was teeming with KGB agents because Dubcek and his associates were about to be returned to Prague, and the KGB was no doubt aware of the possibility of protest. The protesters unfurled their banners—Hands off Czechoslovakia! Shame on the Occupiers! "Throughout my whole conscious life I wanted to be a citizen, a man who calmly and proudly speaks his mind," said one of the protesters, Vladimir Dremlyuga. "For ten minutes on Red Square I was a citizen." Calling the protesters "dirty Jews" and "anti-Soviets," the KGB attacked them physically and dragged them away minutes before the cars bearing Dubcek and his colleagues sped out from the Kremlin across the gray cobbles of Red Square.

Most Russians were indifferent or took a tone of patriotic indignation: we liberated the Czechs from the Nazis, and now look how they show their gratitude. As usual a joke was minted for the occasion: Question: What is the most neutral country in the world? Answer: Czechoslovakia. It doesn't interfere even in its own internal affairs.

Sakharov learned of the demonstration only the next day from Solzhenitsyn, at their first meeting, which, after several reschedulings for security's sake, was set for August 26 at the home of Evgeny Feinberg, a colleague of Sakharov's from FIAN. A sort of intellectual summit, the meeting was held specifically to allow Solzhenitsyn the chance to engage in the "positive criticism" Sakharov had invited. Solzhenitsyn, who had already hailed Sakharov's appearance among the "venal, corrupt and cowardly intelligentsia" as a "miracle," was immediately taken with the man himself: "Merely to see him, to hear his first words, is to be charmed by his tall figure, his look of absolute candor . . . , his warm gentle smile, his bright glance, his pleasantly throaty voice, the thick blurring of the r's to which you soon grow accustomed," and Sakharov's "old school" Moscow intelligentsia dress—a "carefully knotted tie, a tight collar, a suit jacket."

Sakharov was most impressed by Solzhenitsyn's "lively blue eyes and ruddy beard," his "deliberate, precise gestures," and his "tongue-twistingly fast speech delivered in an unexpected treble." Though the evening was so hot that Sakharov would soon loosen his tie and collar, Solzhenitsyn had drawn the curtains and had insisted that the telephone line be cut, at no little subsequent inconvenience to his host, whom Solzhenitsyn had already rebuked for the overly well-set table: "What is this, a reception?" Sakharov himself had noticed that the cabdriver who brought him there had made suspiciously strenuous attempts to engage him in conversation. Solzhenitsyn, the former Zek, favored the covert, whereas Sakharov assumed there was no shaking security, which should be ignored as a matter of principle.

Solzhenitsyn was irritated by the heat, the tension, and the fact that their hosts had not vacated the apartment as he had anticipated. And in the beginning he was also distracted by the "feeling that I could reach out and touch, through that dark-blue suit, the arm that had given the world the hydrogen bomb." But, after congratulating Sakharov for "breaking the conspiracy of silence at the top," Solzhenitsyn passed immediately into his point-by-point rebuttal, which lasted two hours.

Solzhenitsyn's insistence that convergence was impossible afforded Sakharov a private irony since it repeated Slavsky's tirade "almost word for word." Lost in their dream of consumer materialism and lax morals, America and the West, Solzhenitsyn argued, have no real interest in Russia's becoming free. Furthermore, Sakharov had both underestimated the number of Stalin's victims—sixty million had perished—and overestimated the importance of Stalinism. As Solzhenitsyn put it, "there never was any such thing as Stalinism (either as a doctrine, or as a path of national life, or as a state system), and official circles in our country, as well as the Chinese leaders, have every right to insist on this. Stalin was a very consistent and faithful—if also very untalented—heir to the spirit of Lenin's teaching." And to his practice: Lenin had established the secret police, the terror, the Gulag, Stalin only making insanely grandiose use of them. Communism was evil from its inception. It was not reformable, especially by a multiparty system, a form of group egotism with the leaders always betraying their followers. Russia had to follow an extraparty or nonparty path if it was to achieve that spiritual regeneration without which freedom is merely selection or license.

Solzhenitsyn felt triumphant, spent, and humbled at the end of his soliloquy. "It was then, in the two hours during which I made such a bad job of criticizing him, that he conquered me! He was not in the least offended, although I gave him reason enough; he answered mildly, tried to explain himself with an embarrassed little smile, but refused to be the least bit offended—the mark of a large and generous nature."

Though Sakharov's faith in socialism had been shaken by the events in Czechoslovakia and though he was always willing to listen to the voice of intelligent doubt, he was not much moved by Solzhenitsyn's jeremiad. He reiterated the positions he had taken in his essay, but even at that time he realized that despite his respect for Solzhenitsyn, which only grew after the publication of *The Gulag Archipelago,* the two of them were "not at all alike and differ markedly on questions of principle." In fact it would be difficult to imagine two more perfect embodiments of the Westernizer and the Slavophil, the rational, progressive humanist and the defender of the faith that was Russia.

They moved on to more immediate matters: what could be done about the seven who had been arrested for protesting the invasion of Czechoslovakia? Solzhenitsyn saw little hope—so far no one had proved willing to sign a collective letter—but Sakharov felt some action was required. Sakharov knew that he carried much less weight now but still had enough to get through to Andropov. The problem was that, in losing his connection to the Installation, Sakharov had also lost easy access to the high-frequency phone system that linked the leadership. But a few days later he found a way to gain access. Kurchatov had left instructions that Sakharov be allowed into the Atomic Energy Institute without a pass. Sakharov phoned Andropov from there in what would be their last conversation. "Communist Parties in the West are following developments, and it will make matters worse if the demonstrators are tried and sentenced," he told Andropov, who replied that he was so busy with Czechoslovakia that he'd barely been sleeping, that the fate of the demonstrators rested with the Attorney General's office, not the KGB, and that in any case he didn't think the sentences would be particularly severe. No doubt he was truthful about being sleep deprived; the Attorney General was in fact prosecuting the case, though the verdicts would be coordinated with the KGB; and the sentences—terms in a psychiatric prison hospital, exile, and labor camps, variously applied—would not have struck him as unduly harsh.

Coincidentally, or not in the least coincidentally, the trial of another dissident, Anatoly Marchenko, began on August 21, the day of the invasion. Wiry, with close-cropped dark hair and high cheekbones, the son of illiterate Siberian railroad workers, Marchenko had dropped out of school after the eighth grade and worked on construction projects in Siberia and Kazakhstan. Arrested in an indiscriminate roundup after a brawl between Russians and Chechens in which he had taken no part, Marchenko, outraged at the injustice, escaped from the labor camp and tried to flee the Soviet Union for Iran but was rearrested in Ashkhabad. That "treason" cost him another six years. In the Gulag, by fortuitous coincidence, he became friends with Yuli Daniel of the infamous Sinyavsky/Daniel affair. Daniel became Marchenko's tutor and mentor, and he emerged from the camps a writer. *My Testimony,* Marchenko's account of the post-Stalin camps, dispelled the illusions of people like Sakharov, who read the book in samizdat and used it as a source for "Reflections": "The present day Soviet political prisons are just as terrible as under Stalin," wrote Marchenko. "In some things they are better but in others they are worse. It is essential that everyone should know about this—both those who want to know the truth . . . and those who do not wish to know it."

Prisoners were worked to exhaustion and kept so starved that one who was known for periodically slashing himself was asked by another prisoner if could catch the warm blood in a bowl and drink it. The slasher had no objection. For the Zeks, their own bodies were their final zone of freedom, which could be asserted only by flamboyantly senseless acts. One Zek swallowed a whole set of dominos, if only to be the source of trouble instead of its recipient. Defiant tattoos were removed by forcible surgery. "I remember one con that had been operated on three times in that way," wrote Marchenko. "The first time they had cut out a strip of skin from his forehead with the usual sort of inscription in such cases: 'Khrushchev's Slave.' The skin was then cobbled together with rough stitches. He was released and again tattooed his forehead: 'Slave of the USSR.' Again he was taken to the hospital and operated on. And again, for a third time, he covered his whole forehead with 'Slave of the ComParty.' This tattoo was also cut out at the hospital and now, after three operations, the skin was so tightly stretched across his forehead that he could no longer close his eyes. We called him 'The Stare.'"

Once Marchenko was released and wrote his book, he became part of the minuscule opposition, "free thinkers," the term Sakharov preferred to "dissidents." Marchenko of course visited Daniel's wife, Larisa Bogoraz, and quite quickly they fell in love. Daniel's faithlessness with women had already cost him her love, but she would not divorce him even now, so that as his wife she could continue to visit him in the camps.

Marchenko had courted arrest by writing an open letter predicting the invasion of Czechoslovakia a month before it happened, asserting that the Soviet leaders wanted a neo-Stalinist regime in Prague for the best of reasons: "What if suddenly our own leaders should be called upon to account for those deeds which have shamefully been termed 'errors' and 'excesses' or, still milder and undefined, 'difficulties endured in the heroic past' (when it was a matter of millions of people unjustly condemned and murdered, of torture in KGB cellars, of whole peoples proclaimed enemies, of the breakdown of the nation's farming, and other such trivia)?" Marchenko was picked up a week later on a "suspected identity-card violation." From a legal point of view he was not in violation of any regulations, and so even the pretext had its own injustice.

Marchenko and Sakharov threatened the Soviet system from opposite ends of the spectrum, the working class and the scientific elite. Before they even met, Sakharov admired Marchenko as a person of "immaculate honesty." In his account of camp life Marchenko is every bit as hard on his fellow prisoners as on their guards, and as is he is on himself when he fails his own standards of courage.

At his second "summit" with Solzhenitsyn, in the cellist Rostropovich's dacha outside Moscow, Sakharov raised the question of coming to Marchenko's aid. Solzhenitsyn peremptorily refused, saying Marchenko had chosen his fate by attacking the "enemy with a battering ram." Any attempt to help would only harm him and others. Sakharov was "chilled."

* * * *

In the late summer of 1968 Sakharov found himself in the quite unaccustomed position of having nothing to do. He was banished from the Installation and, in the oppressive wake left by the invasion of Czechoslovakia, any political action was futile, incommensurably dangerous. That left science and family.

In September he traveled to Tbilisi, the capital of Georgia, for an international science conference. It had been years since he'd attended one, both because his schedule had been so demanding and because of the fear that his "unsystematic education" would prevent him from following the talks in sufficient detail. And except for travel connected with his work, Sakharov had seen very little of the Soviet Union. Tbilisi, beautifully situated in the high Caucasus Mountains, and Georgia's culture of jovial hospitality only made the prospect more inviting.

The conference was devoted to the theory of gravitation, its cosmological applications and its relation to the theory of elementary particles. Sakharov's fears proved ungrounded, and he thoroughly enjoyed hearing the papers delivered. There was also time for private conversations with some of the scientists—like the American physicist John Wheeler, who had worked with Niels Bohr on nuclear fission in the 1930s, and Wheeler's student Kip Thorne. Wheeler's own thinking on gravitation had been strongly influenced by Sakharov's notion that, in Wheeler's paraphrase, gravity is "an elasticity of space that arises from particle physics . . . the sausage skin only then ceases to be floppy when it is filled with meat!" On a hot dry afternoon Wheeler and Thorne were invited to a lunch of bread, cheese, sausage (presumably filled with the gravity of good meat), fruit juice, wine, and cognac in Zeldovich's hotel room, where for two hours, in Thorne's words, they "munched and talked intensely about the cosmos. . . . What a contrast there was between our two hosts! Zeldovich was a short, muscular, bulldozer of a man, vibrant and impatient in conversation, extroverted and demanding—and exciting to listen to. Sakharov, of whom I had read so much in Western newspapers in recent weeks, was tall and . . . painfully shy. But his hesitant conversation was ladened with fascinating, original ideas. He showed no signs of the stress that he must have been under. . . . It was hard to imagine this mild, unassuming man challenging the Soviet government in the way he had just done. . . . Suddenly in midafternoon, in the midst of our intense conversation, Zeldovich announced that it was time to take a nap. (This was how he managed to keep up his vigorous pace: total interaction intermixed with total relaxation.) Zeldovich then laid down and slept soundly for a half hour while, in deference to his needs, Sakharov, Wheeler, and I sat quietly and read or wrote, awaiting nap's end."

The only conflict during this idyllic interlude of pure science came with Zeldovich, who was on the conference's organizing committee and overrode Sakharov's desire to speak on baryon asymmetry. For Sakharov that subject held particular interest because it was the point at which the subatomic and the cosmological intersected. Baryons are made of three quark particles; protons and neutrons are baryons. The fact that there are more baryons than antibaryons in the universe means that there was a net excess of quarks over antiquarks and goes some way in explaining why in its earliest stages of formation the universe was not annihilated but in fact expanded. Zeldovich had his doubts about Sakharov's work in that area, convincing him to talk instead on a "zero-Lagrangian" gravitational field—that the laws of elementary particle physics dictate the form of gravity. Sakharov was glad to speak on that subject as well, which was more in keeping with the theme of the conference, but he was left with regrets that once again Zeldovich had gotten the better of him. Not that Zeldovich was entirely to blame. Sakharov's own tendency to yield in close confrontations left him wondering if he "should have been more insistent."

There was also good reason to devote time to his family. Klava's health showed no sign of improvement—she continued to lose weight and suffer blackouts from pain. Sakharov arranged for them to spend the month of October at a sanatorium in the south of Russia. The trip almost didn't take place, because the Kremlin Hospital doctors who examined them, while finding Klava in satisfactory health, warned Sakharov that his cardiovascular problems made traveling to a warm climate inadvisable. This time, however, Sakharov was insistent and prevailed over their objections.

He was quickly proved right. Klava had her husband's full attention, the climate was salubrious, her vitality began returning. Soon they were taking longs walk together, as they had done in the days of their courtship. Now they were both close to fifty and had been married to each other for half their lives. Neither of them was in ideal health, and their future was uncertain. Sakharov still drew his Academician's stipend and had saved nearly all the colossal sums that had accompanied his high state awards. But their immediate concern was their elder daughter, Tanya, who was due to give birth to their first grandchild any day. Tanya had also married a physicist, of whom some quipped tartly that he had hoped to marry the daughter of an Academician and instead found himself saddled with a dissident as a father-in-law. Their granddaughter, Marina, was born without difficulty on October 18, freeing Klava from apprehension and giving her additional reason to live.

The month—intimate, restorative, even celebratory—was marred only by two notes that would each prove recurrent. Their only son, ten-year-old Dmitri, was in a nearby sanatorium recovering from infectious hepatitis. During one of their first meetings he took his father aside and whispered a request that from now on his family nickname be "Dima" and not the traditional "Mitya," as Sakharov's father had been called. That headstrong break bothered Sakharov both as a conservative in family matters and for what it seemed to intimate.

Sakharov had probably been allowed access to one of the country's elite health resorts for the simplest of bureaucratic reasons: the proper agencies had yet to be informed that he was off the list. But word got around faster than paperwork—conversations fell to silence at his approach. The void of ostracism had begun forming around him; his charge had been reversed from positive to negative, and now he repelled. But sometimes it was Sakharov who actively put an end to conversation. On a bus ride Sakharov overheard two officials discussing the real reasons why the Crimean Tartars, exiled en masse by Stalin, could not be allowed to return—the Crimean coastline was the vacation resort of the Soviet elite. "But it's their homeland!" interjected Sakharov.

A good month, it ended badly. Klava took a sudden turn for the worse, something impeding the circulation in her hands. But it wasn't critical; she could deal with it on her return to Moscow, the sanatoriums being more suited for rest and recuperation than diagnosis and treatment.

The doctors didn't like what they saw at the outpatient clinic in Moscow and sent Klava back to the Kremlin Hospital. The new year began with the worst possible news: the doctors now detected what all the previous tests and x rays had somehow failed to find: inoperable stomach cancer. She didn't have long. In January 1969 Sakharov signed Klava out of the hospital so she could die at home.

Now their life was entirely filled with the exalted sorrow and little tasks of the final days. Her two daughters grown, Klava's joy was Dima, brown-eyed, his brown hair brushed forward and cut straight across his forehead. His was still the spontaneous love of the child, the mother and son's ease of touch not yet broken off by adolescence.

It was the time of farewell for Andrei and Klava, of last conversations. Klava was adamant that her mother and sister not attend her funeral. He could only agree.

Sakharov looked over at Klava as she watched a women's figure-skating competition on television. Having won a close victory, the young woman on the screen was jubilant. In that moment Sakharov saw Klava bid farewell to life. She gestured that the television be turned off and never asked for it again.

A few days later her pain became so great that she had to be returned to the hospital. Only a miracle could save her. Sakharov flew to Kaluga, where a retired doctor had supposedly found a wonder-working vaccine. The doctor, a forceful woman, refused to accept any money from Sakharov but blasted the idiots at the Ministry of Health for not supporting her work. Sakharov returned at once to Moscow from a desperate journey that was a measure of his grief.

Klava received her first injection on March 7, and that evening she gave out presents to the nurses and orderlies for Women's Day, which was the following day. It was as a mother that she died on Women's Day, her last words, "Close the window. Dima will catch cold."

Her father, Alexei, was informed at once. Klava never had any problem with him and even felt a special solidarity with him, both having been deprived of love by the same woman. But now Sakharov had to choose between violating Klava's wishes or failing to tell a mother that her daughter had died. Dishonored either way, he chose to honor Klava's wishes.

At the funeral, blunt as always, Alexei voiced his unhappiness that his daughter had been cremated, not buried. But, tears streaming down his face, dazed like a sole survivor, Sakharov could no longer be reached.

CHAPTER TEN

VITA NUOVA

HE gave away all his money—139,000 rubles. A fortune. He was supporting his family decently on 750 a month. At that rate the money would have lasted him more than fifteen years.

In late August 1969 Andrei Sakharov made one last trip out to the Installation to clear out his desk, his cottage, and, as it turned out, his bank account. He stipulated that the funds be divided equally among three causes: a children's medical center at the Installation, a cancer hospital, and the Soviet Red Cross for aiding victims of natural disasters. At the end of his official declaration, to forestall any bureaucratic snafus, Sakharov indicated: "In case it proves impossible to use the funds for any one of the stated purposes, the sum should be divided between the other two."

All that was the overmeticulousness of someone who is in fact behaving irrationally—and will regret it later. Sakharov's "fit of generosity," transferring control of his money to the state, would subsequently reduce the help he could give to the families of political prisoners and cause financial complications with his own family. But clearing out a house where you have lived the last twenty years of your life is always a melancholy task, especially when it's a recent widower doing the job. Six months after the death of Klava, Sakharov was still in a state of shock, riches a sin against grief.

Contributing the money to worthy causes had been a natural reaction to tragic shock, the need to commemorate and memorialize, but there was also a political note of defiance, rejection. The sheer quantity of the money, money unspent on any of life's pleasures, was a reminder of a certain poverty in his life with Klava, death revealing self-denial as just another vanity.

But it was not only the loss of Klava that unhinged him. He

had been banished from the place where he had spent the genius and energy of his youth to make his country a nuclear superpower, which had three times awarded him its highest honors. But it was all gone, that life; Klava was dead, and he would never be allowed access to the Installation again. He was between lives, naked of the identity given by love and work.

Work came back first. In May 1969, lobbied by Tamm, Minister of Medium Machine Building Slavsky summoned Sakharov to his office and offered him the chance to return to FIAN as a senior scientist. Sakharov accepted, it being the obvious best solution for both sides. He'd still be employed and at an institution he loved, while for them it was a means of separating him from the Installation "with as little fuss as possible." And who knew? He might yet come up with a Fourth Idea. Neither had entirely given up on the other. And Sakharov "still felt that the government I criticized was *my* government."

Sakharov returned to FIAN in a state of dazed detachment, strikingly disheveled and ill kempt. It was not the old cozy FIAN in central Moscow's tranquil back lanes but a new, imposing neoclassical edifice with stout white columns out on Lenin Boulevard. Like many of the city's grander buildings it had been built by German prisoners of war, some of them laboring on into the fifties, a point of pride for Soviets but one not unalloyed, for it also acknowledged the Germans as the better builders.

One condition of his return was that Sakharov engage in no extrascientific activities at FIAN. "A wolf doesn't hunt near its lair," replied Sakharov, not only accepting the terms, but reversing the normal procedure, shunning those who might be endangered by association with him. Some, misunderstanding, were offended.

At FIAN, Sakharov was again back under Tamm, who headed the Department of Theoretical Physics, but that demotion had more pathos to it than humiliation. Tamm was too ill to come in to work, the old reviver of labor and leisure stranded on the porch of his dacha flanked by an oxygen bottle and a barking dog.

At the time neither science nor politics had any meaning for Sakharov, who was barely able to perform the minimal household tasks and look after his twelve-year-old son, Dima. Klava's last words had been words of concern for Dima. Not only had he lost his mother, he was getting old enough to start to feel the burden of being the son of a great man, the fear you will never equal him.

Clarity of mind returned with October. A colleague showed Sakharov his paper that sought to demonstrate that black holes can't exist. Sakharov disagreed so vehemently that he was moved to write a paper himself, "A Many-Sheeted Cosmological Model," published in 1970 and dedicated to Klava. It was through science that Sakharov began "returning to life," but not to a life of science.

He now devoted himself to catching up with the world. Just before Klava died, Chinese and Soviet troops had exchanged fire on the Ussuri River. Chinese-Soviet relations were at an all-time low, the United States now in a position to play one communist giant against the other. But thanks to the efforts of Sakharov and others, the Soviet Union had achieved nuclear parity with the United States, and the Warsaw Pact could field twice the forces NATO could. Détente, the relaxation of tensions between the USSR and the United States, also suited the American side, which wanted to open up Soviet markets and wring some concessions on human rights from the regime. Suddenly there was a sense of opportunity, even among the dissidents.

In early 1970 Sakharov met Valentin Turchin, a physicist who had an idea that Sakharov found quite appealing—to convince the Soviet leaders on a single point: that the free flow of information is essential to the strength and health of a country's science, technology, economy. Democratization was not heresy but salvation.

Writing to Brezhnev in March 1970, with Turchin and Medvedev, Sakharov stressed the hard, technological basis for his position: "the introduction of the computer into the economy is a factor of decisive importance, radically changing the character of the system of production, and all culture. This phenomenon is justly called the second industrial revolution. Yet the capacity of our computers is hundreds of times less than that of the United States, and as regards utilizing computers in the economy the disparity is so great that it is impossible even to measure. We simply live in a different epoch."

The tone they take is loyal and reasonable. "The source of our difficulties is not in the socialist system." "Democratization must be gradual in order to avoid possible complications and disruptions." "We are seeking to achieve a positive and constructive approach acceptable to the Party-state leadership." But their actual specific recommendations must have caused consternation in the Central Committee. They include an end to jamming, an institute for studying public opinion, amnesty for political prisoners, public control over psychiatric institutions, elimination of the internal passport, independent publishers, multicandidate elections, the return of all nations exiled by Stalin to their homelands.

The letter concludes with a grave and prescient warning: "What awaits our country if a course toward democratization is not taken? We will fall behind the capitalist countries in the course of the second industrial revolution and be gradually transformed into a second-rate provincial power (history has known similar examples). . . . At worst, the results would be 'catastrophic,' at best, 'stagnation.'"

In April 1970 KGB chief Andropov formally petitioned the party's Central Committee for permission *"to install secret listening devices in Sakharov's apartment"* in order to receive "timely information on SAKHAROV's intentions and to discover the contacts inciting him to commit hostile acts." Brezhnev and Suslov, among others, deemed the request reasonable. By the early spring of 1970 Sakharov was a problem at the highest level of state security, a matter considered so sensitive that the KGB typist preparing Andropov's reports was not allowed to know whom they concerned and ordered to leave a blank space for the name, to be filled in later by hand. Sakharov was also given the operational code name "Ascetic."

Andropov's psychological distinction between Sakharov's intentions and acts, the latter supposedly incited by others, in fact reveals the KGB's own psychology: to still be considered salvageable, Sakharov must be viewed as a dupe. Also revealing is the fact that only now had the KGB requested permission to install listening devices. Sakharov had assumed they were in his apartment much earlier, which is why he and Roy Medvedev communicated by writing when discussing the release of "Reflections." The KGB had apparently not been particularly worried about Sakharov the physicist as a security problem. But Sakharov could sometimes be more security conscious than the KGB. He once interrupted a conversation with another physicist to say that those eavesdropping wouldn't have been cleared to possess any knowledge about state atomic secrets and should be spared the discomfort. His wit was at times so dry that it was impossible to determine where sincerity shaded into irony.

The letter from Sakharov, Medvedev, and Turchin was not met with a written rejoinder or an invitation for discussion. To the regime it was at best inadvertent sedition. Yet officials who ran complex organizations, such as Andropov, could not help but see the common sense of it—information has to move as rapidly as possible to where it is most needed. But the question was more hydraulic than philosophic: who controlled the valves and conduits of its flow? The KGB response was to lay a pipeline directly into Sakharov's apartment. Freedom of information, like everything else, belonged to the state.

By the time those listening devices were operative, Sakharov was engaged in two cases in which the state was using psychiatry to punish dissidents, which Solzhenitsyn called "spiritual murder . . . a variant of the gas chamber."

But psychiatric punishment was nothing new in Russia. It had created a famous martyr as early as the 1830s, when the philosopher Piotr Chaadaev was declared insane for professing the idea that Russia would not be a lasting nation but rather would teach the world one great lesson, then vanish from history. The difference was that now the authorities had the chemicals to induce madness while pretending to cure it. For inmates of sound mind it was an added horror to know that their psychiatrists were themselves criminally insane.

In the words of one person who underwent the procedures: "After a breakfast of mush came shock therapy. You're given a large dose of insulin, the sugar disappears from your blood and you go into shock. . . . You're tied to your bed with strips of torn sheet, not ropes. . . . When they're in shock, people go into convulsions. They scream and howl. Their eyes look like they're are going to pop out of their head. Then, after the correct amount of time the nurses come around giving people shots of glucose. A strong shot of glucose will bring a person around in ten or fifteen minutes.

One orderly holds your arm in an iron grip and another winds a piece of rubber around your upper arm, and when the vein is fat and swollen the nurse jabs it with a hypo. The nurse never missed a vein. She was rough but accurate."

In the month of May 1970 two prominent dissenters were forcibly incarcerated in psychiatric institutions. General Piotr Grigorenko was outspoken on a number of issues—Stalin's military blunders in the opening days of the war, the invasion of Czechoslovakia, the right of the Crimean Tartars to return to their homeland, the dangerous tendency to "whitewash" Stalin. The latter was a sensitive point. The Stalinists were in the ascent—Stalin's ninetieth birthday was commemorated in December 1969 and a bust, stern but benign, commissioned for his grave, immediately behind Lenin's tomb, and formally installed in June 1970.

Sakharov learned of Grigorenko's hospitalization from Valery Chalidze, who published a samizdat journal, *Social Problems,* and who asked Sakharov if he would be willing to "cosponsor a complaint" to the Attorney General "protesting Piotr Grigorenko's involuntary confinement in a psychiatric hospital." Sakharov was not only willing but delivered the letter in person.

An even more sensational confinement occurred in late May when Roy Medvedev's twin brother, Zhores, a biochemist and author of a book on the rise and fall of Lysenko, was forcibly detained by three policemen. When Zhores shouted: "Stop, this is a private apartment," he was told by a sergeant: "It belongs to the State and the police have the right to enter any apartment." But Zhores Medvedev's forcible confinement—probably revenge by Lysenkoites working through the local KGB—suddenly became an international event. Even though Sakharov was in ill health—running a temperature of 101 and requiring a hernia operation a month later—he threw himself into the campaign to save Zhores Medvedev, diagnosed as a schizophrenic because he was a scientist but also wrote books on other subjects: "The Grigorenko affair had sensitized me to the political abuse of psychiatry. I had been battling with the Lysenkoites for a long time, and my work with Roy Medvedev on the Memorandum had created a bond between us." On May 30 Sakharov went to the Institute of Genetics, where an international symposium was being held, walked up to the blackboard and, with one hand or the other, chalked the following: "I am collecting signatures in defense of the biologist Zhores Medvedev, who has been forcibly and illegally placed in a psychiatric hospital for his writings. Contact me during the break or reach me at home." He gave his address and phone number.

The message was almost immediately erased and sharply criticized by the director of the institute. A few people came up to Sakharov in the corridor: two or three young geneticists, a scientist who told Sakharov that Zhores Medvedev was an alcoholic, and a Romanian woman who said "such things can happen in any country," to which Sakharov replied, yes, and they must be resisted in any country. Then Sakharov was approached by two men in plain clothes who asked for identification. As usual, Sakharov did not have his papers with him, but this wasn't the old days at the Installation; now he had to vacate the premises. But, in the meantime, he had collected a few signatures and would collect a few more later.

On June 6 Sakharov wrote directly to Brezhnev, protesting Zhores Medvedev's confinement: "psychiatric hospitals must not be used as a means of repression." The only response Sakharov received was the suggestion, made to him at a meeting at the Ministry of Health, that he himself might be a candidate for psychiatric attention, as a person suffering from "obsessive reformist delusions."

The campaign proved victorious. Zhores Medvedev was released in mid-June. But it took another four years for pressure to mount sufficiently to win Grigorenko's release, by which time he was broken in health if not spirit.

After the feverish spring it was an apolitical summer for Sakharov, who underwent the hernia operation in July and then visited the ailing Tamm at his dacha, which was next to Sakharov's. Though confined by his respirator, Tamm still worked, received guests, and had been active in the struggle to free Zhores Medvedev. Tamm now often spoke of Sakharov's tragedy: "that he had to sacrifice his great passion—elementary-particle physics—first to create an atomic and hydrogen bomb," then sacrifice it a second time in the struggle for social justice. But Sakharov still saw nothing tragic in his weapons work, nor any waste in his participation in the social struggle. In fact the two were one, or at least in close symbiosis: he could not be effective in human rights if had not been the "father of the H-bomb." Besides, he was still doing a reasonable amount of science and was about to leave for the International Rochester Conference on elementary particles held that year in Kiev. But he had turned forty-nine in May—there might still be some last flares, but the days of incandescence were over.

Though he always felt most at home in the company of scientists, the world of the dissident intelligentsia attracted him socially for its warmth, eccentricity, stimulation—besides, a good half of the dissidents were physicists and mathematicians, which meant that conversation was not limited to arrests, trials, and sentences. Valery Chalidze, who had invited Sakharov to help defend General Grigorenko, was a thirty-one-year-old physicist whose home—one room in a communal apartment transformed by rugs and daggers into the palace of a Georgian aristocrat—was a natural gathering spot. Black-haired, suavely handsome, nicknamed "Prince," Chalidze would receive visitors reclining on a huge sofa. His faults—adept at the law but tending to legalism, the brashness of youth—were far outweighed for Sakharov by his spirit and intelligence. The conversation at Chalidze's was always lively, the information current and accurate. Sakharov the widower found himself dropping by.

Visiting Chalidze after the elementary particle conference in Kiev, Sakharov was immediately engaged in a high political drama. Two men, Boris Vail and Revolt Pimenov, were arrested for possessing underground literature, Sakharov's "Reflections" included. The trial was scheduled for October.

Chalidze was attempting to arrange for legal counsel for the defendants—at best futile, at worst risky. Sakharov spent even more time now at Chalidze's to stay up with events as he weighed whether to attend the trial, something he had not done before and whose ramifications must be projected as accurately as possible.

Stopping by one day, Sakharov found Chalidze in serious, uninterruptible conversation with a dark and striking woman—her thick eyebrows unusually high, making her eyes even more intent. She smoked heavily, her voice sometimes harsh with nicotine and anger. And she left so dramatically that she never even noticed

Sakharov. Later, Chalidze said with pride: "That was Elena Bonner. She's been helping prisoners nearly all her life."

When Sakharov decided to attend the trial, Chalidze advised him to run it past the head of the Academy so as to ensure admittance. Not long after, Zeldovich appeared with a friendly warning: "I know you're planning to attend Pimenov's trial. You'll put yourself on the other side of the fence."

Sakharov replied that he already was "on the other side."

They left for the pleasant provincial city of Kaluga at 4:00 A.M. by car, not by train, because the theatrical Chalidze wanted Sakharov "to appear like some deus ex machina." After driving a hundred miles through rain and mud, they arrived to find the downstairs hall of the courthouse packed with friends of the defendants, none of whom were allowed access to the second floor, where the trial would be held. The stairs were blocked by police.

As Sakharov approached the policeman, he came into the clear view of Natalya Gesse, who had hobbled on crutches through the Moscow Panic, talked her way into the Red Army on its drive to Berlin, and now, at age fifty-six, was attending the trial of dissidents in Kaluga. She who had seen a lot now saw something new: "We were all there in the lobby, and all of a sudden a person appears, someone I had never seen before. His face was so striking that I just stood there stunned and could not tear my eyes away from that face. I had never seen a person whose face radiated such spirituality and intellect. And he started walking very confidently through the crowd. And when he was passing by me, I just naturally reached out and stroked his shoulder and arm."

The moment was vivid for Sakharov too: "A short, slightly stooped woman standing nearby pressed my arm sympathetically, and her simple, impulsive gesture startled me—people didn't do that sort of thing in my former world."

Announcing himself as Academician Sakharov, he was allowed to pass through the police cordon.

He entered the courtroom and noticed the windowsills lined with identical gray hats, which belonged to the men in identical suits filling the room, ready to shout down any voice, cry out for harsher justice, or catch up on their sleep. It did not take long for Sakharov's worst suspicions to be confirmed. During a break one of the prosecutors asked Sakharov if he agreed that the judge was "thorough and objective."

"The trial is a legal farce."

Natalya Gesse and her friend from Leningrad, Elena Bonner, had taken charge of providing lunch for the support group: milk, kefir, poppy seed rolls, sausage, and cheese. They had placed the food and glasses on a wide windowsill. When Sakharov came down from the courtroom, Bonner, now knowing who he was, offered him food: "Andrei Dmitrievich, take some sandwiches. And what would you like to drink, milk or kefir?" "No, no, no!" replied Sakharov. He took a roll, which he balled up in his hand and ate plain. Bonner thought: "What's he afraid of? The glasses are clean, everything's clean." Then Sakharov excused himself, saying he wanted a hot meal—that hardly being the moment to explain that his horror of cold food required him even to warm up his salad in a frying pan.

Pimenov's final plea was "well reasoned" and lasted three hours. Vail was epigrammatic: "Citizen judges, the verdict decides the defendant's fate and affects his whole life, but it leaves its mark on those who pronounce it as well. Be just."

The verdict was five years of exile for each, which Sakharov considered "bad enough, but mild by Soviet standards." He took some satisfaction in the thought that his presence might have had some influence.

But his contribution had not yet been fully made. Just as he was about to leave the courtroom, Pimenov's wife came up to him and thrust him a green file, nervously whispering: "Hide this and take it with you. It can free my husband."

Sakharov slipped the green file under his coat and walked out of the courthouse to where a car, driver, and bodyguard had been arranged by Chalidze to whisk him to the train station. Sakharov was useful in getting the documents out of the courthouse, but he must not be exposed to jeopardy for any longer than necessary. Two persons, Sergei Kovalyov and Elena Bonner, were dispatched to retrieve the file. They arrived at the train only steps ahead of their plainclothes pursuers, the doors closing in their face. Sakharov handed the file to Bonner and Kovalyov who for some reason did not take the entire contents. That meant that the remainder had to be retrieved, especially now that it had been decided to return the file to the court. The next day Chalidze called to say that Bonner's daughter, Tatiana, would be coming by for it: "You'll recognize her. She looks just like her mother."

Tatiana, twenty, dark-haired, dark-eyed, resembled her mother in coloring, features, and expression—a sharply focused intelligence playing against a passionate intensity. It was, however, Andrei Dmitrievich who confounded visual expectations. He met Tatiana at the door in his undershirt and trousers, an outfit that she associated more with the Soviet proletariat than the Russian intelligentsia, though she realized this was a widower's self-neglect. Sakharov was correct but distant, not asking her in, handing over the last of the green file in the doorway. He may have slipped back into his habitual aloofness, but Academician Sakharov was now moving in decidedly faster company.

* * * *

The regime's initial response to the awarding of the Nobel prize for literature to Aleksandr Solzhenitsyn, announced on October 8, 1970, was also relatively "mild by Soviet standards." A memorandum from the Minister of Internal Affairs to Brezhnev warned against committing the "same glaring errors that we committed with regard to Boris Pasternak" and recommended Solzhenitsyn be published after "painstaking editing" as part of a general policy, "not to execute our enemies publicly, but to smother them in embraces." Solzhenitsyn was not, however, of the sort who submitted either to smothering embraces or painstaking editing. Of his latest novel, *August 1914,* he said: "This book cannot at the present time be published in our native land except in Samizdat because of censorship objections unintelligible to normal human reason and which, in addition, demand that the word God be unfailingly written without a capital letter. To this indignity, I cannot stoop.

"The directive to write God in small letters is the cheapest kind of atheistic pettiness. Both believers and unbelievers must agree that when . . . K.G.B. or Z.A.G.S . . . are written in all caps then we might at least employ one capital letter to designate the highest Creative Force in the Universe."

The Soviet Union was seeking a policy of détente both with the United States and, selectively, with its own internal enemies. General Grigorenko could be confined to a mental institution, but Zhores Medvedev could be released. Pimenov and Vail could be arrested for reading Sakharov in samizdat, but Sakharov, though monitored, was left at liberty. In the case of Solzhenitsyn, the approach proposed by the Minister of Internal Affairs was viewed as too liberal. The prevailing view was that the Nobel committee was using the prize to provoke and embarrass the Soviet Union. Solzhenitsyn would not be allowed to travel to Stockholm.

In early October, Chalidze unexpectedly dropped by to see Sakharov, brimming with an enthusiasm that communicated itself even though Chalidze, out of conspiratorial necessity, confined his communication to writing. His idea was a "Human Rights Committee, a voluntary, non-governmental association that would study and publicize human rights problems in the USSR." Sakharov was attracted but had his doubts. Wouldn't they be duplicating the efforts of the Initiative Group for the Defense of Human Rights, the first such group in the USSR, and the *Chronicle of Current Events,* which had been publishing reports on human rights violations since 1968? Worse, "such a grandiloquently named Committee would attract too much attention and arouse too many false hopes. How were we to respond to the letters, petitions, and complaints that would come flooding in? That we were a study group and not a defense committee? What a mockery that would be!" It was too soon for a yes or a no; he had to think it through.

Andropov was reporting to the Central Committee within days on this latest venture: "*Sakharov* and CHALIDZE . . . are discussing an idea planted by our adversary—the legalization of opposition activities in the USSR. They have in mind the creation of a so-called 'Human Rights Seminar'. . . .It should be noted that this is not the first time that antisocial elements have attempted to carry out the suggestions of Western ideological centers to create legal opposition groups and organizations in our country. All these attempts have been foiled by the Committee for State Security [KGB]." The KGB's psychological image of Sakharov was now shading from the gullible to the haughty. "Moreover, exploiting

Sakharov's ambition, the enemy constantly popularizes his name and his politically dubious actions, thereby deliberately pushing him to commit new antisocial acts." Andropov still felt the solution lay in a "discussion with *Sakharov*" that would "explain to him the illegality of his activities, which bring political and moral harm to the Soviet state."

Though Sakharov was no longer connected with the Installation, it was its director, Yuli Khariton, who relayed Andropov's "urgent" request that Sakharov phone him, possibly for that very "discussion." In early November Sakharov made several attempts to contact Andropov, but none were successful. Either Andropov had lost interest or had merely been reminding Sakharov that his activities were being scrutinized at the highest level.

As soon as Sakharov overcame his doubts about the proposed Human Rights Committee, Chalidze began at once drafting the bylaws, a task he approached with particular zest. They were signed by the three members of the committee—Sakharov, Chalidze, and Andrei Tverdokhlebov, a thirty-year-old theoretical physicist—on November 4, 1970, though the organization was not "formally launched" until a week later at a press conference at Chalidze's apartment. The response of the Western media—Voice of America, BBC, Deutsche Welle—was major and immediate, confirming KGB suspicions of international manipulation.

Some of Sakharov's own suspicions were soon justified. The committee, which met every Thursday at Chalidze's apartment with other prominent members of the human rights movement such as Esenin-Volpin coming by to participate, was not only inundated by letters, petitions, and complaints but the majority of them were addressed to Sakharov in the belief that an Academician wielded greater influence. Yet the effort seemed worthwhile, and there were ancillary pleasures: "Not having been spoiled by an abundance of friends in my life, I prized this opportunity for human contact."

For that reason Sakharov attended Chalidze's birthday party in late November, where he was happy to encounter Elena Bonner again. Though they now had some history together and conversation came more easily, it was still largely confined to arrests and trials. On the day that the formation of the Human Rights Committee was made public, Andrei Amalrik, author of *Will the Soviet Union Survive until 1984?* and the man who had passed Sakharov's "Reflections" to Western correspondents, was put on trial in Sverdlovsk, nearly nine hundred miles from Moscow. Sakharov planned to attend but failed to deal with the logistics in time.

Amalrik had refused to answer questions, contending that the trial was illegal, held "to frighten many, and many will be frightened. But all the same, I think the development of ideological emancipation, having begun, is irreversible." That was not necessarily a good sign, since Amalrik also believed the "liberalization of the post-Stalin period to be a sign of the decay of the regime . . . and no forces in our society . . . are even potentially capable of bringing about a renewal, a democratization of the country."

Sakharov and Bonner met again in connection with the Leningrad "skyjacking" case, in which one of the principal defendants, Eduard Kuznetsov, was a friend of Bonner's. He had been part of a group of sixteen Soviet Jews arrested on June 15, 1970, for planning to skyjack a passenger flight from Leningrad to Murmansk, diverting it from there to Sweden with the hope of reaching Israel. Chalidze and Bonner were both against involving Sakharov in the case, which Amnesty International had refused to consider since the defendants had been prepared to use violence. Sakharov agreed that the skyjackers' behavior had been "hazardous and criminal" but saw beneath the particulars to the principle: "This affair was yet another tragic consequence of the USSR's failure to permit emigration."

Filing an affidavit that she was Kuznetsov's aunt, a fiction to which for some reason the KGB turned a "blind eye," Bonner attended the trial, writing out the day's proceedings from memory each evening, which were then passed on to foreign correspondents in Moscow. When Kuznetsov and the other principal defendant were sentenced to death for treason, the men in identical suits broke into applause, but they fell silent when Bonner rose and shouted: "Fascists! Only fascists would applaud a death sentence!"

Sakharov was shocked by the disproportionality of the sentence —the plane had not been skyjacked, violence had been planned but none carried out. He telegrammed Brezhnev and attempted to get through to him by phone. Learning that Soviet members of the

U.S. National Academy of Science were interceding for the Black Panther Angela Davis, indicted for an armed attempt to free prisoners during a trial, Sakharov decided to write a letter to President Richard Nixon and Soviet President Podgorny requesting clemency for Davis and for the commutation of the death sentences of the skyjackers. He showed the letter to Chalidze, who gave it to Bonner to type. The next day she visited Sakharov at home for the first time to work on a revision of the letter, establishing a working style of cooperative argument. The KGB confiscated Sakharov's letter to Nixon because it gave a "politically incorrect assessment of the decision of the Soviet court," but Chalidze managed to send out a copy through foreign correspondents. Sakharov soon received a telegram from a U.S. deputy secretary for European affairs stating that Davis was accused of "committing a serious crime" but that her trial would be fair and open; Sakharov was invited to attend.

But what probably saved Kuznetsov was chance and Soviet sensitivity to accusations of fascism. Basque terrorists had recently been sentenced to death in Spain, causing an international outcry—cartoons depicted Brezhnev and Franco dancing around a Christmas tree decorated with hanged men. After Franco commuted the Basque terrorists' sentences, an appeal hearing for Kuznetsov and the other defendants was suddenly called for December 30, an event that brought Sakharov and Bonner back together again. Now their conversation was not confined to human rights. He found himself speak freely of his own life and feelings; Bonner told him of her intention to retire at fifty, her right as a wounded World War II veteran, and devote herself to her children.

The following day, December 31, 1970, the death sentences of the Leningrad skyjackers were commuted to fifteen years in a labor camp.

The first year of his new life ended sweetly for Andrei Sakharov. A few minutes before midnight the phone rang. It was Elena calling to wish him a Happy New Year.

* * * *

Elena Bonner was the daughter of a Siberian Jew, Ruth Bonner, and an Armenian father whom she barely knew and who was soon replaced by an Armenian stepfather, Gevorkh Alikhanov, a high-ranking communist. Ruth Bonner was born in 1900, of ancestors sent as forced labor to Siberia, "on foot . . . under the lash," by Catherine the Great. Her father's side of the family "were all against the Tsar and for the poor. The children were split into Socialist Revolutionaries, Bolsheviks, Mensheviks, anarchists, whatever you like, anything but monarchists." A good deal of her childhood and youth was spent visiting relatives in prison. One aunt was sentenced to death for inciting a mutiny of soldiers during the Revolution of 1905, but, as a minor, her sentence was commuted to twelve year's hard labor. During the First World War, Ruth began publishing a magazine at school that exposed local black marketers trading in hashish and opium from Manchuria and China. She even had herself photographed being "hung" from a schoolyard exercise bar. "And there I was—a German war atrocity! By the time I was sixteen I was an all-out Socialist Revolutionary. A real terrorist. I had a fairly elementary program—kill the bad people and let the good ones live. . . . Though I couldn't have defined what 'good' means."

Alarmed by her daughter's radicalism, Ruth's mother decided to send her "little terrorist" to live with her brother, a university professor in Moscow, hoping that this would soften her views and temperament. Ruth arrived in Moscow in early 1917 "just in time for the February Revolution."

After the October Revolution and Civil War, she threw herself into activism, taking part in the famine relief of 1921–1922. In 1921 she first met Alikhanov, a fiery Armenian communist who in 1916 had slapped Beria in the face for the way he treated a young girl and who later fought for Bolshevism in Armenia, proclaiming its victory from a balcony in the capital, Yerevan. They met at the Third Congress of the Communist International, at which Alexandra Kollantai had declared that sex was as natural as drinking a glass of water and of no greater significance. In those heady times it was also considered bourgeois to even register a marriage—love and shared devotion to the cause was enough to sanctify a relationship. It seemed annoying and pointless to pause for such formalities when a new world was being built.

Elena was born two years later, in 1923. Headstrong, "sassy," romantic, handy with tools, with a tendency to get lost in books or the streets of the city, she grew up in the privileged world of the communist elite in both Leningrad and Moscow. Her father became the head of the Balkans section of the Comintern, the organization that dealt with the communist parties in countries outside the Soviet Union and which in the thirties was staffed by such people as Tito, the Italian communist Togliatti, and the famous Spanish revolutionary Dolores Ibarruri, known as "La Pasionaria." Many of these people lived in the same building and were frequent guests in Elena's parents' home. Sergei Kirov, the Leningrad party boss whose assassination in 1934 was the pretext for the Great Terror, gave Lusia, as Elena was known to family and friends, a ride in his car "in front of all the kids." La Pasionaria visited frequently, giving Elena's father a "handsome knit wool shirt, bright blue, almost corn flower in color" and, in general, taking an interest in Elena's father that she found suspect, even following them around to catch them in an intimate moment. Ruth was always working or studying—the Moscow Party Committee, the Institute of Marx-Engels-Lenin, the Stalin Industrial Academy. Sheer exhaustion and the ethos of the times meant that Ruth was not demonstrative to her daughter, who was starved for hugs and kisses.

They lived well by Soviet standards, though everything they had belonged to the state: "It didn't worry me at all (nor my parents) that the furniture did not belong to us and that each piece had a small gold oval with a number attached with two nails." But after the assassination of Kirov, life became shakier, more provisional. The "loud, masterful steps" of soldiers arriving at night to make arrests became a more familiar sound. Everyone looked grim, doomed, especially Elena's father.

He was arrested at work on May 27, 1937. They came for Ruth on December 9, 1937. Elena was there: "Well, that's it," said Ruth, who was wearing the shirt La Pasionaria had given her husband. "She quickly pulled a sweater over it and took her suitcase, packed long ago, from beneath the bed. As she threw her leather coat . . . onto her shoulders, she turned to me. I was astonished by the radiance of her face and by the smile she gave me. She embraced me and kissed me several times . . . and gently pushed me aside."

Refusing to denounce her husband in exchange for a lighter sentence, Ruth received eight years in a women's hard-labor camp in Kazakhstan and eight years of exile as a Family Member of an Enemy of the People. Ruth lost none of her irony or impertinence in the camps. She thought of her aunt, whose sentence for inciting rebellion among soldiers had been commuted from death to twelve years. "A little less than I got. And I didn't start any mutinies." And she didn't pay the guards the proper respect: "Listen to this one," said one guard. "She thinks she's Countess Trubetskaya." Ruth couldn't resist replying: "Trubetskaya was a princess, not a countess."

Ruth learned to survive in the camp by making friends, cutting corners, faking work. But it might have been her love of beauty that saved her. The women were lined up in the subzero darkness for a head count before the day's work began. This was even worse than the hard labor, because at least work allowed you to keep moving or to find a place to hide from the wind. Staring at the huge red sun rising over the frozen steppe, Ruth said: "Look how beautiful it is." The other women thought she had gone mad.

At age fourteen, for all practical purposes an orphan, Elena took her younger brother from Moscow to Leningrad to live with their grandmother. There she finished her schooling, working part-time as a cleaning woman and a file clerk in a factory, and fought her way back into the Communist League, from which she had been expelled as a daughter of Enemies of the People. Elena also found time to study literature and had the first great romance of her life with a young poet, Seva Bagritsky. A fireball of energy, she burned some of it off in gymnastics, track, and volleyball. A photograph of her girls' volleyball team was even published in a magazine that, in some improbable fashion, made its way to the camp in Kazakhstan where Ruth was afforded a glimpse of her daughter.

When the war with Nazi Germany broke out, Elena enlisted at once, serving as a nurse on hospital trains, caring for the wounded after carrying them in from the battlefield. Severely wounded in October 1941 when a bomb exploded near her train, she lay unconscious, covered by earth until she was discovered by chance, after which she was deaf, dumb, and blind for several days. But worse than wounds was the news that Seva had been killed in February 1942. "How did I keep from breaking then? Or perhaps I did break? After all, with him I also lost the treasure that had been entrusted to me by God or by love. And also my rosy, light world view."

Having taken part in mine-clearing operations toward the end of the war, she was discharged as a lieutenant in the medical corps in 1945. She visited her mother in the camp in Kazakhstan before returning to Leningrad, where doctors told her that due to her injuries she would soon go blind and that she should never have children. She began studying Braille.

In December 1945, her camp term over, Ruth returned illegally to Leningrad before beginning her exile: "When the bell rang," says Elena, "and I opened the door, I thought she was a beggar and I offered her some change and a piece of bread. That mistake still hurts."

It was the first December holidays since the end of the war, but Ruth could not celebrate: "My daughter's room was packed with young people, some in uniform, some not. Oh, what young lions they all were. But I was absolutely alienated from all of it and all of them. Absolutely alienated."

Elena enrolled in the Leningrad Medical Institute where she would earn her degree in pediatrics. There one day, perusing some lists on a wall, she heard two young medical students behind her talking, one of whom said: "I'm going to marry those legs." She spun around and replied: "You better ask yourself if they're interested in marrying you."

The student, Ivan Semyonov, was first-generation intelligentsia. He came of peasant stock, his mother illiterate, his father insistent on education for his children. He'd almost not been born in the first place. After bearing five children, his mother decided to take advantage of the easily available abortions, the principle birth control method. The night before she was about to leave for town she was dissuaded by a dream: she was walking down a road and met a priest who was carrying a flowering apple-tree branch and who said to her: "Take this and plant it."

Ivan did marry those legs and the rest of Elena Bonner, who in utter disregard of her doctor's advice gave birth to two children,

Tatiana and Alexei. Motherhood made her no more conservative. She refused to denounce one of the professors in her institute during the Doctors' Plot hysteria and was expelled, only to be reinstated after Stalin's death. In the mid-sixties, Ivan and Elena separated amicably, "at least without any scenes or punches thrown," as she puts it. At around that same time she joined the Communist Party, believing the system could only be reformed from within and intending to do just that.

* * * *

Andrei Sakharov began the new year of 1971 by writing a memorandum to Brezhnev. He prefaced his remarks by requesting a "discussion" of the previous letter on the need for democratization. For the sake of openness, Sakharov also informed Brezhnev of the formation of the Human Rights Committee and enclosed some of its documents: "We hope to be useful to society, and we seek a dialogue with the country's leadership."

Less sharply focused than the letter on democratization, the memorandum is more a list of talking points on "urgent problems." What is new in Sakharov's thinking is that human rights have replaced intellectual freedom as his fundamental premise. He reminds Brezhnev: "The basic aim of the state is the protection and safeguarding of the basic rights of its citizens." Toward that end, he urges an amnesty for political prisoners, open trials, freedom of information ("glasnost") and of religion, the right to leave the country and freely return, and the abolition of psychiatric punishment, the internal passport, the death penalty, jamming, and the privileges for the power elite. He calls for improved education and a struggle against alcoholism, pollution, and cruelty to animals. His work on the Human Rights Committee had sharpened his awareness of the "nationalities problem"—everything from the still dispossessed Crimean Tartars to the Volga Germans wishing to return to Germany, not to mention Ukrainian aspirations for autonomy and the resentment of the Baltic States—Estonia, Latvia, and Lithuania—illegally annexed as part of the secret deal between Hitler and Stalin. He raises that issue in only the most general of terms but calls for a law guaranteeing the right to secede as a "confirmation of the anti-imperialist and anti-chauvinist nature of our policies." On the international front, Sakharov wants the Soviet Union to affirm "our refusal to be the first to use weapons of mass destruction." The Soviet Union's "chief foreign policy problem" is not the United States but China, a view widely shared by the intelligentsia, one of the few points on which Solzhenitsyn and Sakharov would agree. China with its immense population, Siberia with its immense emptiness, combined with the ideological conflicts and territorial disputes, was an equation for war.

The KGB was aware of and worried about Sakharov's memorandum well before it was completed. On February 12, 1971, Andropov reported to the Central Committee: "SAKHAROV is striving to secure a meeting with leaders of CPSU [Communist Party of the Soviet Union] and the Soviet government, and he is preparing a so-called 'Memorandum' which he intends to use for his meeting with the authorities. To judge from a conversation between SAKHAROV and CHALIDZE, one cannot exclude the possibility that this memorandum will be published in samizdat in the USSR (and, possibly, abroad)." It would be "extremely undesirable" if that memorandum were published on the eve of the Twenty-fourth Party Congress, which would open on March 30. "We therefore deem it imperative to decide, without the least possible delay," says Andropov, "the question of organizing a meeting and conversation with him at the Central Committee."

In Andropov's reports and appended material, the psychological profile of Sakharov was becoming more complex. "In daily life, he gives the impression of a typical eccentric scientist, who pays no attention to the minutiae of daily life. His dress is usually untidy, he wears old clothes, and his general appearance is one of neglect. . . . He is described as an honest, compassionate, and conscientious person. He respects intelligent and knowledgeable people; he is principled and courageous in defending his principles; he lives in ideas and in theories; he can think about problems even in the least suitable places. He is absentminded and requires care." But two new broad strokes are added: pride and guilt. Sakharov "proudly asserts that his so-called 'political activity' is currently more important than his work in science, if judged by the worldwide resonance that it has evoked. He declares that he 'has become' famous thanks to the publication of his essay in the anti-

Soviet press in the West and in 'samizdat' within the country, and that the Soviet government, in his opinion, is powerless to do anything with him." But that pride only masks a deeper feeling: "Having made a great contribution to the creation of thermonuclear weapons, SAKHAROV felt his 'guilt' before mankind, and, because of that, he has set himself the task of fighting for peace and preventing thermonuclear war." But Sakharov is not an out-and-out renegade: "In general, SAKHAROV acknowledges the superiority of the socialist over the capitalist system, but he incorrectly perceives many aspects of the real life of Soviet society . . . stubbornly denies that his activities cause political harm to the state and serve as a source of anti-Soviet propaganda for the ideological centers of our adversary."

It was not Sakharov but the other two members of the Human Rights Committee, Chalidze and Tverdokhlebov, who were called in for "prophylactic conversations," that is, warnings. As was the custom before a party congress, the capital was cleared of undesirables—drunks, "parasites" (anyone without officially recognized employment), political and religious dissidents, and activist Jews, who had already demonstrated twice in early March for the right to emigrate. Foreigners also came under close scrutiny. Two Belgians were detained, and a body search revealed that they were in possession of documents they had received from Chalidze. At around eight o'clock on March 29, the eve of the congress, Sakharov was informed by phone that a search was underway at Chalidze's apartment. He rushed over to find Elena Bonner already there.

They were not allowed entry to the apartment, the action, as in a Greek tragedy, occurring off scene. The proceedings followed a time-honored formula. "Independent witnesses"—the building's janitor, neighbors—were present to ensure that regulations were observed, though there was never any question of whose side they were on. The person whose apartment was being searched was not allowed to use the telephone and could not go to the bathroom unaccompanied. "They know how to look and we know how to hide" the Zeks used to say. But what could you hide in one room, and what reason was there to hide when openness was both your watchword and strategy? Every book was opened and held upside down to see what came fluttering out. Not only typewriters were confiscated but used ribbons, address books, diaries, tape recorders, "sometimes even cash, warm apparel, and food, if they suspect that these are destined for prisoners or their families." And, in Chalidze's case, of course, absolutely anything connected with the Human Rights Committee was placed in sacks and sealed.

Around midnight the agents emerged with their sacks. Then Sakharov, Bonner, and a few others sat with Chalidze until nearly three in the morning. Sakharov could see the search had been "devastating" for him. And a search was not an isolated event but part of a fixed, accelerating process—summons, warning, search, arrest, trial, then prison, hard labor, or madhouse.

It was a spring of trials. Sakharov found himself involved with Crimean Tartars, Jewish emigration, Volga Germans, and a case of religious freedom involving a Russian Orthodox sexton: "If I lived in a clerical state, I would speak out in defense of atheists and heretics."

The Party Congress had proceeded without a hitch; the next Five Year Plan, the ninth, was presented, and it was officially announced that the Soviet Union had entered the stage of "real socialism." And, vis-à-vis Sakharov, it had entered a phase of "real antagonism." As Andropov reported: "During the XXIV Party Congress, the Committee for State Security received new information that SAKHAROV has openly switched from a few politically harmful acts of a malcontent to consciously hostile activity." Still, on May 21, 1971, Sakharov's fiftieth birthday, Andropov ameliorated that stance somewhat, reporting a conversation in which Sakharov opposed all revolution: "Now the Communist Party has permeated the whole country to such an extent that any attempt to remove it would involve the removal of living tissue. It would make bloodshed inevitable."

Tamm had died in April, which meant that in the year he turned fifty Sakharov had lost a mentor, protector, collaborator, friend. The little group of friends he kept at FIAN held a party for him, giving him a shortwave radio as a birthday present, Sakharov jesting that a transmitter might have been more appropriate. He also received a painting of Don Quixote, whose image had special meaning for the intelligentsia since the nineteenth century, when the novelist Turgenev had declared that in Russia men of conscience could be only paralyzed Hamlets or foolhardy Quixotes.

Sakharov accepted the painting smilingly and hung it at home. But he had his own doubts too, his black hours. Now it was his turn to confess a father's disappointments to his son when in June he told Dima in a monitored conversation: "I have been removed from dealing with military matters. . . . I am no longer regarded as a person. . . . Earlier I had a pass that said 'Admit everywhere.' Now, I am not allowed in anywhere. Not even to my own installation."

But black moods were rare when he was with Elena Bonner. He pursued relations with a zest that he found pleasantly surprising, as did she, but they both felt the uncertainty that necessarily derives when an acquaintanceship is young and the people involved are not. Sakharov was still unsure whether this was what Goethe had called "the good fortune of a late love," and Bonner still felt their closeness should not exceed the romance of camaraderie.

By the end of the spring Bonner had introduced Sakharov to her mother, Ruth, and her son, Alexei. Sakharov too made sure that meetings were not left to the chance choreography of the cause. Off to a vacation with his daughter Lyuba and son Dima on the Black Sea, he dropped off their dog, Kid, a mix of dachshund and spaniel, to be kept by Bonner at her dacha in the writer's colony of Peredelkino, outside Moscow. Bonner moved in literary circles, and through her Sakharov met the widow of the satirist Yuri Olesha, who told the story of her husband flirting with a waitress, even calling her "my queen." "If she's your queen, then who am I?" asked the wife. "You?" said Olesha. "You are me."

Sakharov settled in on the Black Sea with Lyuba and Dima. Tanya had her own family now—a daughter, Marina, whom Sakharov, an uneasy father but a natural grandfather, particularly adored and who had been born in time for Klava to hold and see. It was Lyuba who had taken Klava's place, caring for both her husband and her son. Not the easiest of jobs. Sakharov had a tendency to drift into a neglect of everything, himself included. He even had to be watched: he would sit on the beach thinking until he was burned red. Dima already felt hurt by life, cheated with a mother gone too early and a father far too great—a father who could pay loving attention but never often enough. That was in part his nature but also a Stalin-era ascetic style that both Sakharov and Klava had shared—children should not be spoiled with attention or affection. His daughter Tanya always had the feeling that they grew in front of him "like grass."

Sakharov walked the beach skipping stones on the water. He had made another unsuccessful attempt to learn how to swim, impressing his teacher with both his courage, going right out into the deep water, and his lack of coordination. That teacher was Mark Perelman, a thirty-nine-year-old theoretical physicist at the Georgian Academy of Sciences' Cybernetics Institute, whom Sakharov knew from conferences in Tbilisi. There, having successful shaken the agents tailing Sakharov, to whom he referred as the "art critics," a lark in itself, they had walked in the woods, Perelman observing that Sakharov took special care not to step on anthills or break branches and was the first to spot a "magnificent white mushroom." Now, at the Black Sea, they talked science and drank tea, Perelman noticing that Sakharov still preferred "cook's tea," apple chunks added for flavor and sweetness. Not infrequently, a question would arise that would cause them both to sit for hours in silence, something with which they were each comfortable. Among other subjects was Sakharov's intention to revise his father's textbook. Lyuba was glad they talked science, not politics. Politics had already cost her father greatly—economically, and in society. And neither of those helped a twenty-two-year-old woman run the life that had been thrust on her.

But it was Lyuba who had to pull her father out of incipient political trouble when he and Perelman, amused by the many pictures of Stalin in the town of Sukhumi, decided with childish glee to look for the most ridiculous depiction of all. They were a little too loud about it and desisted only when an infuriated Lyuba hissed that "they were already starting to attract the outraged attention from supporters of the Great Leader and father of nations."

Then, to add to Lyuba's problems, visitors arrived. Bonner's daughter, Tatiana, and Efrem Yankelevich, the telecommunications student she had married the previous December, also vacationing on the Black Sea, had taken Bonner's suggestion to visit Sakharov. Short, dark-haired, blue-eyed, with a tendency to stammer, Efrem established immediate rapport with Sakharov, who always responded to character rather than personality, though in those circles a person's integrity meant nothing until tested. To Tatiana, Sakharov seemed somewhat ill at ease with his own children, but that may have been an instance of the observer changing the observed, for the entire situation, with its implicit rivalry for Sakharov's allegiance, had to make him uncomfortable. Dima sulked, and Lyuba could not enjoy seeing her father drawn to another woman, another family.

Her father's attraction to Bonner was obvious—he was capable only of strong attraction or of none. But how could he love another woman after their mother?

What was clear to others came clear to them as well, but not until August 24, after a period of separation, did they "confess" their love, the fifty-year-old man and the forty-eight-year-old woman pledging to make each year "count for three."

The next day, listening to an Albinoni concerto at Elena's apartment, Sakharov wept from an excess of joy.

In the fall she took the shockingly untraveled Sakharov to Leningrad to meet her closest friends, three women who lived together in what was known as the House on Pushkin Street, an oasis of free thought and civility. Sloppiness and vodka were not tolerated. Every day was open house. Simple food was served: fried potatoes, salted mushrooms, bread, butter, and jam. Tea was promptly at nine. The police came only once, their search perfunctory, purely to communicate that they were fully apprised of the activities on Pushkin Street but would not demean themselves to harass three old women. One of them, Regina, was bedridden with a heart ailment, and her bedside had become a "confessional." Another woman, Zoya, though up and around, weighed "no more than seventy pounds," which did not prevent her from physically assaulting a huge man who had called her a "kike" in the post office and actually chasing him off. Natalya Gesse was the third.

Glad and proud that he appreciated her friends, Bonner asked him who his friends were.

"Zeldovich," replied Sakharov.

"Zeldovich and that's all?" said Bonner in amazement.

"Zeldovich and that's all."

For Bonner, travel was a passion and an imperative in that Soviet Union of closed borders where the operative quip was "A chicken's not a bird and Bulgaria's not abroad." Camping with Bonner and her son in the woods, Sakharov "boyishly" confessed that this was a first for him. She was astonished.

She'd supply the breadth. That fall Bonner introduced him to the poets, the royalty of Russian literature. He met the balladeer Bulat Okudzhava, whose songs told the story of a generation, Sakharov's, Bonner's. He heard David Samoilov recite his poems in his country house outside Moscow, and he could not help but marvel at it all: "Six months earlier, who would have imagined me in such company."

By early October Andropov reported to Suslov, the gray cardinal still at his post: "Recently, there has been a change in the personal life of SAKHAROV. He has become intimate with a teacher in Medical School No. 2, . . . BONNER (born 1922; member of the CPSU; her parents were earlier subjected to repression, but later rehabilitated.) BONNER supports the negative manifestations and activities of SAKHAROV as a member of the 'Committee' and makes multiple copies of the materials prepared by his fellow collaborators." She was assigned the operational code name "Vixen."

That same October Sakharov and Bonner were further defining their working relationship. Sakharov composed a statement on the right to emigrate, which Bonner typed to his dictation, continuing their tradition of "heated" exchanges. Sakharov was now also proposing marriage. Bonner was reluctant—their marriage might place her children in jeopardy. Sakharov maintained that their union would make them "less vulnerable" until she was finally convinced.

Now the children of each family had to be informed. Bonner's children shared her joy, glad to have Andrei Sakharov in their family. Sakharov's, fearing further loss, were less sympathetic. None of it was any secret and soon reached Andropov's ear: "SAKHAROV's intention to marry BONNER elicited a negative reaction on the part of his daughters, which has caused a tense situation in the family."

Bonner forbade her children to attend the ceremony. Sakharov, hurt by his children's disapproval, did not even inform them of the date, though he knew "such cowardice only makes life more difficult."

But before marrying, they had to attend the trial of the dissident Vladimir Bukovsky on January 5, 1972. Things had changed. Sakharov could no longer penetrate the cordon of KGB agents, who, "insolent and sure of their power," now even questioned his patriotism, asking if he still was a "good Soviet." And Bukovsky's sentence was not "mild by Soviet standards": two years' prison, five labor camp, five internal exile.

On January 7, 1972, Sakharov and Bonner went to register their marriage at their local ZAGS, the civil registry office. His witness was Andrei Tverdokhlebov from the Human Rights Committee, hers Natalya Gesse. But the KGB had also sent "its own witnesses—half a dozen men in identical black suits." Bonner's daughter, Tatiana, dashed in breathlessly at the last moment, angering her mother, who wished to keep her children and this marriage as separate as possible.

That same evening, to gather support for clemency in Bukovsky's sentence, Sakharov and Bonner flew to Kiev, man and wife.

Andrei Sakharov's paternal great-great grandfather.

Andrei Sakharov's paternal great-great grandmother, 1860s.

Dmitri Sakharov, Andrei's father, in student uniform, 1909.

Dmitri Sakharov, c. 1949.

Dmitri Sakharov, c. 1950.

Ekaterina Sofiano,
Sakharov's mother, 1909.

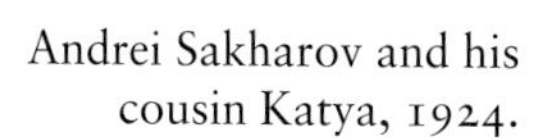

Andrei Sakharov and his
cousin Katya, 1924.

The children of the courtyard at No. 3 Granatny Lane. Andrei is second from left; Grisha Umansky is seated, sixth from left.

Alexei and Zinaida Sofiano, Sakharov's maternal grandparents, at their dacha, c. 1925.

Andrei Sakharov, 1943.

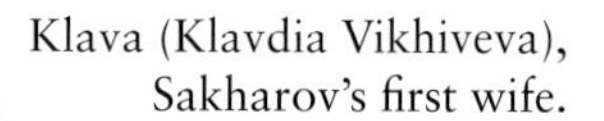

Klava (Klavdia Vikhiveva),
Sakharov's first wife.

Sakharov with Klava and their daughter Tanya, 1948.

Andrei Sakharov, 1950.

Andrei Sakharov's daughters, Lyuba and Tanya, 1956.

Andrei Sakharov and Igor Kurchatov at the Atomic Energy Institute, summer 1957.

Andrei Sakharov and his son Dmitri (Dima), 1970.

Elena Bonner as a child with her mother (left), Leningrad, December 4, 1927.

Elena Bonner, Leningrad, 1949.

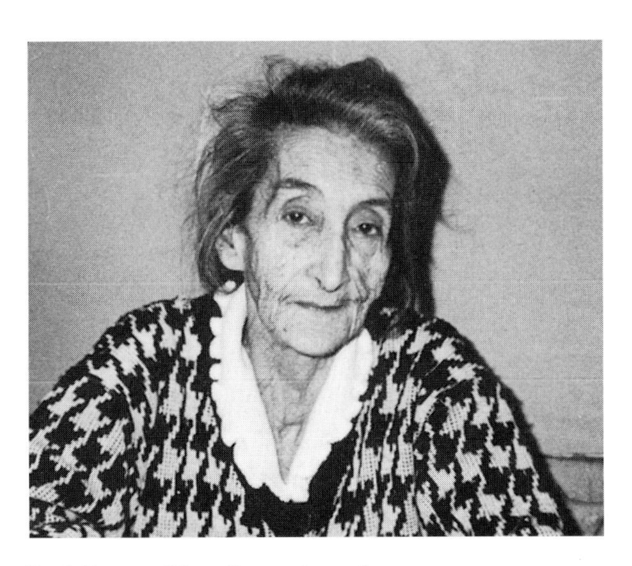

Ruth Bonner, Elena Bonner's mother.

In Vilnius, Lithuania, on December 10, 1975, the day Sergei Kovalyov was sentenced. Efrem Yankelevich is third from the right.

Sakharov and Bonner with Jewish refusenik leader Vladimir Slepak, 1978.

Sakharov outside the Moscow courthouse during the trial of Yuri Orlov in May 1978.

Police cordon outside the courthouse during Orlov's trial.

Sakharov and Bonner in Gorky, January 1980.

Sakharov on a balcony in Gorky, 1984.

Bonner photographed by Sakharov, Gorky, 1985.

Sakharov returns to Moscow, December 23, 1986.

With the grandchildren, Sheremetevo Airport, Moscow, June 1987. Sakharov holds Anya and Matvei.

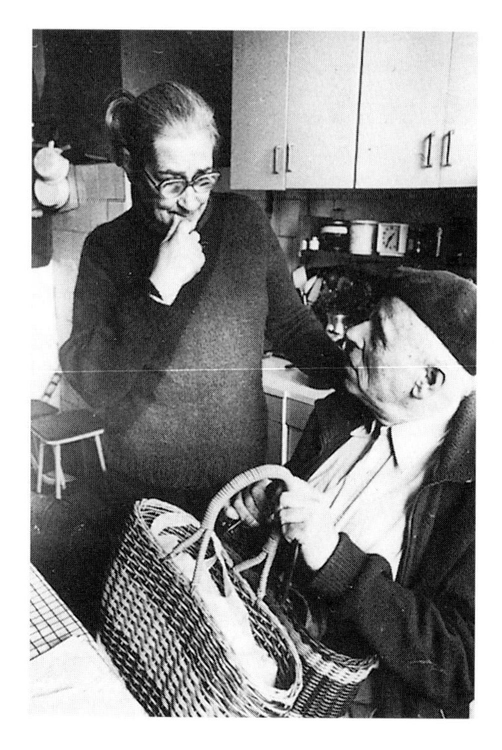

Sakharov and Bonner back in their Moscow kitchen, autumn 1987.

Andrei Sakharov and Tatiana Yankelevich, Newton, Massachusetts, November 1988.

With Edward Teller,
Washington, D.C.,
November 1988.

With President Ronald Reagan at the White House, November 1988.

In Spitak, epicenter of the Armenian earthquake of December 1988.

Sakharov speaking at the First Congress of People's Deputies. Gorbachev is in the background.

CHAPTER ELEVEN

ESCALATIONS

LOVE led Andrei Sakharov into the labyrinth of the "apartment question." In principle, as the Soviets liked to say, his own situation was quite good. The dacha in Zhukovka, the elite settlement outside Moscow, remained in his use—the authorities had wished to rescind the privilege but, to their consternation, discovered that the original documents had been signed by Stalin himself. And he still had the roomy apartment in the plushly wooded part of Moscow where Dima and Lyuba lived. Since it was out of the question to move Elena in there, Sakharov had begun living with her after they had declared their love in August. The problem was that Bonner's mother, Ruth, her son, Alexei, her daughter, Tatiana, and Tatiana's husband, Efrem, were already living in that two-room, 325-square-foot apartment. Every inch was packed; as one visitor noticed, "skis were stored next to the flush bowl in a tiny toilet. Ice skates dangled closely overhead."

A solution was found: Ruth and Alexei slept in one room, Elena and Andrei in the other, Tatiana and Efrem on a couch in the kitchen. Tolerable, but not a situation any of them wanted to continue for very long. The apartment question, which had not existed for Sakharov since the early days of his marriage with Klava, became all too real again. Now he would have to enter that maze of dingy offices, ill-tempered clerks, poorly printed forms that had to be filled out perfectly, plus a more shadowy zone of favors, deals, swaps, and bribes.

When an apartment was obtained for Efrem and Tatiana in January, all their friends pitched in to help them move, paint walls, wash floors. Vladimir Gershovich, the "Jewish Hercules," carried a refrigerator up seven flights on his back. Bonner, known for her cooking, made a roast with potatoes that the team washed down with vodka.

The movers were dissidents—the people who protested, helped the families of prisoners, gave blood to get a day off (which they would spend standing outside courtroom doors blocked by KGB), and published samizdat. Even Hercules was a refusenik.

Even though moving in with Bonner was the obvious choice for Sakharov, it was still read as emotional treason by his own children, who believed he had signed a written promise to Klava never to marry again. The logistics dictated neglect for Dima, now fifteen. In March Sakharov and Bonner decided to take a real honeymoon. It was to be a family event, Alexei coming with Bonner and Sakharov inviting Dima, but he, "owing in part to his sisters' influence," refused. And so it was only Sakharov, Bonner, and Alexei who saw the ancient Silk Route cities, the scimitars of Arabic script on the minarets, the brown-gold domes of the buildings resembling camel caravans, the reflecting pools touched by willows, the teahouse where men in blue silk turbans with small tattoos under their eyes recline on large bedlike structures as the long vowels of an Uzbek love song pour from a scratchy PA and an hour passes in the wink of an eye, like the whim of a caliph.

While they were in the neighborhood they flew over the Pamir Mountains to Dushanbe, the capital of the Tadzhikistan Republic—just west of China and north of Afghanistan—on human rights business, which was quickly reported to the Central Committee by the KGB. Every year the KGB had a new reason to be especially vigilant: 1971 had been the hundredth anniversary of Lenin's birth; 1972 was marked both by the fiftieth anniversary of the USSR's official founding in 1922 and by heightened activity in détente. President Nixon, who was still fighting Vietnam and would soon be fighting Watergate, visited Beijing in February and Moscow in May, the order of his itinerary signaling precedence and priority. The Soviet leaders and the Soviet dissidents were united by a fear of China.

Not liking tension on two fronts, the Soviet leaders had begun seeking a détente with the United States in the very late sixties. Though détente in time would result in cooperation in space, agreements to limit the deployment of nuclear weapons, grain sales, and Jewish emigration (35,000 in 1973, over 50,000 in 1979), some Soviet dissidents were doubtful from the start. Amalrik wrote: "Americans apparently also believe that the gradual improvement in the standard of living, as well as the spread of Western culture and ways of life, will gradually transform Soviet society—that the foreign tourists, jazz records and miniskirts will help to create a 'human socialism.' It is possible that we will indeed have a 'socialism' with bare knees someday, but not likely one with a human face." Henry Kissinger, one of détente's initial architects, also had his own reservations: "We cannot be indifferent to the denial of human liberty, but we cannot, at the same time, so insist on transformation in the domestic structure of the Soviet system as to give up the general evolution that we are hopefully now beginning."

In fact, from the Soviet point of view, the "denial of human liberty" was by 1972 an integral element in their understanding of détente. No new Budapests or Pragues would be tolerated, especially at home. The first target was the *Chronicle of Current Events,* which had been publishing since 1968 and whose mission was to print impartial, factual accounts of arrests, trials, sentences. Its motto was Article 19 of the Universal Declaration of Human Rights: "Everyone has the right to freedom of opinion and expression; this right includes freedom to hold opinions without interference and to seek, receive and impart information and ideas through any media and regardless of frontiers." The Soviet authorities held quite a different opinion, one that they enforced with a wave of nationwide arrests in the first half of the year. The campaign against the *Chronicle of Current Events* was so intense that the journal ceased publication in summer 1972, a great victory for the KGB but only if it proved final.

In June, Piotr Yakir was arrested. A tragic glory reflected on Yakir, whose father had been one of the several generals executed by Stalin on the eve of World War II and who had grown up in the Gulag, spending his fourteenth through his thirty-first years in "various forms of detention," which he described in his book *A Childhood in Prison.* Shaggy-haired, shaggy-bearded, a cigarette always in his mouth and a glass usually in his hand, Yakir had every reason to want every freedom.

Remembering that Minister of Machine Building Slavsky had been close to General Yakir when they both served in the cavalry, Sakharov decided to put all past bitterness aside and try to convince Slavsky to intercede. Sakharov had not been to the ministry for some time; how much it had remained the same only made him feel how much he had changed. Even though General Yakir had been safely rehabilitated, Slavsky refused to honor his memory or to help his son who, he told Sakharov, must be "as anti-Soviet as you are." That would soon prove tragically untrue. Yakir was already buckling under interrogation and would be the first dissident to recant and inform.

The next violent shock in a violent year came not from the secret police but from terrorists. In early September at the Munich Olympics, Palestinian terrorists massacred eleven Israeli athletes. The Jews of Moscow organized a silent protest in front of the Lebanese Embassy because that country gave haven to a Palestinian terrorist organization called Black September. Bonner was ill that day, but Sakharov decided to attend and was accompanied by Tatiana, Efrem, and Alexei. The first wave of demonstrators had already been arrested and taken to a drunk tank, which tended to be empty by day. Sakharov, and Bonner's two children and son-in-law, were taken to the drunk tank in the second wave, processed, and released. Even as he was going through the formalities, Sakharov had a keen sense that it "had been a mistake to let them come with me."

* * * *

But there were also peaceful interstices of science such as seminars at FIAN, which was protective of Sakharov. Even the scientists who were in the institute's party organization did what they could to deflect the inane directives that came down on them. Sakharov had especially good relations with one of them, Boris Bolotovsky, a physicist by training and a communist "by conviction," that is, the belief that socialism was the only cure for anti-Semitism. One day Bolotovsky asked: "You write a lot of letters, Andrei Dmitrievich. Did you ever get an answer even to one?" "No," said Sakharov, "I never received a single answer. I'm not surprised though because I think for a very long time before I write those letters, and so they must still be thinking how to best answer me." Bolotovsky could never quite decide to what exact extent Sakharov was joking.

Sakharov also made time to attend conferences in Baku and Yerevan, having now caught some of Bonner's wanderlust. In Samarkand billboards of Lenin and construction cranes rose over the mulberry trees and the emerald tomb of Tamerlane. In Baku he saw the temple of the fire worshipers, the caravansary where Marco Polo tarried on his way to Cathay noticing a black substance oozing from the ground (runoff from an underground lake of oil), prehistoric sculptures of animals, a five-foot-long stone instrument that "emitted a pleasing tone when struck."

Even when Sakharov was doing science in exotic settings, politics was never far from hand. He had already begun collecting signatures for an appeal for an amnesty for political prisoners and an end to the death penalty in honor of the fiftieth anniversary of the founding of the Soviet Union, a card that could be played by both sides. But no one was signing in Baku.

The political atmosphere was more severe than it had been in quite some time, and people were taking no chances. Some refused on truly held convictions—Kapitsa for instance, whose courage in the worst of times had already been amply demonstrated: "There are more important things than a few political prisoners. Mankind is facing major challenges. The gravest danger is the demographic explosion, the continuing population increases in the developing nations—*that* threatens millions with death by starvation." Solzhenitsyn declined to sign, saying it could interfere with his work; even Chalidze hesitated before finally signing. The KGB discounted the signatures that Sakharov did manage to collect, such as that of the cellist Rostropovich, as those of "known . . . participants in various provocations" while also assigning a new role to Bonner—the instigator "who systematically fans the flames."

Sakharov also attended the Yerevan conference on gravitational theory in October, Bonner hiking the hills while he listened to papers and attended colloquia. But in their free time she showed him Armenia's "quasi-Biblical landscapes and architecture" and the balcony where her father had proclaimed the victory of Red power. The idyll came to an abrupt end when Sakharov and Bonner received word that Tatiana had been expelled from Moscow University. They flew back to Moscow at once.

Tatiana, a senior, who had finished her course work and needed only to complete her dissertation to receive a degree in journalism, was a night student, which required her to hold a day job in her field. She was in fact a junior editor on an outstanding science magazine for schoolchildren, but her name did not appear on the masthead. First she was deemed "not employed," then "not suitably employed." No one was the least deceived by those clumsy niceties—Tatiana was being punished for attending the demonstration at the Lebanese Embassy and even more so for being Bonner's daughter, Sakharov's daughter-in-law. For the first time it was plain to Sakharov and to Bonner that "the children . . . had become hostages."

In an effort to get Tatiana reinstated, Sakharov had several meetings with the rector of Moscow University, Academician Ivan Petrovsky. Sakharov had entirely positive feelings for Petrovsky, considering him a "distinguished mathematician" who had "displayed great courage and determination in defending the entire academic community" during the anti-Semitic campaigns of the early 1950s. Now, to Sakharov's surprise and chagrin, he found Petrovsky "less than honest," unwilling to admit that Tatiana's expulsion was unjust or even that he "was powerless to reverse the decision." At their last meeting Sakharov lost his temper and pounded the table twice with his fist. Later that day Petrovsky dropped dead from a heart attack, and in some quarters, including the Academy, Sakharov was considered complicit in Petrovsky's death. "To this day," said Joseph Shklovsky, who had been the leader of Sakharov's train car on the journey to Ashkhabad during World War II and was now a leading astrophysicist, "I can't forgive Andrei Sakharov for the sharp rebuke he delivered to the poor rector."

The conflict was sharpening, tempers fraying. At a shoving match between dissidents and KGB outside a courthouse, Sakharov was among those whose arms were twisted behind their back. Elena Bonner "marched up to the senior KGB officer, who was commanding from the sidelines, and slapped his face." For that act, as a member of the Communist Party, Bonner was summoned to appear before the Moscow Party Committee on November 9. Expecting to browbeat her into recanting her act of "hooliganism," the committee was shocked when she placed her party card on the table as a sign of her willingness to resign then and there. One party member who knew Bonner well and was genuinely concerned for her said: "What are you doing? Think of your children!"

"Stay out of it! What's this got to do with the children?"

"Why are you so hostile to the Soviet system?" said another member, taking a more conventional approach. "It's given you everything, an education, interesting work."

"No one gave me anything. I fought in the war, nearly lost my sight, I worked day and night," replied Bonner.

"It's all because you're so bitter," said a KGB type. "You keep on saying your father was executed, and it just isn't so."

Resisting the temptation to discuss this question of vital importance to her, Elena Bonner left the room, her party card still on the table. The system could not be reformed from within.

Also in November, succumbing to pressure, Chalidze emigrated in circumstances that proved divisive. Solzhenitsyn condemned him: "The greatest of simpletons will agree that before you get a visa to go and lecture foreigners on the rights of man in the USSR you will need to talk things over with the KGB." Sakharov was more forgiving, even though Chalidze's lack of candor had "soured" their relationship. Sakharov and Chalidze clashed over whether Chalidze would be stripped of his citizenship upon his arrival in the West. Sakharov thought it "practically inevitable." Chalidze vehemently denied it. It took only three weeks for Sakharov to be proved correct. But, unlike Solzhenitsyn, Sakharov saw no need for an overt deal—Chalidze was glad to avoid arrest, the authorities were glad to be rid of him. Before emigrating, Chalidze resigned from the Human Rights Committee, and the other founding member, Andrei Tverdokhlebov, Sakharov's witness at his wedding, did the same, in part because of his disapproval of Sakharov's "attitude toward Chalidze's departure." Though it would continue for a couple of years, the Human Rights Committee had begun its slow demise.

The year ended badly, and the next one would be worse.

* * * *

In 1972 the *Chronicle of Current Events* was the target; in 1973 it was Sakharov. The initial blow was glancing, oblique. In February the *Literary Gazette,* the newspaper of the Writers' Union, ran a review of a new book by Harrison Salisbury, managing editor of the *New York Times,* who had written the introduction and notes for Sakharov's "Reflections." That essay had never been mentioned in the Soviet press in the five years since its publication. But now the reviewer of Salisbury's book took the opportunity, still without mentioning it, to ridicule Sakharov as a naive utopian who wants "people to be like flowers" and who envisions a "world government to which everyone will be subject—Soviet workers and collective farmers and Texas oil barons, Pentagon generals and the heroic patriots of South Vietnam, American 'Black Panthers' and racists from the 'John Birch Society.'"

Though Andropov considered "Reflections" an "ideologically pernicious essay" and thought the review attacking Sakharov deserved a "positive mark," he considered that any further attacks could only serve the "next anti-Soviet campaign." For that reason he informed the Central Committee that "it would be expedient to ban mention of SAKHAROV's name in official publications of the Soviet press." A short while later the report was stamped: "The consent of the Central Committee secured."

Sakharov was startled and puzzled by the attack coming so long after the publication of his essay—"just why the conspiracy of silence about me should have been broken at that particular moment I have no idea." Ultimately, however, it wasn't the timing that was important but the public nature of the attack, the deliberate escalation. The article was followed in the press by a letter from Soviet musicologists and composers, including Khachaturian and Shostakovich, expressing their indignation at Sakharov's "actions directed against the Soviet Union's policy on the relaxation of international tension, as well as his slanderous statement about socialist reality." Ruth Bonner could not help but remember the drumbeat campaigns of the thirties. It was already a matter of family discussion that Elena Bonner's children would have to leave the country at some point if they were ever to have a semblance of a normal life. By the spring of 1973 this was no longer a matter of principle but of opportunity: Tatiana, her husband, Efrem, and her brother, Alexei, all received invitations from the president of the Massachusetts Institute of Technology (MIT) to spend a year studying there, an obvious prelude to emigration. But to Sakharov and Bonner, "the thought of their departure, with little hope of ever seeing them again, was difficult to face."

Sakharov also knew how to escalate, giving a wide-ranging interview to Olle Stenholm, a Swedish radio correspondent, which was broadcast on July 2. He spoke of the direction his own life had taken: "I began by confronting global problems and only later on more concrete, personal, and human ones." Usually dispassionate in analysis, Sakharov voiced animus against the West for using the Russian dissidents as "small change." His pessimism, though bitterly realistic, was free of despair. "It seems to me that almost nothing can be done. . . . Because the system has a very strong internal stability." Still, people could speak out, protest, set an example, and create ideals, "because if there are no ideals then there can be no hope." But he drew a very distinct line between moral imperative and expectations of success: "if a man does not keep silent it does not mean he hopes necessarily to achieve something. These are not the same question. He may hope for nothing but nonetheless speak because he cannot, simply cannot remain silent."

Stark in its criticism of the Soviet Union as a society both rigid and "on the decline," the picture Sakharov now presented was free of any of the ritual linguistic obeisance that marked his former statements. The Soviet Union was a closed, totalitarian society ruled by a small privileged class. Fearsome if beheld from without, the USSR was in fact unhealthy, suffering from "tiredness, apathy, cynicism." The economy, burdened by central planning, stifling initiative, could not deliver goods and services, making life "much more boring and dreary than need be." Education and medical care were free, but it was common knowledge that the best presents got you the best doctors; a ruble under the pillow ensured that the nurses changed your sheets, and cost-free root canals did not include anesthesia. And, without the free flow of information "intellectual life . . . simply does not exist." The balance had shifted for Sakharov: "on the whole our state has displayed more destructive features than positive ones. . . . I am skeptical about socialism in general. I don't see that socialism offers some kind of new theoretical plan, so to speak, for the better organization of society."

The last question the Swedish interviewer asked was about fear. "I have not feared for myself, but that is, you might say, part of my character and partly because I began from a very high social position. . . . But now I have grounds for fearing such forms of pressure, which are not directed against me personally but against members of my family, members of the family of my wife. That is the most painful thing, because it is very real and is coming closer to us."

It had taken the *Literary Gazette* five years to respond to "Reflections"; it needed only two weeks to respond to Sakharov's Swedish interview. The article's title, "A Purveyor of Libel," alluded to criminal action since the statute most commonly applied to dissidents was slander of the Soviet state, a point driven home by references to Sakharov's "Western 'clients.' "

The "Purveyor of Libel" article was not an isolated event but part of a process. In early August Bonner's son, Alexei, was refused admission to Moscow University, and on August 16, 1973, Sakharov was invited for a conversation with Deputy Prosecutor Malyarov. "This conversation is intended to be in the nature of a warning," said the deputy prosecutor. Referring to the Swedish interview and to other instances of actions that were "harmful and openly anti-Soviet [in] character," the deputy prosecutor made the same point several times: "By the nature of your previous work, you had access to state secrets of particular importance. You signed a commitment not to divulge state secrets and not to meet with foreigners. But you do meet with foreigners and you are giving them information that may be of interest to foreign intelligence agencies. . . . Any state has the right to defend itself. There are appropriate articles in the Criminal Code, and no one will be permitted to violate them."

Sakharov replied that he saw nothing illegal in meeting with foreign journalists, which wouldn't even be necessary if he could be published in the Soviet press as he would prefer. Though Malyarov was not at all interested in an exchange of views, views were exchanged, on the subject of Western democracy. "You seem to like the American way of life," said the deputy prosecutor, "even though they permit the unrestricted sale of guns, they murder their Presidents, and now they've got this demagogic fraud of the Watergate case. Sweden, too, is proud of her freedom, and they have pornographic pictures on every street. I saw them myself. Don't tell me you are for pornography, for that kind of freedom?"

"I am not familiar with either the American or the Swedish way of life," replied Sakharov. "They probably have their own problems and I would not idealize them. But you mention the Watergate case. To me, it is a good illustration of American democracy."

The deputy prosecutor snorted in indignation, saying it was all "just a show. All Nixon has to do is show a little firmness, and the whole thing will come to nothing."

Sakharov reiterated that he did not consider himself in violation of the law. "In particular, I cannot agree with your statement that my meetings with foreign correspondents are illegal or that they endanger state secrets." To demonstrate his sincerity and to defy that warning, Sakharov held a major press conference five days after his conversation with the deputy prosecutor, about thirty correspondents packing his apartment on August 21. Sakharov began the conference with a brief disquisition on the nature of warnings: those that precede arrest or permission to emigrate, those that are given to witnesses before trials, and the blanket warnings such as the ones that appeared for every new issue of the *Chronicle of Current Events*—"appropriate persons would be arrested and those already under arrest would be sentenced to longer terms." In his own case, the interview he had given the Swedish correspondent had "evidently been the last drop that filled the cup to overflowing." The recent attacks on him in the press were an attempt at intimidation, as were the attacks on his family. In addition to Alexei's being refused admission to the university and Tatiana's expulsion, her husband,

Efrem, had now been fired from his job. The pattern was classic, the message clear.

Yet Sakharov refused to accept the very assumptions on which the warnings were based. "I like to believe that the loyal character of my activities will ultimately be understood." "Loyal to what?" he was asked by one correspondent. "Loyal in the literal meaning of the word, namely lawful." In Sakharov's opinion, both the foreign correspondents and the Soviet authorities labored under a similar illusion—that there was any such thing as a democratic movement in the USSR: "I have viewed it essentially as an expression of concern for the fate of particular persons who have become the victims of injustice. . . . It is certainly not a movement of any kind. It is normal human activity that cannot be regarded as political . . . as pursuing a particular political goal, such as a struggle for power. . . . So that the authorities really have no grounds for concern, especially not for any repressions. If there are any grounds for concern, these must be within their minds."

Now it was Sakharov's turn to issue a warning: the West was playing a dangerous game by accepting détente without insisting on a concurrent democratization of Soviet society. A country that does not respect its own constitution can hardly be expected to respect any other agreements except in the interim when they serve its self-interest. The result will be the "cultivation and encouragement of a closed country, where everything that happens may be shielded from outside eyes, a country wearing a mask that hides its true face. I would not wish it on anyone to live next to such a neighbor, especially if he is at the same time armed to the teeth." Sakharov went on record as supporting the Jackson Amendment, then pending in the U.S. Congress, which linked easier trade with the Soviet Union to unrestricted emigration. On September 14, 1973, he restated this position in greater detail and with greater force in a letter to the Congress of the United States. Expressing his open support for the amendment and criticizing those who viewed it as dangerous to détente, Sakharov warned that failing to pass it would be "tantamount to total capitulation of democratic principles in the face of blackmail, deceit, and violence."

Retribution was swift. Eight days after the press conference, the August 29 edition of *Pravda* printed a letter signed by forty Academicians denouncing Sakharov, who "went so far as to oppose the Soviet Union's policies aimed at reducing international tension. . . . His utterances align him with highly reactionary circles that are working against peaceful coexistence among nations. . . . A. D. Sakharov thus makes of himself an instrument of propaganda hostile to the Soviet Union and other socialist nations." Some of his former colleagues at the Installation as well as Khariton, the director, had signed. Zeldovich had, mercifully, not been asked to contribute his signature, and Kapitsa with his usual forthrightness had simply refused. The new head of the Academy found a very Russian excuse—too drunk to come to the phone. When reading this and later such letters, Sakharov would express a complex compassion both for those who had signed and for those who had not. The former had disgraced themselves and would incur their children's enmity; the latter would lose privileges. As one Russian put it: "I know writers who will sign any statement, make any denunciation of Sakharov or whomever the authorities want, to get something published or to get a trip abroad. I know a scientist who will stop at nothing for a trip to Japan."

The soon to be infamous letter of the forty Academicians was followed by volleys from every quarter of society—filmmakers, factory workers, war veterans, and collective farmers who "unanimously support the outstanding Soviet scientists who have denounced Sakharov, and express our keen indignation at his actions and words." There was no escape from the assault. Even when walking on the beach during a brief vacation on the Black Sea, where Alexei was trying to teach Sakharov how to swim, he overheard people describing him as a "traitor." Rumors were floated that Sakharov was a Jew who had changed his name from Zuckerman ("zucker" and "sakhar" both meaning "sugar").

Back in Moscow, in a statement distributed to foreign correspondents, Sakharov defended himself against the charge that he was an "*enemy of détente,* working against the most precious of prizes, peace, whose price was counted in the blood of millions of Soviet citizens." Western response by such figures as Bruno Kreisky, the chancellor of Austria, and Günter Grass, the novelist, was strong and immediate. Philip Handler, president of the U.S. National Academy of Science, applied more than moral suasion to his Soviet counterpart: Soviet-American scientific cooperation was, he said, at stake.

By mid-September the cannonade subsided. But it had already accomplished a good deal—in millions of minds the name of Sakharov was now associated with treason. And since press campaigns were often only a prelude to something worse, the silence of the guns was more ominous than their roar.

The press campaign was standard, but the next move was not. On October 18, a Sunday morning, only Sakharov, Bonner, and Alexei were home, Ruth having gone to visit Tatiana's first child, a boy, Matvei, born only a few weeks ago. The bell rang. Sakharov answered the door and found himself facing two men, Arabic in appearance, one tall, the other short and stocky. Against his better judgment he let them in. The tall one did the talking. His Russian, a product of Lumumba University, was "accented but correct" as he announced that they were from Black September: "We want an immediate statement in writing from you admitting your complete lack of competence in Middle Eastern affairs and disavowing your statement of October 11" (referring to an interview in which Sakharov had placed the principal blame on the Arabs for the wars with Israel). When Bonner reached for her cigarette lighter by the telephone, the short man "sprang like a cat and blocked her."

"I am not about to write or sign anything under duress," said Sakharov.

"You'll be sorry."

"What can you do to us—kill us?" burst out Bonner. "You're not the first to threaten that!"

"We could kill you, of course, but we could do worse: you've got children, a grandchild. . . ."

That threat, the most painful, was also the most revelatory. Matvei had been born on September 24, three weeks and one day before. That, ordinarily, would not have been information available to a Middle Eastern terrorist group, which meant that though those men might indeed be Palestinians and sincere in their demands, this was nevertheless a KGB charade.

When the doorbell rang, startling everyone, the short one herded Sakharov, Bonner, and Alexei into the back of the apartment as the tall one yelled: "Black September acts without warning. We've made an exception with you this time, but there won't be any more!"

After a minute, Sakharov, Bonner, and Alexei rushed out and found the apartment empty, the phone line cut. A short while later their surprise visitor, the novelist Vladimir Voinovich, drove Sakharov to a nearby police station to report the incident, a pro forma protest. When asked if he had been frightened, Sakharov thought for a long moment then replied, "I guess not."

The KGB was not only staging charades; it was considering Sakharov's long-term fate as well. In the public mind, both in Russia and abroad, and in the KGB mentality, the names of Sakharov and Solzhenitsyn were now inextricably linked. On September 18, 1973, Andropov reported to the Central Committee: "The hysteria stirred up lately in the West around the names of Sakharov and Solzhenitsyn . . . represents the product of a prearranged and coordinated program. . . . In contrast to Solzhenitsyn, Sakharov is more restrained in his futile attempts to prove 'the unacceptability of the Soviet system.' Nevertheless, he is definitely degrading into anti-Sovietism." This, wrote Andropov, "forces one" to consider "more radical measures to terminate the hostile acts of Solzhenitsyn and Sakharov," but the differences between the two required different solutions. Andropov still advocated a meeting between Sakharov and a Soviet leader, but more for the sake of appearances now than with any hope of converting him. Andropov proposed a new solution to the Sakharov problem—deprivation of state awards, exile to a closed city—but this struck the Central Committee as too radical, premature.

Solzhenitsyn had linked his name with Sakharov's in an August 23 interview with Associated Press and *Le Monde,* and in a September 5 article, "Peace and Violence," Solzhenitsyn proposed that Sakharov be awarded the Nobel prize for peace. Solzhenitsyn lauded "Sakharov's indefatigable, devoted (and for him personally dangerous) opposition to systematic state violence." Knowing that the linear Western mind might not be able to fathom all the "paradoxicality" of the Russian psyche, Solzhenitsyn cautioned the Nobel committee against being put off by Sakharov's work on thermonuclear weapons: "for the human spirit's admission of its previous errors, its cleansing itself of them, its redeeming of them, contains the higher meaning of humanity's sojourn on Earth."

Sakharov and Solzhenitsyn met a few times in the fall of 1973. Each of the two men still held enormous respect for the other, though every meeting made their differences the more apparent. Sakharov found that even when he agreed with Solzhenitsyn in general he was disturbed by the "peremptory nature of his judgments, the absence of nuance, and his lack of tolerance for the opinion of others." In Solzhenitsyn's heartfelt mystical patriotism, his belief that Russia's salvation lay in the virgin lands of Siberia away from the filth and clamor of modern civilization, Sakharov could detect only an anti-Western bias, a flight from reason. The two men clashed intellectually, their wives with greater heat. Learning that Bonner's children were considering emigration, Solzhenitsyn's wife told her that she should worry more about the Russian people, none of whom had such luxurious choices. "Don't give me that 'Russian people' shit!" yelled Bonner. "You make breakfast for your own children, not for the whole Russian people!"

But what particularly incensed Sakharov was Solzhenitsyn's attitude toward him and Elena—that she was a dominating hysteric with the poor naive Sakharov "under her heel." That view, with roots in Slav machismo, was shared by the KGB and by some of Sakharov's colleagues, who preferred "quiet diplomacy" and were offended by her "sharp tongue" and "explosive temper" or shamed by her courage.

* * * *

Andropov made his next move in November. On the thirteenth Bonner was summoned to the KGB investigation department at Lefortovo Prison. The cause, and pretext, was the Western publication of *Prison Diaries* by Eduard Kuznetsov, the Leningrad "skyjacker" whose aunt Bonner had pretended to be to gain access to the courtroom, where she had made a transcript of the proceedings. The *Diaries* had come into her possession when a woman walked into their apartment, placed a "tiny package carefully sewn into a piece of cloth" on the table, and walked out, without saying a single word. The cloth contained a manuscript on cigarette papers, the letters so minute that Sakharov and Bonner could not read them even with a powerful magnifying glass. Better, younger eyes were found and the manuscript typed and ferried abroad. Two people had already been arrested in connection with transmitting the manuscript when in September Bonner issued a statement, claiming sole and complete responsibility. At Lefortovo Prison she was questioned by a master interrogator, a "brilliant performer" who could pass from civility to rage without missing a beat. She was summoned nearly every day for the next two weeks. Sakharov would accompany her to Lefortovo and wait for her downstairs. Finally, Bonner decided to refuse any further summonses. She had stated her refusal to testify many times already, and it was clear that the interrogations had no real relation to the case but were only attempts to inflict psychological damage. When Sakharov intercepted one summons, saying Bonner was ill and he would accept responsibility for her failure to appear, the summonses ceased. But Bonner was furious with him—she could fight her own battles!

The year of battles had wearied them both, and in December they checked into the Academy Hospital. Her pulse was racing at 120 per minute, a thyroid condition having flared up, and it was time for him to have the cardiac tests that had been long advised and long postponed. They had a private room where they could rest and work in peace, Sakharov writing, Bonner typing, editing. They received guests, her literary friends, his scientific colleagues. But there were some unsettling notes as well. The doctors at the Academy Hospital seemed "quite indifferent" to Bonner's thyroid condition, and it was only later in Leningrad that a physician advised an operation. Natalya Gesse knew the perfect surgeon, a good friend of hers. But, in a single, telling instance of the effectiveness of the anti-Sakharov campaign, that surgeon declined; he was about to defend his dissertation and feared that he would be denied his degree if he practiced his art on the renegade's wife. The membrane of support around Sakharov and Bonner was thin indeed if even a friend of Natalya Gesse feared to help.

In early January Solzhenitsyn's thirteen-year-old stepson arrived at Sakharov's apartment with a copy of the first volume of *The*

Gulag Archipelago concealed under his shirt, in such a hurry to deliver the book that he even refused the customary tea. Ten minutes later Sakharov and Bonner were "devouring that masterpiece," captivated by his voice, "angry, mournful, sardonic" as it evoked a "somber world of gray camps surrounded by barbed wire, investigators' offices and torture chambers flooded with merciless electric light, icy mines in Kolyma and Norilsk." It was Solzhenitsyn at his best—the artist as a hero of history.

The Gulag Archipelago, a history of Soviet crimes against the Soviet people, now itself became part of that history. One woman from whose home a copy was confiscated hung herself. When *Gulag* was published abroad, the Soviet press trained its sights on Solzhenitsyn. Despite his growing disaffiliation from Solzhenitsyn, Sakharov had a clear sense of the value and importance of this work and immediately spoke out in its defense.

But Solzhenitsyn's fate had already been crafted. On February 12 Sakharov learned by phone that Solzhenitsyn had been arrested. Fifteen minutes later he and Bonner were at Solzhenitsyn's apartment, where his wife recounted the details of the arrest to each new arrival and then returned to sorting papers, burning some. People kept coming, two teakettles were constantly steaming on the stove, the phone never stopped ringing, Sakharov took a few of the calls; "the shock of events untied my tongue, and I spoke clearly and forcefully."

The next day Sakharov, Bonner, and several others gathered in their kitchen and wrote the "Moscow Appeal," demanding Solzhenitsyn's release and an "international tribunal to investigate the crimes described in *The Gulag Archipelago.*" Neither would occur. The Soviet authorities had decided to inflict on Solzhenitsyn the fate Pasternak had so dreaded. Just as Stalin had deported Trotsky from Odessa, a city associated with Jews, the authorities now chose to deport Solzhenitsyn to Germany, land of the traditional enemy. The Politburo calculated that the storm of controversy the expulsion would rouse would be more sensational than substantive. Solzhenitsyn had not been imprisoned or shot—it was hard to feel sorry for someone being accorded a hero's welcome, traveling to Zurich, as Andropov put it, to "touch his millions being held in Swiss banks." And in a display of Soviet humanism, Solzhenitsyn's wife, children, mother-in-law, and archives were being permitted to join him in the West. Sakharov attended that farewell party, lugubrious and exalted, where "many a fine Russian song was sung."

The solution to the Solzhenitsyn problem allowed the KGB to concentrate on Sakharov but, since the writer's expulsion had caused such an uproar, Andropov recommended "no public statements against SAKHAROV during the next five or six months."

It was a fortuitous respite, because Sakharov and Bonner had problems closer to home. Another, braver surgeon having been found, the operation on her thyroid was successfully performed in late February but had the unforeseen consequence of severely aggravating her glaucoma, itself a result of the concussion she had suffered during the war that had left her temporarily blind, deaf, and dumb. What was not initially clear was whether this unforeseen consequence of surgery would be lasting or transient.

In the early spring Sakharov felt unbesieged enough to accept a commission from the *Saturday Review* for a special issue on "The World in Fifty Years"—or, as reflected in the magic mirror of the KGB mentality, he wrote the article "on orders of bourgeois propaganda organs." In his vision of 2024, the world is divided into densely populated work areas, including flying and subterranean cities, and large ecological reserves where people restore their inner balance. Fusion energy is available, and asteroids are blasted closer to Earth by nuclear bombs to make their mineral wealth more easily exploitable. People communicate on a global videophone system and a precursor of the World Wide Web, which he called the Universal Information System (UIS). Opening his article with anxiety over the dangers posed by his country to the world, he concludes it with a declaration of his faith in the fundamental health and intelligence of the species: "I believe that mankind will find a rational solution to the complex problem of realizing the grand, necessary, and inevitable goals of progress without losing the humaneness of humanity and the naturalness of nature."

In the bizarre ecology of the Soviet system, Sakharov's "libelous fabrications" earned him five hundred dollars, which in a hard-currency store he immediately converted into canned meat and other food for prisoners in the camps. A similar conversion occurred with the 150,000 francs Sakharov was awarded on May 2, 1974, by the French Cino del Duca Foundation for his contributions to "modern humanism." Sakharov was glad to give a portion of the money to Bonner, who had long desired to set up a fund for the children of political prisoners for, as she said in her statement: "I was fourteen when my parents were arrested in 1937, and it's been my fate to know the bitterness of loss and that strange half-orphaned state and the press of financial deprivation."

By late spring, however, these honors were overshadowed by Bonner's deteriorating vision. Neither her usual medicine nor stronger medications much relieved the pain; she could no longer distinguish a friend's features from across a room. "Mother's going completely blind," Tatiana told Sakharov. In early June, Bonner checked in to the Moscow Eye Hospital for an operation, but weeks passed with no surgery scheduled.

On June 27 President Richard Nixon, indicted as a co-conspirator in the Watergate burglary cover-up, which would compel his resignation on August 8, arrived in Moscow for a summit with Brezhnev. Though this summit was largely symbolic, if not an exercise in denial, Sakharov took the opportunity to stage his first hunger strike to "call attention to the plight of political prisoners . . . ethnic Germans imprisoned for demonstrating for the right to emigrate; and prisoners in psychiatric hospitals." With Tatiana interpreting, Sakharov gave interviews to the journalists in Moscow covering the summit. An attempt at transmitting a televised version through the Ostankino relay station in Moscow was blocked, the several minutes of blank screen as eloquent as anything Sakharov could say about the lack of free expression in the Soviet Union.

Worried about the effect of the hunger strike on Sakharov's health, Bonner would slip the hospital watchman a ruble, hail a cab, and spend a few hours a day with him at their apartment. The KGB was solicitous about his condition as well, Andropov reporting to the Central Committee on July 1: "in case the hunger strike becomes dangerous to SAKHAROV's health (this, according to doctors, may occur in ten to twelve days), he will be hospitalized and force-fed." Feeling that his point had been made and noticing a deterioration in his health, Sakharov broke off the hunger strike after six days.

After weeks of inaction at the hospital, one of the staff told Bonner in confidence: "We don't know what they're planning to do with you, but you ought to get yourself out of here as fast as you can." Medical murder was a tradition in the Soviet Union since Commissar of War Frunze had died during routine surgery in 1925. There was no choice but to seek medical treatment abroad. Still, Sakharov and Bonner were reluctant to take that step. But once the decision had been reached, they both felt it to be "irrevocable; there could be no turning back."

Now it was a race of sight against time. Bonner immediately contacted friends in Italy, who took care of the necessary paperwork on their side. But that was nothing compared with all the perfectly filled out forms she needed to present to OVIR, the Department of Visas and Registration. First her papers were rejected because Bonner had not included the address of her ex-husband—no, it couldn't be filled in by hand, she had to go home and type it in. Then they were rejected because Bonner had not filled in "father's place of death." It took Bonner another two days to obtain a copy of his death certificate, the sort granted to rehabilitated victims of the Terror on which neither place nor cause of death is indicated. Finally, her papers no longer rejectable, they disappeared into the cloud bank of the bureaucracy.

Sakharov continued to press the American Embassy and Henry Kissinger to intervene for Bonner's children and son-in-law to travel to the United States. Senator James Buckley, a conservative Republican from New York, visited Sakharov in November, after first meeting in a Moscow synagogue with Jews wishing to emigrate, all of which raised the hackles of the KGB. This time, however, Sakharov intervened on behalf of ethnic Germans who had settled along the Volga in the time of Catherine the Great and who now wished to return to Germany, pressing a list of some six thousand on Buckley to pass on to the German government—"not a difficult thing for him to do, perhaps, but how many others are willing to undertake such tasks?" He and Sakharov had a good conversation, whose most important result was the frustration it left Sakharov with. He felt unable to articulate, even to himself, the changes time and experience had wrought on his worldview. Bonner suggested putting his thoughts down in the form of an open letter to Buckley. Sakharov liked the idea of writing, but not the genre.

The cold weather came, and the KGB went on the attack. On November 27 a search was conducted at the apartment of Andrei Tverdokhlebov, one of the original three members of the Human Rights Committee, and among the materials confiscated were *The Gulag Archipelago* and "declarations and appeals by SAKHAROV." In December it was the apartment of Sergei Kovalyov who, along with Tatyana Velikanova, had in May 1974 revived the *Chronicle of Current Events* after more than a year and a half of silence, publishing four issues within two weeks. The resumption was accompanied by a brave and provocative statement:

"The reason for the interruption in the publication of the *Chronicle* lies in numerous and unambiguous threats made by the KGB that every issue would be answered with arrests of people suspected of publishing or distributing either old or new issues of the *Chronicle*. The kind of situation in which people are forced to make a decision affecting others as well as themselves does not require any explanation. However, continued silence would have supported, albeit passively and indirectly, the tactic of taking hostages—a practice incompatible with law, morality, and human dignity. Therefore the *Chronicle* will renew its publication, trying to adhere to the style and position of former issues."

Ever since the days of the green file, Sakharov had great regard for Kovalyov's intelligence and courage, considering him a "great friend." Blond, blue-eyed, with the sort of skin that looks tan even in winter, Kovalyov, though thirty-four, seemed boyish to Sakharov. Kovalyov worked in an area that interested Sakharov, "neural networks and other topics in electrophysiology on the borderline between biology and information science." Kovalyov's internal passport had been confiscated during the search, and he had been told to retrieve it on December 27, which usually meant saving the state the expense of sending an officer out to make the arrest. On the evening of the twenty-sixth a few of his friends gathered in the Sakharovs' kitchen on Chkalov Street and had eaten and were already drinking tea when Kovalyov, certain he would be arrested the next day and with plenty to do, finally arrived. He did take advantage of his pecarious situation to ask a favor of Elena Bonner: "Give me one last taste of your cabbage soup."

For Sakharov that winter, warnings turned to threats. In December while the U.S. Congress was debating the Jackson Amendment, Sakharov found a typewritten note from the "Central Committee of the Russian Christian Party" saying: "This debate was provoked by your activities. If you don't cease them, we'll take measures—starting, as you might expect, with the Yankeleviches, senior and junior." That threat was reiterated in person to Efrem Yankelevich when he was throwing out the garbage: two men told him that he and his son would end up in the garbage themselves if Sakharov didn't halt his "shenanigans." Everyone was very disturbed, especially by this second threat against Matvei, now fifteen months old. And Alexei, helping a blind man find his way, was attacked by a youth gang in a side street, the blind man having suddenly vanished.

Sakharov was having problems with his own children, though of a different sort, and dealing with them with his standard avoidance. He saw more of Bonner's children than his own children if only because of their living arrangements. And it would have been difficult to deny that he felt more comfortable with Bonner's children than with his own, whether because they were more his sort or because he had no past or bond of blood with them. With his own there were squabbles over money, all the more painful for being petty, and for what they masked. In 1973 Sakharov transferred almost fifteen thousand rubles to his children. Sixteen now, Dima was not doing well in school and displaying a taste for bad company. The relationship between his children and Elena's was hardly ideal. Since both he and Elena had daughters named Tatiana, Tanya for short, the press inevitably mixed up the two, creating additional ill will. Sakharov's children did not win Elena's heart, nor she theirs.

Only with Bonner did Sakharov feel at home. And for that reason he loved the double portrait of them painted by Boris Berger in that bad December. Impressionistic, as if they were in a darkness lit by a snowfall of light, the painting catches their essential selves—Sakharov is at once inward and adoring; Bonner, looking dead-on at the painter, the viewer, is "poised with a cigarette, ready to rush into action to help someone." More important, the artist saw what Solzhenitsyn had been blind to, the bond between them. The painting gave Sakharov the "strange and exciting feeling that although the physical, material existence of that moment has faded into the past and will continue to recede farther and farther after our deaths, there is at the same time something deep and lasting there, fixed for all eternity."

* * * *

The year 1975, dramatic even by Sakharov and Bonner's standards, began in relative tranquillity, though one edged by anxiety over her visa. While continuing to speak out on general issues and in defense of individuals, Sakharov devoted the first half of the year to voicing all he had failed to say in his conversation with Senator Buckley. Once again the writing was slow and arduous. Like his father, he agonized over every word, but his burden was greater—all his statements had worldwide impact both because they were his and because they were heard as the voice of many. Even the title gave him trouble, as the KGB's taped conversations indicated: "While discussing this opus with his wife E. G. BONNER, SAKHAROV listed the following versions for its title: 'The Prospects for Détente,' 'An Open Letter,' 'A Letter to Friends in the West,' 'Thoughts on Peace' and 'Thoughts on the Present and the Future.'"

Finally, he settled on "My Country and the World." Some seventy pages long, the new essay is, like its predecessor "Reflections," tightly constructed: introduction, five main themes, conclusion. But the tone is different. This time Sakharov includes no "optimistic futurology" or any rhetorical flourishes about the Soviet Union as a "breakthrough society." Sober and somber, the essay is certainly anti-Soviet from the official point of view; he depicts a Soviet Union that is sluggish, repressive, dangerously opaque. But of course it was the Soviet Union that was anti-Soviet, to use the formulation of Russia's best jester, Vladimir Voinovich.

Still advocating convergence and gradual change, Sakharov

would have felt remiss had he not included a list of reforms needed "to bring our country out of a constant state of general crisis." The reforms—partial economic denationalization, amnesty for political prisoners, the right to leave the country and return, freedom of speech, a multiparty system, even a convertible ruble—are still offered with a genuine intent to be of assistance to his country. And they were still the last thing Soviet leaders wanted to hear.

In his essay Sakharov is waiting to see how the United States and the Soviet Union will relate to the most basic of rights: to chose one's country of residence. That will show whether détente is a "profound, comprehensive process of historical significance, which includes the democratization and greater opening of Soviet society, or an unprincipled political game." Reducing the nuclear danger remains the overarching priority. But how dependable are agreements with a society in which it is against the law for political prisoners even to refer to themselves as such? In fact, in the USSR they are all political prisoners. "Only when he has all rights is a man free. If you are locked in your house, you feel like a prisoner, even if you have no need to go outside and you are not beating your head against the door day and night. That's the sort of prisoner we all are now."

Used to having the charge of naivete flung at him, Sakharov now flings it himself. Distance, a lack of any similar experience, a glut of information, a tendency toward left liberalism—all numb and immunize intellectuals in the West against sensing what is so "strange, terrible and monstrous" about his country, where people have disappeared by the millions and are still disappearing, one by one.

Finally, in late April, OVIR summoned Bonner to its central Moscow office. Accompanied by Sakharov and foreign correspondents, Bonner was informed that sufficient treatment was available in the Soviet Union and her visa application had therefore been denied. Sakharov and Bonner gave a press conference on the spot, which "startled several Soviet bureaucrats who were waiting to pick up travel documents," and called another for May 9, when the Soviet Union celebrates its victory over Nazi Germany, a victory in which Bonner had taken part. If no visa were received,

Sakharov would begin a three-day hunger strike. At the eleventh hour, OVIR made a counteroffer: treatment at the ophthalmological institute of her choice, with the possibility of foreign specialists brought in at government expense. But, no longer willing to entrust her sight to Soviet medicine, Bonner refused.

Now their last hope was world opinion. Queen Juliana of the Netherlands and West German chancellor Willy Brandt brought up the issue when visiting the USSR, and the Federation of American Scientists warned Brezhnev that scientific cooperation was at stake. For the KGB the question was which commotion was worse —the one from not letting her out or the one she'd make if they did.

By late July tests had shown that the "necrosis was approaching the yellow spot, the retina's most sensitive element." OVIR called again, but only to reiterate their position. Cursing, Bonner replied: "It's your fault I'm going blind, but I'm not going to any of your doctors."

It had been one last test of her resolve. Late the next day OVIR called again to say she should come there at once; her visa had been approved.

"I won't be able to get there before you close," said Bonner.

"No problem. We'll wait for you."

Orders had to have come down from the very top, one call on a high-frequency phone. It was a political decision in a season of high political symbolism. The Soviet Union had won a cosmic propaganda victory in mid-July when its *Soyuz* and the American *Apollo* spacecraft linked up in Earth orbit, showing the Soviets to be the equal of America in projecting power and grandeur. The signing of the Helsinki Final Act, scheduled for August 1, would stress human rights and nonintervention but, more important for the Soviets, it would ratify the boundaries created by World War II. Sakharov and Bonner could be depended on to exploit the occasion to highlight her situation. Bonner would be less of an embarrassment if she were abroad during the signing.

Accepting the political decision, the KGB nevertheless registered its displeasure in acts of malice disguised with the ambiguity that was their "signature." Bonner was to leave on the Moscow–Paris train on the evening on August 9 and would then continue to

Siena, Italy, for the operation. That morning her grandson Matvei, one month shy of his second birthday and the target of previous warnings by Black September and the Russian Christian Party's Central Committee, went into convulsions. He was taken to the hospital "unconscious and rigid, frothing at the mouth, with his eyes rolled back." The night was spent without knowing whether he would live or die.

When, the next morning, it was clear that Matvei would survive, Sakharov, tears in his eyes, said to Efrem: "Now you understand that you all should leave." It would never be known what exactly had brought on the convulsions, though a likely suspect was a poisoned cookie from a stranger. Sakharov refused to blame the KGB publicly without some hard evidence. Bonner postponed her trip for a week, and on the day of her next scheduled departure she and Sakharov received an envelope stuffed with photographs of mutilated eyes. Postmarked Norway, the envelope had originally contained a copy of an appeal to Brezhnev by a Lithuanian emigrant whose wife had been refused permission join him.

But Sakharov and Bonner were used to the KGB's postal tricks and the thug language of warning. Hardly about to be intimidated into blindness, Bonner left Moscow on the Paris train on August 16. She and Sakharov would be apart for the first time, and with the Soviet border between them. That she would be allowed to return to the country was not a zero risk.

None of this was good for Sakharov's heart. That June, after an alarming EKG, he had been confined to bed rest. Fifty-four, balding, hair graying, he now found it hard to climb stairs.

September brought joy and relief. The operation performed on Elena in Siena was successful in halting the glaucoma, though not in restoring any of the vision already lost. (Sakharov had previously received word that the operation had been a failure but knew this to be "another KGB 'joke.'") Bonner's daughter, Tatiana, gave birth to her second child, Anya, in September, and Bonner's son, Alexei, now married, became a father for the first time in October when his daughter Katya was born. But the joy was tinged—new grandchildren, new hostages.

On October 9 Sakharov received a phone call from a friend, Yuri Tuvim, with an appealing invitation. Aware that Sakharov's apartment was always filled with dissidents and petitioners, most of whom stayed too long and talked too much, draining energy from Sakharov, who was still weak after his recent heart problems, Tuvim suggested that Sakharov drop by for tea and a few hours of peace and quiet. Sakharov responded with alacrity and, wearing a suit and tie, arrived with his mother-in-law, Ruth, within half an hour. Yuri Tuvim was just putting on the teakettle when the doorbell rang. Bearing a bunch of flowers and a bottle of vodka, the writers Lev Kopelev and Vladimir Voinovich pushed past Tuvim, who objected that the last thing Sakharov needed was guests and commotion. They paid him no attention, excitedly informing Sakharov that he had been awarded the Nobel Peace Prize for 1975. Tuvim thought Sakharov took the news calmly, though to Voinovich he seemed at a loss. Then the bell rang again, and Tuvim opened the door to a corridor jammed with dissidents and correspondents. Collected now, choosing his words carefully, Sakharov said : "I feel I share this honor with our prisoners of conscience—they have sacrificed their most precious possession, their liberty, in defending others by open and non-violent means." Flashbulbs, questions, congratulations. So ended Sakharov's quiet tea at Tuvim's.

The following day Andropov contextualized the event for the Central Committee: the Nobel committee had awarded Sakharov the peace prize "for provocative purposes, in order to support his anti-Soviet activities and on that basis to consolidate hostile-minded elements within the country." He proposed a series of countermeasures, including an open letter by prominent academicians to be published in *Izvestiya.* Quite soon seventy-two scientists proved willing to sign; six were not. Zeldovich wriggled out by saying "he was thinking of preparing an individual letter" (by phone Zeldovich urged Sakharov to renounce the prize, using stock phrases that would appeal more to those bugging Sakharov's line than to Sakharov himself); Khariton, director of the Installation, who had signed the letter of the forty Academicians, felt he had already done his share; and Kapitsa thought Sakharov should be called in for a talk at the Academy.

In the belief that the word should serve the state, Andropov also proposed that "the editorial board of the newspaper *Trud*

[Labor] should publish a satire portraying the award of the Nobel Peace Prize to Sakharov . . . as a reward from Western reactionary circles for his continual slandering of the Soviet social and state order." Quite soon the newspaper published "A Chronicle of Life in High Society." Its tone was one of venomous self-righteous sarcasm: "We are delighted that Mme Bonner has at last found eye surgeons befitting her social status." Sakharov was depicted as bowing to Senator Buckley "till our intellectual's nose was scraping the floor." And for good measure a dollop of anti-Semitism and treason was added: "Sakharov has been promised more than $100,000. It's difficult to find an exchange rate to compare that sum to the thirty pieces of silver received by Judas. Mrs. Bonner, who is well versed in such matters, is probably the person best qualified to answer." Her response was to wonder why an atheistic state was suddenly so fond of Biblical allusions.

In a masterstroke of hypocrisy Andropov suggested "using the channels of the KGB, [to] promote in the West articles showing the absurdity of the decision to award the Peace Prize to the inventor of a weapon of mass destruction." In the odd bedfellows department, this was a view shared by some dissidents, including Zhores Medvedev, for whose release from a psychiatric institution Sakharov had battled. In no event, said Andropov, should Sakharov be allowed to travel abroad to accept the award, as he was in "possession of state and military secrets." Sakharov, fully expecting to be rejected, made application for a visa on October 20.

While waiting for OVIR's reply, Sakharov began work on his Nobel lecture, and for once the "writing came easily . . . a delightful task." Though the writing itself went well, the process was continually interrupted—by journalists seeking interviews and by others seeking his aid in the redress of injustice, whose number only grew after the awarding of the prize. One of them, a young man with "childlike blue eyes," visited Sakharov at his apartment, complaining of KGB harassment and forcible confinement to a mental hospital. But it was not a clear enough case for Sakharov to pursue, especially since there had already been instances of a defense mounted for persons who in fact proved insane. The young man accompanied Sakharov on the train out to his dacha, continuing to beseech his help. After Sakharov disembarked, the young man was thrown off the train and killed. Though the incident was murky, Sakharov was inclined to believe that this was a "murder designed to make clear to me that my public activities could only have tragic consequences."

Later in November Sakharov's daughter Lyuba went into labor, and her first child was stillborn. Encountering his colleague Evgeny Feinberg on the stairs at FIAN, the usually undemonstrative Sakharov startled him by throwing his hands up in the air in a posture of Old Testament lamentation. "A terrible thing has happened," he kept saying, "a terrible thing."

On November 14 he was summoned to OVIR and informed that his application for a visa had been denied him as "an individual possessing knowledge of state secrets." Though promising to contest the decision, Sakharov realized that there was little probability of success and in a telephone conversation with Bonner, then in Rome, asked her to represent him at the Nobel ceremonies in Oslo. A few days later, fearing KGB reprisals, specifically that she might not be allowed back into the country, Sakharov reversed his decision. "I understand," she replied with disappointment. "But you're making a mistake."

Her opinion was seconded by Efrem: "I don't think you're right, Andrei Dmitrievich," said Efrem, arguing that only she could represent him. It took Sakharov thirty minutes of inward struggle to reverse his position. "I shudder to think of what would have happened had it not been for Efrem."

Though it was no easy thing to contest Sakharov in a moral decision affecting his personal life, Efrem, for all his tendency to stutter, had found the words to compel Sakharov to reconsider. Now taking Bonner's place as Sakharov's editorial typist, Efrem also proved an "exacting" first reader. And on December 5 he was at Sakharov's side as a small group of dissidents, matched in number by foreign correspondents and KGB agents—a perfect image of détente at that stage—removed their hats on Pushkin Square and observed a minute of silence. And on December 8 he was with Sakharov as they boarded the train for Vilnius, the capital of Lithuania, where the trial of Sakharov and Bonner's "great friend" Sergei Kovalyov was to be held. Kovalyov had been arrested nearly a year earlier, after responding to a KGB request to come in and "pick up" his passport. So certain was he of his fate that he had asked Bonner: "Give me one last taste of your cabbage soup." Now, the "coincidence of the Nobel Ceremony and the opening of the trial" was, in Sakharov's opinion, "almost certainly no accident."

They arrived in Vilnius by train on what struck Efrem as a "gloomy Baltic winter morning." Lithuania can have vile winters—it was there that the retreating Napoleon had to slide down an icy hill on his bottom, losing all dignity—but there was only a little snow on the ground in that early December. It was the Lithuanian people who were cold—to anyone speaking Russian. The Lithuanians obeyed but with hatred and disdain.

The days when Sakharov could enter a courtroom were long over. Still, he decided to lodge a request with the presiding judge and prosecutor and strode right past their secretaries into their offices, accompanied by Efrem, who "managed to tape their revealing and flustered responses," which came down to a refusal. Sakharov would have to stand outside with everyone else, the courtroom's benches packed with people who needed no signal to jeer with gusto. The KGB had particular enmity for Kovalyov, who had resumed publishing the *Chronicle of Current Events,* thereby depriving them of a victory they thought already won.

Sakharov and the others standing vigil had no illusion that they could influence the verdict. They had other purposes, both spiritual and practical. They were there to bear witness, to show solidarity and, since no foreign correspondents were allowed to attend, to serve as the sole reliable conduit of information concerning the trial, the light that deprives the criminal of the cover of darkness.

Cut off from the trial itself, Sakharov was also cut off from Elena. All his calls to her in Oslo were blocked. But on December 10 he was again linked to her by "airwaves" and "elation" when, huddled by a radio, he heard a flourish of trumpets and the sound of her footsteps, so familiar to him, as she mounted the podium to read the Nobel prize acceptance speech. His words, her voice.

With some fifteen Lithuanians attending, they celebrated the prize that evening with toasts, a good meal, and a "fantastic Lithuanian cake." The celebratory mood lasted until Kovalyov's wife arrived trembling to report that Kovalyov, who had been defending himself, had demanded that Sakharov and some of the others be allowed inside the courtroom. Laughingly refused, the normally even tempered Kovalyov called the court a "herd of swine" and demanded that he be removed from the courtroom. The judge was more than willing to oblige, and now the trial would proceed unimpeded by the presence of defendant or counsel.

On the day before Kovalyov's sentencing, Bonner delivered Sakharov's Nobel lecture. "It was Antigone who stood there. Proud and beautiful, she was at that moment the incarnation of resistance to the tyrant," observed the Norwegian painter Victor Sparre, as she read: "Peace, progress, human rights, these three goals are indissolubly linked; it is impossible to achieve one of them if the others are ignored." The reduction of the nuclear danger was "inconceivable without an open society." Otherwise, détente was "simply another version of Munich." The signing of the Helsinki Final Act had ratified postwar borders to Soviet satisfaction but had not resulted in any "real improvement" concerning human rights within the Soviet Union. Sakharov reiterated that he was sharing the award with all "prisoners of conscience and all political prisoners" in his country and, unafraid to bore that gathering of the tuxedoed and gowned, he listed all their names, dozens upon Slavic dozens, including one with the not uncommon name of Gorbachev.

Not only do peace and human rights depend on an open society, but progress does as well, he said, in a subtext that continued his quarrel with Solzhenitsyn: "any attempt to reduce the tempo of scientific and technological progress, to reverse the process of urbanization, to call for isolationism, patriarchal ways of life, and a renaissance based on national traditions, would be unrealistic . . . and lead to the decline and fall of our civilization."

Sakharov concluded with a vision that arched from the prehistoric to the universal: "Thousands of years ago human tribes suffered great privations in the struggle to survive. It was then important not only to be able to handle a club, but also to possess the ability to think intelligently, to take care of the knowledge and experience garnered by the tribe, and to develop the links that would provide cooperation with other tribes. Today the human race is faced with a similar test. In infinite space many civilizations are bound to exist, among them societies that may be wiser and more 'successful' than ours. I support the cosmological hypothesis which states that the development of the universe is repeated in its basic characteristics an infinite number of times. Further, other civilizations, including more 'successful' ones, should exist an infinite number of times on the 'preceding' and the 'following' pages on the Book of the Universe. Yet we should not minimize our sacred endeavors in this world, where, like faint glimmers in the dark, we have emerged for a moment from the nothingness of dark unconsciousness into material existence. We must make good the demands of reason and create a life worthy of ourselves and of the goals we only dimly perceive."

CHAPTER TWELVE

DUEL

AT a meeting of the KGB Collegium, Andropov now declared Sakharov "Public Enemy No. 1." Andropov, too, was sincere. His cynicism did not exist in a void but was always in the service of the state, the end that justified any means. And it justified them because it was only in the Soviet Union that someone like him could rise from the working class to become ambassador to Budapest, not to mention chief of intelligence. (It had been his guile and zeal in suppressing the Hungarian Uprising that had singled him out for the power elite, becoming KGB chief without ever having served in the organization.) The vaunted rights of the capitalist world were meaningless because, for all the millions of workers without work, "their lives do not improve one iota as a result of this," he said with harsh conviction to a meeting of voters. The workers of the Soviet Union needed only one party, the party of the workers, the Communist Party. It was the party that had made his own rise possible; he came up through the Young Communist League, whose ranks had been fortuitously thinned by Stalin's purges.

Born in 1914, Andropov was old enough to remember the romantic aura of the Civil War, Chekists in leather jackets dispensing justice from Mausers. Ascetic, aloof, sensitive to language and nuance, Andropov always preferred to call communists "Bolsheviks," and KGB agents "Chekists." General Dmitri Volkogonov, who knew him, wrote: "A Marxist idealist, Andropov had a profound belief in the virtual messianic role of the KGB and in the inexhaustible vitality of Leninism."

With the Russian bent for literature, which apparently knows no bounds, Andropov, like the founder of the Cheka, Felix Dzierzhinsky, tried his hand at poetry. The results make for something of a unique literary genre—the brevity of life as mourned by one of its abbreviators:

> We are fleeting in this world, beneath the moon.
> Life is an instant. Non-being is forever.
> The Earth spins in the universe,
> Men live and vanish. . . .

Declaring Sakharov Enemy No. 1 was a high-stakes game for Andropov, who was now obliged to solve the problem. Unlike Solzhenitsyn, Sakharov could not simply be put on a plane and whisked out of the country. Sakharov possessed state secrets and, though no doubt a fool of honor, could always reveal something inadvertently because certain issues, such as disarmament, always touched directly on the classified. His anti-Sovietism should not be underestimated and might even induce him to go to work for the West, affording them yet another technological advantage. And again unlike Solzhenitsyn, who was quickly alienating the West's liberal intelligentsia, Sakharov would fit in much better and could become a focal point of resistance from abroad.

Under no circumstances could Sakharov be killed or injured; Andropov gave explicit orders to that effect. The only feasible solution was internal exile, an idea reactivated in Andropov's mind by Sakharov's Nobel prize. In December 1975 he had requested the Central Committee to banish Sakharov to Sverdlovsk-44, a nuclear weapons installation located east of the Urals and closed to foreigners. But that request was denied, the Central Committee confident that the costs of the Sakharov problem were absorbable. Sakharov's press conferences protesting violations of the Helsinki Final Act could also be used to convince the West that the Helsinki accords were being honored. He should remain at liberty, but nearly everyone else could be picked up.

Andropov had no choice but to bow to the collective wisdom of the Central Committee, of which he was also a member. The Central Committee might set the limits on the KGB's treatment of Sakharov, but within those limits Andropov held absolute sway. A KGB major who defected said of Andropov that "he directs all operations personally, and everything that happens in the KGB is done only with his knowledge and under his direction."

* * * *

Andropov had the might and weight of the state behind him; Sakharov had his kitchen. The center of life in every Soviet home, it was, in the apartment on Chkalov Street, an outpost of freedom in a society the Zeks called the Big Camp, the whole country at best a minimum-security facility.

People came for liberty, conversation, cabbage soup before prison.

The company was excellent—the skeptical Amalrik, the comical Voinovich, and Kopelev, huge with a huge beard, speaking of German culture or the Gulag, both of which he knew well.

Anything could be discussed there except, of course, tactical matters, the children's magical slate with gray liquidy paper and wooden stylus having proved especially suitable for such communication.

Sakharov was hardly of the sort to hold court; in fact he spoke so little that he struck some as being a bit boring, as he had at the Installation, though, as there, he would at times sparkle with wit or flash with indignation. He elicited universal admiration, but not everyone accepted his specific proposals. Amalrik said: "I agree with Solzhenitsyn that Sakharov is a poor tactician—though God forbid he should follow the tactics of Solzhenitsyn. But Sakharov is a great strategist, and that cancels out his weakness as a tactician."

They argued fiercely, one of the pleasures in a society of leaden unanimity. And they had the rhapsodic consciousness of belonging to the chosen few, the few who had chosen to live freely where that could cost you your freedom.

In a poem whose description of the atmosphere is accurate even in its too high notes of exaltation, Vladimir Kornilov wrote:

> Evenings in the kitchen
> At Andrei
> Dmitrievich's, evenings in the kitchen. . . .
> Although winter, fiercely chilling,
> Had begun its age-old work . . .
> All those not yet locked up in camps,

All those in exile, half locked up,
All those who are teaching peace from
New York professorial chairs,
Or going blind in strict discipline barracks,
Like the words in one poem,
Are irrevocably included
In the Age of Andrei Sakharov
The best in the country's history.

But sooner or later—usually later, Sakharov reading Bonner and himself to sleep with English mysteries at two or three in the morning—those evenings in that "inn of merry beggars," as she called it, would come to an end, and the guests would have to return to "Soviet reality." Yet there was always time for one last toast with cognac: "To the success of our hopeless cause!"

There was scant cause for hope when the prosecution always got what it demanded. On the day after Bonner had delivered Sakharov's Nobel lecture, the verdict on Sergei Kovalyov was read in his absence in the Lithuanian Supreme Court in Vilnius: seven years' labor camp plus three of exile. Standing outside, Sakharov and Efrem heard the courtroom erupt in applause. Efrem went pale, and Sakharov wept as the KGB clowned and cavorted.

But there were some happy exceptions. For Efrem, Bonner's "triumphant return from Oslo to Moscow was a high point of those years, possibly the only bright and joyous occasion I can recall." Met by a great crowd of family, friends, and correspondents, she and Sakharov gave an impromptu press conference. "The KGB, who wanted to break it up, had thought of nothing better than to send a dozen cleaning ladies, who started to wipe the floor simultaneously and in every possible direction. Actually, a very funny scene." Then they all returned to the city in a long cortege of the sort usually associated with visiting dignitaries or the power elite.

Andropov was now employing a new weapon—timing. Kovalyov's trial had coincided with the Nobel ceremonies, and the next two major trials were scheduled for the same date, April 6, 1976. Andrei Tverdokhlebov, a cofounder with Sakharov of the Human Rights Committee and of the first Amnesty International group in the USSR, would be tried in Moscow while Mustafa Dzhemilev, a leader of the Crimean Tartar movement, would face trial in Omsk, Siberia. Certain that the Moscow trial would be well attended, Sakharov and Bonner flew to Omsk. The cause of the Tartars had moved Sakharov as early as his Installation days when, on a bus for the elite, he had interrupted a conversation to defend them. Branded a traitor nation by Stalin, the entire Tartar population was herded into freight cars in the middle of the night in March 1944 and resettled in Siberia, Uzbekistan, Kazakhstan, 46.2 percent of them dying in the first year and a half. "Rehabilitated" by Khrushchev at the Twentieth Party Congress, where he denounced Stalin, they had been trying to get home ever since.

Arriving in Siberia, Sakharov and Bonner learned that the trial of Dzhemilev, the Tartar activist, had been postponed. They flew back from Omsk to Moscow, but they couldn't attend Tverdokhlebov's either for it too had been postponed, and by the time it started they had already flown back to Omsk for Dzhemilev's trial.

Dzhemilev, on a hunger strike, barely able to stand on his feet, was greatly encouraged when his sister informed him Sakharov was outside. For that his sister was expelled from the courtroom, as was his mother the next day.

"Let his mother in!" Sakharov shouted to the KGB agents blocking the door, jeering and shoving. Bonner slapped the face of the plainclothesman running the operation, and Sakharov went for his assistant. The police moved at once; it was over in seconds.

Sakharov was not able to return to Moscow before Tverdokhlebov's sentencing, five years internal exile. Precisely because he felt closer to Tverdokhlebov, Sakharov knew he should attend Dzhemilev's trial. Sakharov had been present for "several depressing hours" at the search that preceded, in standard fashion, Tverdokhlebov's arrest. Dzhemilev was of a small people and being tried in a distant city. These were not in the least pleasant choices, and they inevitably caused, as Andropov put it, a "negative reaction of their friends."

Partially in response to the dual trials of April, a group of human rights activists founded the Moscow Helsinki Watch Group to oversee the implementation of the Helsinki Final Act. One of its initial members was the firebrand Anatoly Shcharansky, who fought for human rights and Jewish emigration, still finding time to act as Sakharov's English interpreter, making him a triply attractive target to the KGB. Sakharov declined to become a member of the Helsinki Watch Group, preferring "the freedom of speaking out as an individual."

And it was as an individual that Andropov next attacked him. Gallant, old-fashioned, Sakharov was known to be touchy on the subject of his wife, and so an article on Bonner titled "Sakharov's Evil Genius?" was published in *The Russian Voice*, an obscure Russian-language newspaper in New York, which could then of course be quoted as a foreign press account. It was no secret that some of Sakharov's friends and colleagues already thought she was radicalizing him, while others such as Zeldovich blamed Bonner for diverting Sakharov from science. Hawking, he would mutter, never let anything distract him.

Not that Sakharov was neglecting science. His connection with FIAN, though limited, was vital to him. At the dacha and on vacation especially, he found time to write, and he was still publishing papers. He attended the annual Rochester Conference on Elementary Particles held that year in Tbilisi. Georgia—for all its famed mountains, subtropical Black Sea beaches, its lamb-and-red-wine hospitality—also had the feel of being a place apart, not quite in the Soviet Union. And the conference itself was international, itself offering some escape from the Soviet Union, as claustrophobic as it was vast.

Possessing physics-and-detective-story English, Sakharov attended the lectures by his American and British colleagues. There were exciting reports—the existence of the charmed quark had been confirmed. And he had his own revelations: "In Tbilisi I became a true believer in quantum chromodynamics and Grand Unified Theory."

Sakharov's own wit, scientific and otherwise, was also in fine form. At a ceremonial dinner one physicist, Lev Okun, in his version of table talk, tried out his favorite problem on Sakharov:

"One end of a rubber band one kilometer long is attached to a wall, and you are holding the other end in your hand. A beetle begins to crawl along this rubber band from the wall toward you, at a velocity of one centimeter per second. While it is crawling that first centimeter, you stretch out the rubber band by one kilometer; while it is crawling the second centimeter, you stretch out the rubber band another kilometer; and so on, each second. The question is, does the beetle manage to crawl all the way to you, and if so, then how long does it take?"

It took most people a day to solve the problem, though one had done it in an hour. Sakharov wrote out the solution on the back on an envelope, "then and there, without comment. The elapsed time was about a minute."

And walking through the streets of Tbilisi after the banquet with another physicist who noticed that pairs of young men, clearly not science types, were walking both ahead of and behind them, Sakharov dismissed the subject with a wave of his hand, saying not to worry, they were only there to protect him; the last thing the government needed was for something to happen to him. The other physicist wasn't quite sure how to take this, for it was both accurate and absurd. Some ironies were for Sakharov alone.

* * * *

A bit of a tightwad except when impulsively giving away fortunes, Sakharov was always appalled at the state funds spent on him. In August 1978 he again had occasion to observe the wasteful largesse of the KGB.

Andrei Tverdokhlebov, whose trial Sakharov had been unable to attend, had been banished to five years' internal exile in a small, remote settlement in Yakutia, from where he sent his parents photographs of himself. The photos were passed on to Sakharov and Bonner at the dacha in Zhukovka. Peering at them with the intensity of those whose sight is endangered and do not waste a single look, Bonner saw something very wrong in Tverdokhlebov's expression. "We've got to go see him," she wrote to Sakharov on a piece of paper. He nodded. They packed their bags, took the train into Moscow and a cab to the airport—actually two, the first being smashed from behind by a car with diplomatic plates, causing both Bonner and Sakharov whiplash.

Called as big as India and as empty as Iceland, Yakutia is the largest region of Siberia. Sakharov and Bonner flew first to Mirny, a new diamond-mining city. As the plane approached, Sakharov peered with curiosity at the blue mounds of slag around the mine itself, a downward-spiraling cone of blue-gray kimberlite where trucks chugged beetle-like. Looking up at the endless green taiga, Sakharov was reminded of his ongoing polemic with Solzhenitsyn: "So this is the region Solzhenitsyn regards as an untouched reserve for the evolution of the Russian people, the nation's 'saving grace.' We're still a long way from being able to make this swampy soil really productive, unless of course we're prepared to do as Stalin did and send millions of forced laborers to their graves."

For the KGB, which could reschedule trials, juggling a domestic flight was child's play. Sakharov and Bonner were stranded in Mirny for the next twenty-four hours. They walked around the city in the evening, by chance encountering some of the few other independent souls in Russia, drifters always a step ahead of the law; but Siberia, perennially short of workers, was always more interested in hands than in papers.

Sakharov and Bonner spent the night in the terminal sleeping on benches, then flew the 375 miles farther northeast to Nyurba the next day. There they were informed that the bus that covered the 15 miles between Nyurba and Nyurbachan, their final destination, had been canceled, no detail too local. The police were unwilling to provide any assistance and unable to contain their laughter at the very idea.

Though it was getting dark, Sakharov and Bonner decided to walk, the KGB still having no jurisdiction over their legs. The forest path was moonlit, the air fresh, a Siberia of stars above the trees. They stopped for bread and cheese, sipping coffee from a thermos. In a lovely twist, all the KGB's machinations had only afforded them hours of happiness.

Arriving at daybreak, they encountered fear. In that small settlement no one would even tell them where Tverdokhlebov lived. But Bonner, who had looked hard at the photograph, now recognized the house.

What disturbed Bonner in the photograph was also present in their reunion, an underlying coolness, a turning away they could neither fail to feel nor understand. And time would prove that feeling right: "the rift between us deepened until we lost contact with him entirely."

To make matters worse, Sakharov badly injured his leg in a posthole, and on the return journey, now unimpeded by the KGB, who wanted them out of there, he suffered such severe heart pain at the airport that he had to lie down.

By October, after a period of recuperation, he and Bonner were appealing to the UN secretary-general, the Security Council, and the president of Lebanon in connection with the "tragic situation of the wounded, of children and women in the besieged Palestinian camp of Tel-Zaatar. . . . Use your great authority and influence to save the dying."

Although his relations with the Palestinians were uneasy at best, Sakharov always felt free to speak out not only against their atrocities but also those committed against them.

On December 5, accompanied by Efrem and Tatiana, Sakharov attended the now traditional gathering on Pushkin Square to remove his cap and observe what Andropov termed "the so-called 'minute of silence.' " No sooner did Sakharov remove his cap then a KGB agent, guised as an irate citizen, poured a bag of snow and mud over his head.

But Baptists threw red carnations over the KGB's heads to Sakharov. And détente had resulted in greater freedom for the foreign press, which meant that the act did not go unwitnessed. Among the correspondents present on Pushkin Square was the AP's daring George Krimsky, who not only reported on dissidents but socialized with them. Sakharov had formed a friendship with him for which Krimsky paid with harassment and slashed tires. Their talk was frank; in one interview Krimsky asked Sakharov if his activity hadn't fallen off somewhat since the Nobel prize. Sakharov replied that "one year is never the same as another," but he also felt free enough to also say: "I must admit that I am beginning to feel the strain after so many years of public activity for which I was psychologically ill prepared. . . . Nevertheless, I do not feel that I have given up. . . . I live day by day, doing what life requires of me."

The New Year's holidays brought some respite for Sakharov, a tired man. Their little clan—Sakharov and Bonner, Tatiana and Efrem Yankelevich along with their two children, Matvei and Anna, and Bonner's mother, Ruth—went out to the dacha in Zhukovka, located on a piney bluff beside a meandering river. Their living situation had become increasingly complex. After Matvei's mysterious convulsions, the family decided that it wasn't safe for the Yankeleviches to be living alone in their new apartment in a relatively isolated area. The Yankeleviches would move into the dacha to relieve the pressure that has to build when three generations of women are sharing one kitchen as happened when all seven lived together in the apartment on Chkalov. Or else Sakharov and Bonner would move out to the Yankeleviches' apartment, whose isolation now proved beneficent, allowing Sakharov some peace from the never ending stream of those seeking his aid who, even if they could not be helped, had to be heard out. He could even do a little science.

The ideal solution would be to swap the two small apartments they had for one large one where they all could live safely and with some privacy. That entailed both a search for suitable candidates and the inevitable quests for official permission. Even among all the problems they had, Sakharov still considered housing "probably the major problem in our life."

At the dacha the children danced outdoors around a decorated New Year's tree, watched over by a scientist dressed as Grandfather Frost. Gifts, poetry, Bengal lights. Sakharov took to the role of grandfather, knowing just what fanciful stories and drawings would delight Matvei and Anya, the woman in his life now, as he liked to joke. The child in Sakharov was alive and well, which Bonner sometimes found endearing, sometimes "infantile." Sakharov's own granddaughter, Marina, occasionally had the impression there were no grown-ups in the room when she was left alone with her grandfather, who would be heating grapefruit on a radiator.

The holidays ended in explosion. On January 9, 1978, a bomb killed seven and wounded more than thirty in the Moscow subway. Bonner was absolutely opposed to Sakharov's making any public allegations about KGB involvement—"They have long memories"—especially when there was no hard evidence, but he was unmovable: "I cannot shake off the deep sense that the explosion in the Moscow subway with its tragic deaths is a new provocation of the agencies of repression," he told Krimsky in an interview. It could well be an attempt to discredit human rights activists, who drew their inner strength and derived their moral authority from the "total, principled rejection of force or the advocacy of violence." It was all too much like the Reichstag fire of 1933 and the assassination of Kirov in 1934, which touched off the Terror.

The immediate consequence for Sakharov was his second summons to the Attorney General, an idea that originated with Andropov, who also recommended that Krimsky be expelled from the country. On January 25, at the Attorney General's, Sakharov was presented with a document to sign:

> WARNING
>
> Citizen A. D. Sakharov is hereby warned that he has issued a deliberately false statement in which he alleged that the explosion in the Moscow subway was a provocation designed by government agencies directed against so-called dissidents. Citizen A. D. Sakharov is warned that the continuation and repetition of his criminal acts will result in his liability in accordance with the law of the land.

Sakharov refused to sign. The next day a press campaign that threatened to be as ferocious as that of 1973 opened with an *Istvestiya* article: "A Slanderer Is Warned."

Watched by KGB agents in fur hats with scarves covering their faces, Krimsky was expelled in February 1978—not, since this was Russia, without farce. The young border guard at passport control would not allow him out of the country, because his last name had been misspelled "Krinsky" on his exit visa. Laughing, Krimsky told him to confer with a superior at once if he wanted his career to outlast the day.

But when Sakharov, kissing Krimsky on the lips Russian-style, presented him with a copy of "Alarm and Hope," Krimsky, not wishing to cause AP any more trouble, had a failure of nerve. The gift, he said, was better sent through other channels. "Sakharov looked crestfallen. I will regret my decision till my dying day."

None of this prevented Sakharov from initiating a correspondence with newly elected U.S. President Jimmy Carter, saying it was "our duty and yours" to "defend those who suffer because of their non-violent struggle for an open society, for justice, for other people whose rights are violated." Carter replied with staunch reassurance: "Human rights is a central concern of my administration. . . . We shall use our good offices to seek the release of prisoners of conscience." Sakharov was elated. Brezhnev was not, denouncing the exchange of letters as "outright attempts by American official agencies to interfere in the internal affairs of the Soviet Union." The issue was also argued on the op-ed pages of American newspapers, where Carter was both chided and praised. Arthur Schlesinger, Jr., called the move a "considerable and very serious success . . . human rights is evidently one of those ideas whose day has finally arrived." Others, such as James Reston, criticized the act as "an innocent mistake of inexperience."

In Sakharov's reply to Carter's letter, the extent of the correspondence, he informed the president that four members of Helsinki Watch Groups had been arrested. Would the defenders of Helsinki be defended?

In late February and early March Sakharov had also somehow found time to write an essay whose title, "Alarm and Hope," exactly reflected his mood. He still viewed convergence as the only ultimate guarantee of safety in a nuclear world. The problem was that while Western societies were evolving, becoming what, "with a few exceptions, can be called 'capitalism with a human face,' " he was "less certain of a reciprocal evolution of totalitarian socialism toward pluralism."

Human rights was an integral component of any negotiations—what good is an agreement with a nation that breaks its social contract with its own people? The moral internationalism of the human rights movement is reflected in the essay's epigraph from Martin Luther King, of whom Sakharov felt neglectful in not acknowledging in his Nobel speech: "Injustice anywhere is a threat to justice everywhere."

* * * *

By late April, wearied and dispirited by the latest wave of arrests and the failure to secure permission for the apartment swap, a "major blow," Sakharov, Bonner, and Matvei took a vacation in the Black Sea resort of Sochi. Sakharov spent little time on the beach, the climb back up now too steep for him. But he did enjoy celebrating his fifty-sixth birthday on May 21, 1977, the three of them in a dockside restaurant, toasting the occasion with Pepsi-Cola, the very beverage of détente. But that same night they were woken by a phone call from Moscow—Tatiana was facing criminal charges in a case that also involved Efrem's mother. Cutting their vacation short, they were back in Moscow the next day.

Efrem had also been summoned to the Attorney General's and informed that criminal charges might be brought against him as well. The severity of the attacks was mounting incrementally; at some point something irrevocable would occur. By the summer it was clear both that Elena would require another eye operation and that, for everyone's sake, the children must emigrate. The decision could no longer be deferred.

The subject had come and gone over the last years, Bonner the first to take the most forceful position, telling Efrem he was heading for prison. Sakharov had not shared her certainty until Matvei went into convulsions. But even with these latest menacing developments, Efrem still argued that he could be of more use to the cause by staying. But then, "through the joint efforts of the Sakharovs and the KGB, I was finally persuaded to leave the country." When summoned to OVIR and told he had permission to emigrate, "I said to the OVIR director, a conspicuously well-fed and well-dressed man, who probably got fat on Jewish bribes or just bribes from visa applicants, and to the KGB officer present, that I will go nowhere until Elena Bonner is allowed to go to Italy for eye treatment. . . . It worked. She got her permission in the next day or so. . . . In a week or so, I used the same tactics to secure visas for my mother and her parents."

It was indeed a rare instance of the KGB and the Sakharovs sharing a common goal. On July 20, 1977, Andropov reported to the Central Committee that the KGB found it "expedient not to oppose BONNER's trip. One can expect that BONNER will attempt to use her presence abroad for antisocial purposes. However, the very fact that she was permitted to go to Italy will significantly diminish her level of activity. This decision will put SAKHAROV himself in a difficult position, since he will find it difficult to explain the authorities' treatment of his wife. One must also take into account the fact that rejection of a trip for medical treatment will entail large costs in terms of propaganda."

On September 1, 1977, around one hundred people gathered at the dacha in Zhukovka for a farewell party. As one of Sakharov's friends put it: "Those sendoffs were always noisy and lively, but everyone was heavy at heart."

At Sheremetevo Airport, on September 5, Efrem expressed a last regret to Sakharov: "I am sorry, Andrei Dmitrievich, I could not help you as much as I wanted." Kissing him, Sakharov said: "We do what we can do."

That left one hostage, Bonner's son, Alexei, who, having been denied entry into Moscow State University, was now, on November 5, expelled from the institute to which he had gained admission. Unemployed and unemployable, he was immediately subject to the draft. The Red Army was notorious for its brutal hazing, resulting in numerous suicides and deaths.

A sickly child who had become athletic, introspective but fiery when aroused, Alexei had even shamed Sakharov, who had advised him to attend the Lenin class at his school that led to automatic membership in the Young Communist League, saying it was only a "minor formality." Alexei shot back: "Andrei Dmitrievich, you allow yourself to be honest. Why do you advise me to behave differently?" Alexei had always been against leaving the country, at fifteen exclaiming: "Psychologically, I'm more prepared for a Mordovian labor camp then for emigration!"

The crisis had not taken the expected form and soon led to another. Twenty-one, married, and a father, Alexei had fallen in love with a slim Mongolian beauty, Liza Alexeyeva. Now he would have to ask his wife, Olga, for a divorce and leave the country without Liza.

In the middle of that dramatic December, Alexei got the chance to see those Mordovian labor camps for which he considered himself psychologically prepared. He, Sakharov, and Bonner, back from a rather unsuccessful eye operation, set out on a long train journey for the Dubrovlag camps in an effort to visit Eduard Kuznetsov of the Leningrad skyjacking affair, whose death sentence she and Sakharov had fought to commute. All Bonner's previous attempts at visiting him had failed—the camp authorities could cancel the annual meeting on a moment's notice. But Sakharov's presence might force a meeting.

Bonner had visited her mother in a camp, and at the Installation Sakharov had seen the lines of Zeks bent over shovels, but neither he nor Alexei had ever actually entered the Zone, where all color suddenly fades to brown and gray. Spotlights, guard towers, barbed wire, ground plowed to show footprints, voices "vicious and grating," the barking of the German shepherds that hardly seemed "man's best friend . . . when the man is dressed in a standard-issue gray quilted jacket or the stripes of a special-regimen prisoner."

They stayed in a cold, ratty hotel for visitors that also served as a dormitory for unmarried camp staff, whose leisure was devoted to TV soccer games, vodka, and fistfights. Learning of their arrival, Kuznetsov immediately announced a hunger strike, to continue until he was allowed to see his visitors. But Andropov was not about to allow Public Enemy No. 1 to defeat him in a test of wills.

In the ten days they spent at the camp hotel, Sakharov, Bonner, and Alexei came to know two of the guards. One, Kolya, took a defiant pleasure in recounting the punishments he had inflicted. The other, Vanya, countered him with a rebuke and a tale of mercy on his part, winning Sakharov's heart. A few days later he and Bonner chanced to be in the television room with Vanya. The program was a reading by the poet David Samoilov, to whom Bonner had introduced Sakharov in the early days of their courtship. Samoilov's poem was addressed to Pushkin in a time of epidemic: "Thank God you are free, in Russia, at your estate in Boldino and in quarantine."

Learning that Sakharov and Bonner were friends of the poet startled Vanya with the realization that there truly was another, better life, and the better part of him wanted to know more of it. Sakharov experienced a moment of hope that justified the otherwise futile journey, for Vanya was proof that "this wretched, downtrodden, corrupt and drunken people—no longer even a *people* in any real sense of the word—is not yet entirely lost, not yet dead."

That December Sakharov had also accepted an invitation to speak in Stockholm at the Amnesty International Conference on the Abolition of the Death Penalty, which was of course even less of a possibility than visiting Kuznetsov in the camps. But, glad to be linked to his grandfather Ivan, the gadfly lawyer who compiled the anthology *Against the Death Penalty* after the failed revolution of 1905, Sakharov wrote an essay condemning capital punishment as a "savage, immoral institution which undermines the ethical and legal foundations of a society." Only an evolved, humane society can reduce or eliminate crime and its punishments. "Such a society is still no more than a dream. Only by setting an example of humane conduct today can we instill the hope that it may someday be achieved."

Sakharov's positions never neatly coincided with anyone's. He was a staunch advocate of nuclear power, which he saw as cheaper and safer than coal and gas, and essential to the economic independence of any country that did not wish to be blackmailed by the energy producers and, in some cases, their Soviet sponsors, which struck him as a more likely problem than radiation, theft, and "do-it-yourself" atomic bombs.

December of 1977 was shaped by three nonevents: the failure to meet with Kuznetsov, the inability to travel to Sweden, and Sakharov's absence from the traditional silent gathering on Pushkin Square. "I'd never been all that enthusiastic about these demonstrations, which smacked of 'revolutionary' party rallies, and I disliked the role of 'opposition leader' into which I was thrust."

Sakharov was always adamant that he was not the leader of the movement, because no such movement existed. No decision could be more individual than that to risk your liberty and your life. It was only natural that like-minded individuals would join forces to defend a fellow human being. But that was not a movement with its "rallies," which Sakharov found suspect and distasteful.

But the dissidents did not only quarrel with the world; they quarreled with each other. Sakharov and General Grigorenko, who by then had been released from psychiatric confinement but would soon be stripped of his citizenship while traveling abroad, clashed over moving the date of the Pushkin Square demonstrations from the fifth, Soviet Constitution Day, to the tenth, the day the United Nations had adopted the Universal Declaration of Human Rights. Sakharov was, paradoxically, against both the change *and* attending the demonstration.

In rejecting the role thrust upon him, Sakharov was of too lucid mind not to know that others perceived both a movement and a leader—the foreign press, foreign governments, ordinary Soviet citizens seeking justice, the other dissidents. As Andrei Amalrik said: "Whenever we were drafting a statement or setting up a committee, the first thing we asked ourselves was whether Sakharov would sign the statement or head up the committee." Amalrik sometimes saw him as "a solitary monk under a leaky umbrella" with "an infallible sense of good and a constant readiness to oppose evil. There is something of the saint in him." Father Sergei Zheludkov, who fought in World War II, then became a priest and finally a dissident, also saw "certain traits of personal saintliness" in Sakharov, "to use the Old Russian Church term, his constant *sorrowing* for each one, each prisoner of conscience in our country." One day, "being young and foolish," Efrem had asked Sakharov "which one of Christ's commandments he finds more difficult to observe. 'I have no difficulty with any of them,' " Sakharov answered, "without hesitation, seriously, and seemingly neither disturbed nor surprised by the question."

The KGB was, however, doubt free: there was a movement, and there was a leader. The movement was made up of many movements—Tartars, Jews, ethnic Germans, nationalists of every stripe, priests and sectarians, urban intellectuals—but there was only one overall leader to whom they all turned. And it was time-honored practice to destroy the movement first, the leader last.

* * * *

There was no send-off when it was time for Alexei to emigrate. He preferred to say his farewells individually—to Ruth, his mother, Sakharov, Liza. On March 1, 1978, Alexei stopped on the way to the airport to lay three red carnations at Pushkin's statue. Liza moved into the apartment on Chkalov, replacing Alexei as hostage.

Nothing could be immediately done about her situation, and in the meantime two important trials had been scheduled on the same day, May 15, one in Moscow the other in Tbilisi. Both were Helsinki Watch trials, the KGB having decided to destroy this organization. Dual trials prevented people from attending both, thus also fracturing the movement, a collateral benefit.

Sakharov decided to divide his time between the two, though he had declined to join the Helsinki Watch Groups. But his decision to divide his time was obviated when one leader of the Georgian Helsinki Group dramatically recanted. "Human strength has its limitations, and there's a tendency to overestimate one's staying power," said Sakharov. "All the more reason to admire those who stand firm."

That left Sakharov free to attend the Moscow trial of Yuri Orlov. A renowned physicist, Orlov had risen from the peasantry and had worked as a laborer in a tank factory. Red-haired, tough, Orlov was believed by the dissident community to be among the last arrested because no KGB officer wanted the unenviable task of trying to break him. Sakharov had thought that the authorities would hesitate to arrest Orlov or at worst sentence him to internal exile. But the prosecution was demanding the maximum—seven years' labor camp and five years' exile. The KGB was hitting harder than he expected, bearing down as they had in the early seventies when they shut down the *Chronicle of Current Events*. The trial was rougher than usual—a heavy uniformed presence, police barriers keeping everyone fifty feet from the courthouse, Orlov's wife and sons roughed up during a search for tape recorders, his lawyer ejected from the courtroom and locked in an office.

On the last day of the trial, outraged that the defendant's friends were not even allowed in when the verdict was read, Sakharov moved through the crowd, adamant in his call for them to be let in. But not all the people milling around the barriers were there to show support. A fight broke out with plainclothes KGB, Sakharov and Bonner in the thick of it. Sakharov slapped a policeman in the face. As they were being dragged away, Bonner punched the local chief of police by accident, for which she later apologized at the station, though still maintaining: "I was right to hit the KGB agent, and don't regret it."

In his report on the incident to the Central Committee, Andropov focused on similar episodes in the past, proving a pattern of violent antisocial behavior on the part of Sakharov, who had now "slapped the face of a policeman." And that was a slap in the face of them all.

* * * *

That same spring Sakharov's younger daughter, Lyuba, gave birth to her first child, a son, Grisha. Fifty-seven that May, Sakharov was a grandfather again. Relations with his own family were still tinged with their resentment and his guilt. He was supporting his younger brother, Yura—who, always eclipsed by Andrei, had now slipped into paranoia, persecuted for Andrei's acts. Having tried his hand at physics, Sakharov's son, Dima, had left the university and was adrift, as in fact he more or less had been since his mother died, and was drinking now. The foreign press corps was aware of Sakharov's complicated relations with his first family but did not always do the legwork to check their impressions, one correspondent reporting Sakharov shunned by both his children and his first wife.

But there was little time to savor spring and new birth, for the KGB had scheduled the next two trials for July 10, one in Moscow, the other in Kaluga, some ninety miles away. Sakharov and Bonner decided to "cover both."

The Moscow trial of Anatoly Shcharansky was already the focus of worldwide attention, since the charge was the capital crime of espionage and since the atmosphere of anti-Semitism surrounding his case was odiously reminiscent of the Doctors' Plot, which had been foiled only by Stalin's death, "the Jews' luck" as it was sometimes called. And it was Alexander Ginzburg's bad luck that he was to be tried in a provincial city when most attention would be focused elsewhere. Shcharansky was facing prison and the camps for the first time, but Ginzburg had been in trouble since 1960, when he published the first underground journal, *Syntax*, for which he was arrested that same year. He had helped compile the white book on the Sinyavsky-Daniel trial and spirit it away to the West, for which he was also arrested. And, until his most recent arrest, he had been managing the Russian Fund to Aid Political prisoners established by Solzhenitsyn in 1974 in Switzerland.

Having already served multiple terms, Ginzburg, now forty-two, "looked like an old man," as Amalrik said.

But Ginzburg's trial had particular meaning for Sakharov. Ginzburg had been the first individual Sakharov had ever stood up for publicly and the first for whom he had personally suffered, losing his position as head of the Department of Theoretical Physics. And Ginzburg's trial was being held in the same Kaluga courtroom where Sakharov had first spoken with Elena Bonner, refusing her offer of cold food.

Luck would prove on Ginzburg's side. A mere eighteen months later he would be among a batch of dissidents swapped for two Soviet spies at JFK Airport in New York, whereas eight years would pass before Shcharansky breathed free air again.

Anatoly Shcharansky, like many Russian Jews, had to re-create his own Jewish identity from scratch, so thoroughly had the twentieth century and Soviet society bleached out religion, language, tradition. Israel's lightning victory in the 1967 War had transformed the Russian Jews' image of their people, and themselves, from victims to victors. But not everyone could rise to the heroic mode. One of the trial's uglier anti-Semitic aspects was that Shcharansky's principal accuser was also a Jew, Sanya Lipavsky, a helpful doctor, friend of the dissidents, a "sweet and slightly unctuous man with a trim moustache." The message to the Soviet people was, not only can't you trust the Jews, they can't trust each other.

Shcharansky had been imprisoned for interrogation from March 1977 until July 1978. Under particular pressure because espionage was punishable by death, Shcharansky, twenty-nine, did not break under interrogation but emerged tempered, self-assured, even ardent. Able to think many moves ahead like the chess player he was, Shcharansky was also aided in the one-on-one of interrogation by practices designed to strengthen the spirit. In his cell he performed the relaxation techniques taught by the refuseniks, composed a prayer in his "primitive" Hebrew, all the more powerful for that, and found his own ways of using a prisoner's natural tendency to slip into reveries to infuse himself with the élan of hope and courage. He also had the image and photograph of the woman he loved fiercely, Avital, whom he had married on the day before her departure to Israel. Despite the KGB, the Central Committee, and the Gulag, he lived with the certainty that he would be reunited with her in Jerusalem. No hope could seem more naive, yet for Shcharansky naivete was "an essential component for the person who rejects the spiritual slavery of his society and struggles against a powerful regime. Perhaps it guarantees that you won't be frightened to death or paralyzed by fear."

Defending himself, Shcharansky was by far a better lawyer than any the court would have allowed or appointed. For all the time the state had to build its case against him, it proved surprisingly flimsy, which only added to the aggressive meaninglessness of the trial. And it wouldn't have been Russia without some comical grotesquerie: a witness for Ginzburg's trial in Kaluga had somehow ended up at Shcharansky's in Moscow and had rattled off half of his testimony before anyone noticed.

But the comic relief only sharpened the pain. On the pretext that she would be called as a witness, Shcharansky's mother was not allowed to attend the trial. After a struggle, Shcharansky's brother was allowed in and took notes on the palm of his hand. The prosecution did not ask for death but twelve years' camp, three prison. For reasons that can only be imagined, Shcharansky was not given a total of fifteen but thirteen, which his cellmate, a swindler, assured him was a lucky number; he even had it tattooed on his arm.

Shcharansky was finally able to meet with his mother after the sentence and asked her if she had met Sakharov. "Met him?" she replied, answering one question with two. "Who do you think was standing with me every day?"

* * * *

After the twin trials of July, the summer of 1978 was, as Bonner put it, a "bit less loaded down than usual," and Sakharov began writing a memoir, which he had begun as an aide-mémoire. When Bonner had left for her first eye operation in 1975, Sakharov kept a journal of what happened in her four-month absence. And he had done the same during her second trip abroad. In May of 1978 Bonner began trying to convince Sakharov to write a memoir, if only for her alone. He resisted the idea. They argued, bluntly, hotly, with no attempt to hurt the other's feelings, but none to spare them either. He said she knew all about him, and she said she didn't and he'd better write his memoirs before he turned into a "hopeless sclerotic." While agreeing that something should be written, he countered that maybe she was the one to do it. Or maybe they should write it together.

Sakharov had also gotten into the habit of keeping a diary even when Bonner was not out of the country. He made his entries late at night and sometimes would read them to Bonner, who'd already be in bed. And they argued about that as well. You keep a diary for yourself, she objected. But that at long last gave Sakharov the opportunity to use Olesha's line he had so loved on first hearing: "You are me."

By the time they left in September for the Black Sea resort of Sukhumi, he had already written the first chapters which, in a special memorandum to the Central Committee, Andropov characterized as a "book of an autobiographical nature" that, in discussing his work at the Installation, divulged "state secrets." The September trip to Sukhumi was marred by an incident—Bonner suffered a hemorrhage in her eye of such obvious severity that they would clearly have to fight for another visa to Italy—and a non-incident. Mates Agrest, the "Soviet rabbi" who had had his son circumcised at the Installation and who, when expelled, had been saved from homelessness by Sakharov, was now living in the vicinity of Sukhumi, still engaged in classified work. Agrest was categorically forbidden at work to meet with Sakharov. "To this day . . . my heart is pained and my face reddens with shame that I did not have the daring to defy my superiors."

In the fall Sakharov concentrated his efforts on securing Bonner's visa, phoning officials, writing an appeal directly to Brezhnev in mid-November. On the twenty-ninth of that month Sakharov and Bonner were out, leaving Liza and Ruth at the apartment. But Liza and Ruth also had to leave to put a scheduled international call through to Tatiana, Efrem, and Alexei, who were living in Newton, Massachusetts, a suburb of Boston. For an hour, in violation of their own security regulations, the apartment was left empty.

They had left the apartment locked, and it was locked when they returned. But Liza couldn't find her robe, and Sakharov noticed that some odd items were missing: the old pair of pants he liked to wear around the house, a favorite blue jacket Klava had given him, his glasses. Items of obvious value had been left untouched. It was only the next day that he discovered that documents had also been stolen, including the memoirs.

It was clear that the KGB, which continually monitored their movements, had seized on that hour's window of opportunity; what wasn't clear was whether this was the search that precedes arrest—in which case it was usually done openly—or simply confiscation of material deemed especially harmful.

Still having received no response to Bonner's visa request, they announced a hunger strike to begin on January 3, 1979, and began stocking up on Borzhomi, Georgian mineral water that helps maintain the body's electrolyte balance when no food is being taken. On December 14, 1978, Andropov informed the Central Committee: "The Committee for State Security has again decided to allow Bonner the trip to Italy. Such a decision is tactically justified because the fact of permission arouses surprise and envy on the part of her and Sakharov's accomplices, and this leads to greater discords and hostility within their milieu."

* * * *

Bonner flew to Italy on January 15, 1979. She had tried, and failed, to mitigate Sakharov's sharp response to the Moscow subway bombing and would not have proved any more successful when in late January he learned that the defendants, Armenian terrorists, had been tried in secret and sentenced to death, their families rushed in to Moscow for one final meeting. No sooner had Sakharov drafted his appeal to Brezhnev for a stay of execution and further investigation than he learned that the executions had already taken place. Still, he lodged his appeal with the Supreme Soviet to shame them.

But, in the authorities' view, the shame was his. Now Sakharov was not only attacked in the press—"Shame on Those Who Defend Murderers"—he was attacked on the phone, on the street, in his home.

He argued with his callers, but they always had an answer:

"Why weren't the defendants' families present at the trial?"

"For fear of reprisal by the victims' relatives."

In one of the many letters he received, the writer vowed to cut off Sakharov's head and impale it in front of the American Embassy.

When he flew to Tashkent for a trial, he was accosted in the street by a "relative of a victim": "If you don't leave for Moscow today, I won't be responsible for my actions. I've done time, and I'm willing to do more!"

And back in Moscow two KGB thugs with a taste for theatrics arrived at Sakharov's apartment—he let everyone in. They too were relatives; they too threatened violence, and there even was some; a little melee broke out, Liza getting a sharp jab in the stomach.

Andropov was now, for all practical purposes, first among equals. A vain and doddering embarrassment, Brezhnev slurred his words when reading speeches, mixing up the order of the pages. Since 1975 in special memoranda, Andropov had been informing the Central Committee of the country's incipient crisis. The USSR was depleting its gold reserves to buy wheat from the West. Agriculture remained a low-grade disaster. Harvests had been poor and, no matter whether good or poor, a third of the yield was always lost to waste and spoilage. The Soviet Union had ten times more farmers than America yet bought food from them, a humiliation in itself. If it weren't for the food people grew on "private plots," 1 percent of the land producing more than a quarter of the country's food, the situation would have resulted in worker protests like those in Poland. Industry was at a standstill because more than half the state's money went into the arms race, the stakes soaring at the speed of technology.

Andropov also kept track of the exact number of times Sakharov visited foreign embassies, the ideal place for information to be transmitted, and how often he received foreign guests such as Zbigniew Romaszewski, a member of the Polish Workers Defense Committee (KOR), which was forging bonds between the intelligentsia and the working class in Poland, where the visit of the Pope in June 1979 brought out people in such huge numbers that an atomized country suddenly saw that it was still a nation.

The question of foreign contact was an enduring one. If Soviets were allowed to travel abroad, some would defect, bad press even when a famous ballet dancer was not involved. If large numbers of foreigners were allowed into the country, as would be the case when the 1980 Olympics were held in Moscow, there was a high risk of contagion. Sakharov was in favor of the Olympics for that very reason and thought boycotting the games would "spoil" their spirit.

But the KGB took positive pleasure in spoiling the spirit of a moment and punctured a tire on Sakharov's car when Bonner returned to Moscow on April 15. Two French correspondents helped them change the tire only to soon find all of theirs punctured.

Bonner had been gone three months, flying to Boston after her operation to visit her children and grandchildren. Adaptation was not proving easy. The invitations from MIT were pro forma. Alexei had, however, gotten into Brandeis University, but Efrem had not succeeded in finding work in telecommunications, his specialty. Even looking for work was difficult when he remained involved in human rights in the Soviet Union and would soon be impossible when he was officially appointed Sakharov's representative abroad.

A period of relative calm ended in November with the arrest of Tatyana Velikanova. A forty-six-year-old mathematician, a grandmother, in 1969 Velikanova had been a founding member of the Initiative Group for the Defense of Human Rights in the USSR; the first of its kind, it protested human rights violations to the United Nations and distributed copies to Western correspondents. By the time of her arrest none of the other founding members were at liberty. But it was the part she had played in bringing the *Chronicle of Current Events* back to life that had proved unforgivable.

Sakharov had met her back in the old days of the parties at Chalidze's, when the dissidents were "young in spirit and pure of heart" and few enough to fit together in one room. Later, when the movement had been infiltrated by informers and "professional dissidents," attracted more by the style and risk, she exclaimed: "How did all this filth creep in?"

The movement was tainted, the authorities were more active

and severe, détente was almost dead, the decade ending badly. At the very end of December the Politburo sent Soviet troops into Afghanistan. As Marshal Ustinov, head of the military, put it, a "limited contingent will remain a year or a year and a half in Afghanistan until full stabilization is achieved," meaning the emplacement of a government sympathetic to Moscow. The soldiers also had illusions, some believing they were marching off to "their Spain." Perhaps it was the desire to both distract and unify the populace in a "short, victorious" war that blinded the Soviet leaders to the fact that Afghanistan had never been conquered. But that was a paradoxical Soviet Union, sluggish yet aggressive, decrepit yet confident.

Whatever the illusions and motivations involved, Sakharov viewed the invasion as "expansionism." And he reversed his position on the Olympics to accord with the changed circumstances: "The USSR should withdraw its troops from Afghanistan," Sakharov told a correspondent from *Die Welt* on January 1, 1980. "If it doesn't, the Olympic Committee should refuse to hold the games in a country that is waging war." A similar interview he gave the *New York Times* on January 3 was broadcast back into Russia by the Voice of America. But there was no immediate, discernibly hostile response from the authorities. In fact, on January 7, 1980, Elena Bonner's mother, Ruth, was granted permission to travel abroad and visit her grandchildren and great-grandchildren. KGB agents gathered by Sakharov's apartment house when ABC came for a taped television interview on the fourteenth, but there was nothing unusual about that, though Sakharov did sense "something peculiar in the air—a mixture of hostility and gloating." A week later he was awakened at one in the morning by a phone call from a friend who had learned from party connections that Sakharov was to be stripped of his state awards and exiled from Moscow.

But in that society it was always difficult to know what was fact, what was rumor, what was warning. "A month ago, I wouldn't have taken it seriously," Sakharov responded, "but now, with Afghanistan, anything's possible."

All he could do was go about his business. The next day, January 22, was a Tuesday, and the weekly theoretical physics seminar would take place that afternoon. Sakharov ordered an Academy car from the pool and left a little early, taking a shopping bag with him to pick up groceries at the Academy commissary on the way. Sitting democratically in front as was the custom, Sakharov still did not notice that the Krasnokholmsky Bridge was unusually empty for that time of day. Suddenly a traffic patrol car was beside them, signaling them to pull over. As the driver was showing his license, Sakharov heard the rear doors open and turned to see two men getting in, flashing red IDs. They ordered the driver to follow the patrol car to the Attorney General's office on Pushkin Street. Citizen Sakharov was under arrest.

CHAPTER THIRTEEN

THE BLESSINGS OF EXILE

STUNNED by sudden arrest, Sakharov did not have his wits about him when apprised by the deputy prosecutor that he had been stripped of his state awards and would be exiled to the closed city of Gorky; Bonner, under no sanction, could accompany him. And he did not bother to protest the obvious: exile could result only from a verdict passed by a court of law, and his was therefore unlawful.

In a KGB minivan with curtained windows he was driven through the streets of Moscow to the airport. Bonner arrived later, having the foresight to bring their shortwave radio. For a time they thought that they might not be going to Gorky at all; maybe he'd only been told that to make things go smoother. But even that didn't matter. "We were too relieved at being reunited to worry about where we were headed—we didn't care if it was to the ends of the earth. In some strange fashion, we were actually happy."

But it was to Gorky, by special plane, accompanied by twelve KGB agents, good food even served, which, by then quite hungry, they ate and enjoyed.

Transferred in Gorky to a KGB minivan, Bonner asked: "Where are we going?" Grinning, one escort replied: "Home."

Home was an eleven-story slapdash tan brick building on Gagarin Boulevard, where the city ended in rubbish and woods. Home was the four-room ground-floor apartment, one room of which would, from time to time, be occupied by the "landlady." The furnishings were Soviet generic, and the place had a provisional air, as if it had been used as a hostel. No telephone. A police observation post directly opposite controlling the flow of anyone attempting to visit them. A policeman by their front door twenty-four hours a day. A jamming station set up for them alone. Home was a punishment designed by Yuri Andropov.

The best line of resistance was in fact to make the place their home, by continuing to live as they always had. The first order of business was to draft a statement, which Bonner, free to travel, brought back to Moscow with her five days later. Though Sakharov concentrated on his own illegal exile, he cited other examples of injustice and again condemned the invasion of Afghanistan before concluding: "I am prepared to stand public and open trial."

What he instead received was a visit the following day from two patriotic "drunks" who waved guns and made threats to "turn this apartment into an Afghanistan! You won't be here long. They'll take you to a sanatorium where they have medicine that turns people into idiots."

At the end of his statement on exile Sakharov had thanked those who would be coming to his defense. Now it was his turn to be written up in the *Chronicle of Current Events*. Now it was time for his scientific colleagues and fellow Academicians to show their true colors. Now it was time for appeals to be made for the maker of appeals, and letters to be written for the writer of so many. Anatoly Marchenko, temporarily at liberty, wrote a letter to Kapitsa requesting intercession for Sakharov, reminding Kapitsa that he was getting old and "now is the time to think of your soul."

Referring to Socrates and Lenin, Kapitsa wrote to Andropov, "greatly disturbed" by Sakharov's fate. It seemed counterproductive, radicalizing him while depriving society of the benefit of his scientific genius. "Wouldn't it be better," concluded Kapitsa, "simply to put the engine into reverse?" Andropov replied that since Sakharov was a "subversive" who slandered the Soviet Union, passed information to foreign embassies, and defended the terrorists of the Moscow subway bombing (Bonner was right: that did stick in their mind), any "reverse" was out of the question.

Despite quiet domestic diplomacy, despite foreign protest and scientific boycott, Sakharov would continue to be confined to the city limits, trailed constantly, none but the most casual contact permitted. It had been decided to exile Sakharov not to a Siberian installation but to a pleasant city in "European Russia" which, because of its large military industry, was closed to foreigners. It was a punishment "mild by Soviet standards," but Sakharov chafed against his "gilded cage."

Andropov had created an impermeable bubble of isolation around Sakharov, depriving him of all contact, which he had only lately come to prize. His information flow was regulated but not totally controlled. His shortwave was jammed in his apartment, but he was not prevented from talking walks with it, standing in the wind at the cemetery (where the reception was good) to hear his own case discussed. He was receiving scientific materials by mail, delivered by a good-natured woman assigned especially to the task by the post office. Bonner was free to travel and could ferry his statements to Moscow; in that of late February, Sakharov was no longer saying he was "prepared" for trial but insisting on public trial. All the positive elements of his situation could be used as propaganda and, since their situation had been crafted by the KGB, inherent was the threat that they could alter it for the worse at any moment.

The rhythm of life that Sakharov and Bonner established in Gorky was twofold and distinct. When they were together, he worked on science, his memoirs, the occasional statement or appeal. Every few weeks Bonner would leave for Moscow to make contact with the outside world and to stock up on the provisions available only in the capital. Those partings were always risky, superstitious, both of them remembering, sometimes even quoting, Osip Mandelstam's lines: "Who can know at the word 'separation' / Just what kind of separation is in store?"

While she was away, Sakharov would at first live on the food she'd prepared for him, then revert to shopping and cooking for himself, warming up everything, salad and cottage cheese included. He enjoyed doing the dishes, so soothingly quotidian. He worked and took walks, not to the woods out past the raw high-rises connected by mud and planks, but to listen to the radio. When Bonner was there, they explored the city by car, which pretty much could be seen in a day. Gorky had a Kremlin, a fortress around which old Russian cities were built, but it was nowhere as grand and sumptuous as Moscow's. Even the Gorky Kremlin's one cathedral was, though lovely, modest and bare. Like all the other townspeople and visitors, Sakharov liked to stroll on the high bluff that overlooked a great swath of river and plains shining off into the distance. It was beautiful there in any season. The city still had a number of old Russian wooden houses, made from thick logs and seeming to sink slowly into the ground, torpor of the provinces.

The city had been known as Nizhny Novgorod until 1932, when it had been renamed for its most illustrious native son, the writer Maxim Gorky. But it had been renamed for a pseudonym—the writer's real name was Peshkov; he had assumed the nom de plume of Gorky, meaning, "bitter." It was the right name for a place of exile, though for most Russians it had quite positive associations: Maxim Gorky was a symbol of success whose works appealed to the common man, proving that the Soviets were capable of creating a culture uniquely their own—working-class, tough, heroic, a little sentimental, but in the right way. The model for socialist realism, the officially enforced aesthetic, was Gorky's novel *Mother,* which had been written under unusual circumstances. Sent to America to raise funds after the failed Revolution of 1905, Gorky outraged New York society by arriving with his mistress. He was scorned by Mark Twain, who remarked that Gorky holds out one hand to us and slaps our face with the other. Ejected from his hotel, Gorky was given refuge by an American millionaire, at whose upstate home he wrote the mother of all socialist realist novels. In the early thirties, when Stalin began enforcing his will in earnest, Gorky grew disillusioned with Soviet Russia and, now inconvenient, he may have been poisoned. After his death the old trading city at the junction of the Oka and Volga Rivers was renamed in his honor as was the main street of Moscow itself. There was a joke that Gorky was embarrassed by all this but was told that the name could not be more appropriate for an era that would be "bitter to the max."

Science filled Bonner's absences. Sakharov was realistic about any contribution he might yet make while still hoping for "a flash of inspiration." He derived great satisfaction from studying the latest developments in science, not "greedy" for any further achievement or renown. In fact he would publish three papers in that first year of exile on particle physics and cosmology. But he was increasingly aggrieved that none of his colleagues or fellow Academicians had yet spoken out in his defense. A few were, however, working behind the scenes. His friends on FIAN's party committee slyly thwarted efforts to expel him from FIAN by requesting the necessary legal documents, of which, since his exile was illegal, there were none. Vitaly Ginzburg, the head of FIAN's Department of Theoretical Physics, had a grudging admiration for Sakharov and successfully lobbied for him to be allowed visits from his colleagues at FIAN. Fearing that he would be turned into a "one-man sharashka" (a gulag for scientists), Sakharov was suspicious about those visits, the very first of them, on April 11, 1980, coinciding with the arrival of foreign scientists in Moscow; their protests could be countered with the verifiable information that Sakharov was receiving his FIAN colleagues in the comfort of his own four-room apartment. The policeman moved away from his post at the door while the scientists were visiting and the jamming was suspended, but only until the moment they left.

The only scientist who dared visit him through "chance" encounter was Misha Levin, his university mate, with whom he had strolled the streets of Moscow discussing Pushkin and physics. On a dank, raw March day the two of them strolled the streets of Gorky, Sakharov voicing his disgust at the careerism of the Academicians and confessing that his great dream was to live until everything was known about the lifetime of the proton. But Pushkin did not go unmentioned either—Sakharov, again at work on his memoirs, saying that it was shame that Pushkin had burned his. On the stairs Levin noticed how bad Sakharov's heart was.

Sakharov's other unpermitted guest was Natalya Gesse. When forbidden by a KGB captain to travel to Gorky, she replied: "I'll go anyway."

"Sakharov insulted both us and the government very deeply."

"When was that, when he gave you the H-bomb?"

Natalya knew the tricks. A friend buys a train ticket to Gorky thirty days in advance. On the day of departure he boards the train with her bag. At the last second she gets on, he gets off. It would be beneath the dignity of the Moscow KGB to stop the train to drag her off, and the Gorky KGB would be equally reluctant to drag a tiny woman in her midsixties from a table in the café where she had just happened to encounter Sakharov and Bonner.

When Sakharov wasn't doing science, he worked on his memoirs which, now more than a challenging intellectual project to fill the long days, had become an element in the struggle—the KGB had made them an issue by stealing the early chapters from him in Moscow. He also wrote an essay-length statement, "A Letter from Exile," that was divided into four parts: "World Problems," "Western Problems," "Domestic Problems," and in the final section "Some Words about Myself," his own problems in exile. The essay restates his basic beliefs but also displays the inexorable evolution of his thought. If in recent statements he had accentuated his skepticism about socialism, he now took a forcefully positive view of the United States as "the historically determined leader of the movement toward a pluralist and free society, vital to mankind." Because of the overarching nuclear danger, the United States must find a way of reaching agreements with the USSR even though "everything is as it was under the system created by Stalin." The second round of Strategic Arms Limitation Talks (SALT II) proved that agreement could be reached, though now, with the invasion of Afghanistan, the United States had withdrawn from the talks as well as the Olympics—as well they should, thought Sakharov. He was a believer in parity as a precondition for serious arms negotiation, but there was increasingly less parity in his vision of convergence. The more the USSR, "a closed totalitarian state with a largely militarized economy and bureaucratically centralized control," became like the West, the better. And, abandoning the leaden lexicon of politics, he spoke with plain ardor of his persuasion: "I believe in Western man. I have faith in his mind, which is practical and efficient and at the same time aspires to great goals."

The article was published as "An Alarming Time" in the *New York Times Magazine* on May 4, 1980. There it further influenced American public opinion, which was already pro-Sakharov, in no small part due to the unremitting labors of Efrem, Tatiana, and Alexei. Sakharov would have been immeasurably pleased if his own colleagues had shown a tenth part of the activism of the

American scientists who spoke out for him. Zeldovich had begged off any public statement, saying that, as it was, he couldn't travel any farther west than Hungary. After that, Sakharov was no longer sure he could shake the man's hand.

* * * *

They passed the long winter evenings in Gorky with good meals cooked by Elena, music, television, hours of conversation, reading "until you can't stand it."

Once in a while they went to the movies, braving a blizzard for Belmondo.

But there was always too much time, and it would quickly fill with anxiety and the pain of separation. All Bonner's family, her mother, children, and grandchildren, were now in America. Liza had been barred from Gorky, pistols waved in her face. By counting the number of special food parcels delivered to the observation post across from his apartment, Sakharov estimated that thirty-five agents had been assigned to them. No one could get through.

After the first winter of exile, the first spring of exile was especially welcome. They walked in the woods, or on the high bluff above the river. She planted flowers in pots on their prefab cement and brick balcony and taught him about them, for he would be responsible for their care in her absence, and what interested her interested him.

That summer the workers at the Lenin shipyards in Gdansk, Poland, went on strike and seized control of the yards, where they quickly organized food distribution, a newspaper, masses. Wiry, foxy, with a machine-gun delivery, Lech Walesa was emerging at the forefront of the movement. Sakharov, in tune with the Polish August as he had been with the Prague Spring, himself experienced a resurgence of political heart. He wrote to Brezhnev requesting that Liza be allowed to join Alexei. There was a reasonable chance this request might be granted, the other children having been allowed to emigrate. He wrote to China's State Council chairman asking that the sentence of a Chinese human rights activist be reconsidered. He wrote an appeal for Tatyana Velikanova, the arrest of his good friend still paining him. And finally, outraged, heartbroken, disgusted by ten months of silence from his fellow

Academicians, on October 20, 1980, he wrote an open letter to Anatoly Alexandrov, president of the Soviet Academy of Sciences, which put the matter straight to him: "Is the leadership of the Soviet Academy of Sciences prepared to demand my immediate return to Moscow and an open trial which will determine my guilt or innocence . . . ?"

Though placing a question mark at the end of his sentence, Sakharov was all too aware of the probability that it was rhetorical. He had turned fifty-nine during that first spring in exile and saw little likelihood of not turning sixty there either, though who could have predicted that the strikes would spread across Poland and that the workers in that workers' state would establish their own union, well-named Solidarity. As he knew, "the mole of history burrows unnoticed, and change often comes when not expected."

Facing a future of exile, he had sources of strength and reasons to hope. The children, except for Liza, were safe, if heartbreakingly distant. He and Bonner could enjoy each other's company for sinfully long stretches. He had his work on the memoirs, which were now taking on size and shape, the barest outline of his life—H-bomb, human rights, exile—causing even him some pause at the mismatch of personality and fate. But there was always the question about the stability of their own situation. With some increased depredations, it still remained structurally unchanged. But anything could cause it to worsen—a shift in the world situation, power struggles to replace Brezhnev, whose condition was now so hopeless that the famous healer Dzhuna was rumored to be treating him. Or Sakharov himself could issue a statement that would strike the wrong nerve.

Only retrospect could show how idyllic that first year of exile had been, and that retrospect was quick in coming.

* * * *

"Teeth are the bane of humankind," Sakharov had once remarked to a colleague with an appropriate grin and grimace. That bane, age's indignity of bridgework, had accompanied him into exile.

But in Gorky even a dental appointment posed a danger. From the very start, Sakharov and Bonner knew the KGB had access to their apartment; there were signs, and the only question was exactly what role their intermittent landlady played in those incursions. One day, forgetting her cigarettes, Bonner went back to the apartment to discover the KGB tearing the place apart. Caught by surprise, they immediately fled through the window in the landlady's room. This arrangement was apparently to keep the police out of the loop, and indeed the policeman at the door seemed shocked both by the incursion and the disorder. But the forgotten cigarettes proved a stroke of luck. When the landlady sat for hours in her room, her presence could never be entirely forgotten. Now, her cover gone, she was gone with it.

But that did not mean that the KGB was denied access to the apartment. The police could be brought into the loop, nothing easier. And that in turn meant that when Elena was in Moscow, Sakharov had to carry a satchel of manuscripts weighing some twenty-five pounds with him whenever he went out, whether to the market, where he was just another customer with a string shopping bag and a beret or, as he did on March 13, 1981, to the dentist, having been notified that his bridgework was ready. Perhaps it was because Bonner was away, or because he was distracted by a disturbing error he had found in one of his scientific papers, or because nothing seems more banal than a dental appointment—whatever it was, it was enough to lull Sakharov, who accepted without question that his bag could not be brought into the "sterile area," even though the area was clearly not in the least sterile, and who also accepted the assurances of the nurse, who promised to watch the bag, saying: "Don't worry, nothing ever gets lost here."

Her statement was, within its own limits, quite accurate. While Sakharov was in the chair, that bag containing notes on science and current events, his diaries, the three thick notebooks in which he wrote his memoirs, and even some literary efforts—essays on Faulkner and Pushkin—were snatched, either by design or happenstance. Even worse was knowing it was his own fault. Bonner came back from Moscow to find him "in a state of shock, literally trembling."

But once he emerged from the shock, Sakharov responded to the theft of his writings by writing furiously. Four days after the theft Anatoly Marchenko was arrested for the sixth time. One of the counts against him was the letter he wrote to Kapitsa in Sakharov's defense. For that reason, and because of the spiritual kinship he felt with Marchenko, Sakharov, working with Bonner, wrote an appeal calling upon "all honest people of our country and the world to do everything in their power to defend and help him." But, at war in Afghanistan, the state was increasingly aggressive at home, and Marchenko's sixth term was ten years' strict regimen labor camp, five years' internal exile. "If the state believes the only means of answering people like me is to keep us behind bars," declared Marchenko, "then I have no objections. It means I'll be behind bars until the end of my days, I'll be your eternal prisoner."

The theft had also included an essay for a conference honoring Sakharov at Rockefeller University. Eleven days later, he had rewritten it. He addressed the audience as fellow members of international science, "the only real worldwide community which exists today." Speaking of the plight of Marchenko and others, he also spoke of his own—the theft of his papers, a KGB attempt "to rob me of memory." Liza, who had not received any word on her visa application for a year and a half, was a "hostage of the state." Sakharov appealed to his fellow scientists to act as befits a community of reason: "Western scientists face no threat of prison or labor camp for public stands; they cannot be bribed by an offer of foreign travel to forsake such activity. But this in no way diminishes their responsibility."

He also wrote an article "How to Preserve World Peace," in which he reminded the outside world that the Soviet Union had resumed jamming after the interlude of détente and that nothing of substance could change until "public opinion in the West realizes the seriousness of the totalitarian threat and achieves greater psychological readiness to meet it."

The KGB did not let this volley go unanswered. On May 12, 1981, Liza, who had already attempted suicide by pills, was summoned to OVIR and informed that she lacked sufficient grounds for emigration; she had no relatives abroad and, moreover, her father, a retired military officer, was opposed. On May 14 and again on May 19 she was summoned by the police in connection with possible implication in a criminal case and threatened with being thrown out the window. The timing was hardly accidental. On May 21 Sakharov was to turn sixty, and his name would be honored throughout the world. The KGB could not prevent the conferences, the festschrift to be published by Knopf, the greetings from U.S. President Ronald Reagan calling Sakharov "one of the true spiritual heroes of our time," but they could attempt to spoil the mood by striking hard at Liza.

But celebration was itself a form of resistance. Sakharov and Bonner dressed for dinner and toasted the occasion, the clink of their glasses reverberating in the echo chamber of their apartment.

Five days after he turned sixty, Sakharov wrote a detailed letter to Brezhnev protesting his illegal exile in Gorky, restating his willingness to bear responsibility for his own acts under the law, but concentrating on Liza: "the KGB's use of my daughter-in-law as a hostage for purposes of revenge and as a means of pressure on me is particularly ignoble." Meanwhile, in the West, Efrem, Tatiana, and Alexei were themselves taking action. They discovered that two American states allow for proxy marriage. Standing in for the slim, Mongolian beauty at the marriage ceremony in Butte, Montana, on June 9 was Ed Kline, a happy bear of a man. A successful New York businessman, Kline was also a serious student of Soviet affairs and the owner of Chekhov Publishing, which issued, among other things, the *Chronicle of Current Events* in English.

The proxy marriage was a stroke that yielded double consequence. The Soviets, signatory to agreements recognizing the validity of such unions, were deprived of the objection that Liza and Alexei were in no way related. And Liza's father, now convinced that Alexei was sincere in his love, wrote to Brezhnev withdrawing any of his previous objections. There were no longer any legal impediments to her emigration, and Liza was now therefore detained in the country illegally.

Exchanging notes in September 1981, Sakharov and Bonner decided that a "hunger strike was a moral imperative and our only practical course of action." They chose late November, when Brezhnev would be in Bonn for talks with Chancellor Helmut Schmidt. Once the commitment was made and the date chosen, there was no longer any reason for silence. But Sakharov and Bonner refrained from making their decision public and official in order to allow matters to be settled quietly, as was always possible, and to allow Bonner to travel twice to Moscow, bringing his new diaries and the new draft of the memoirs for safekeeping there. She also collected Borzhomi mineral water to balance their electrolytes during the hunger strike and contacted their own community to make their reasons plain.

On October 21 Sakharov sent telegrams to Brezhnev and Andropov setting the date for the hunger strike at November 22, a month's notice, ample time to avoid direct confrontation. The KGB's reaction was quick and crude—they stole their car. "Whenever the authorities did not like something, it was our car that suffered," Bonner said with sympathy as for a comrade who has also suffered "KGB persecution." The theft deprived them of transportation and signaled that the KGB was prepared for violent action.

Word was abroad of the impending hunger strike, and world pressure was being brought to bear. But, at war in Afghanistan, with Poland out of control, its Olympics boycotted, an anticommunist president in the White House, and an anticommunist pope in the Vatican, the Soviet Union was not terribly concerned about its image. On November 15 Bonner made her last trip to Moscow, the hunger strike now a week away but still no sign or word of any change from the authorities.

But then, on November 16, OVIR, after a long delay for consultation, accepted Liza's application for a visa. In itself that meant nothing. The paperwork must be ready for any eventuality; the KGB would be running a variety of scenarios, and retreat was always an option.

What Bonner learned in Moscow, and Sakharov would learn from her, was that their own community was not wholeheartedly with them. As Revolt Pimenov, a dissident Sakharov had defended, put it, Sakharov and Bonner were risking their lives for Alexei and Liza's right "to argue, to make up, to fall into bed." Sakharov was indignant. What better reason than to help unite people in love? Liza was vilified for not dissuading Sakharov and Bonner, as if she had that power. Besides, there were other dimensions to their decision that did not concern Liza directly. They were affirming the basic principle of a person's right to travel freely, they were protesting Sakharov's illegal exile, and perhaps most important: "Our fortunes in Gorky had hit bottom. A victory was badly needed." They were also contesting the dissident community for building "something of a 'cult of personality' " around him, which in practice meant that he was "regarded as a means to an end rather than a human being." And it was as a human being who makes his own decisions that he would now assume the pains of hunger and the risk of death.

On November 20, two days before the date chosen, Sakharov received a letter pleading with him to desist: "this absurd action will prove fatal." As further inducement, the writer also included a newspaper account of the "death agonies" of the hunger strikers in Northern Ireland's Maze Prison. But Sakharov was contrarily inspired to keep an account of his own:

November 21.

Lusia [Elena's nickname among family and friends] arrived, wearing a new hat. . . .

Tomorrow, the first day of our hunger strike, film footage taken of Lusia and Liza on November 16 will be shown on Western television together with the film Lusia shot of me here.

Tonight at midnight, after supper, Lusia and I clinked our glasses of Carlsbad salts [a laxative]. The hunger strike has begun. Our plan is to massage each other's backs every day, and to take a five-minute warm bath. Lusia got hold of a notebook to record our weight, blood pressure (she has a sphygmomanometer), pulse, water intake, and urine output.

November 22.

The first day of our hunger strike.

We're settling into a pattern. We watch a lot of television. Beethoven's Fifth Symphony was on last night. . . .

There's a lot about our hunger strike on the radio.

November 23.

The second day, and we're both slightly dizzy.

Brezhnev's visit to Bonn was covered extensively on television. It's clear that they have not, and could not, come to agreement about medium-range missiles. . . .

On the evening of the 23rd, Lusia reminded me of Goethe's dictum:

> He alone is worthy of life and freedom
> Who each day does battle for them anew!

November 24,
The third day.

. . . It seems the KGB is not going to let Liza leave under the cover of Brezhnev's trip. I said to Lusia: "Let's go back to our basic idea. If the KGB doesn't want us to die, sooner or later they'll let Liza go. And if they want us to die, they have plenty of ways to get rid of us without the hunger strike."

November 25.

The fourth day of the hunger strike.

. . . Lusia's legs have started to swell. That worries us. Her heart's not good either. I'm worried because it's her first hunger strike. We received an anonymous telegram: "Andrei, have pity on Lusia."

November 26.

. . . In France, even the left socialists have come out for Liza.

November 27.

The sixth day of the hunger strike.

Lusia has lost little weight, which doesn't make sense, but there is less swelling. The hunger strike continues to be harder on her. She has a headache. For the second day we walk back and forth on the terrace for about an hour in the afternoon sun. We take measured steps toward each other, pass each other like ships at sea, and then turn at either end. . . .

Our lungs are clear. Lusia's blood pressure was 140. That's not good—usually pressure falls during a hunger strike.

In the evening we heard on the Voice of America (other stations also picked it up): "According to information received from a scientist with access to Sakharov, his health has taken a turn for the worse." We were infuriated: we feel fine. . . .

November 28.

The seventh day.

... I ... sent a message to Liza and our relatives and friends, assuring them we were "feeling fine. All signs and symptoms according to the book. We're determined to stay the course."

... I don't like the fact that Lusia's weight has hung at 132 pounds for three days. She didn't feel well today, but it's hard to know just how bad, since she never complains. Maybe the crisis has begun. ...

November 29.

Lusia asked me if I share her conviction that we have made the right decision. I do. Our whole past existence is fading into the distance. A sense of composure, of the absence of doubt. ...

... We read Mihajlo Mihajlov's *Kontinent* article, "The Return of the Grand Inquisitor," about the danger of Solzhenitsyn-style nationalism and warning against muddling religion, nationalism, the state, and politics all together. A serious article, beautifully written, to the point. I am 100 percent in agreement with it.

November 30.

... A summons was delivered requesting us to meet with detective Rukavishnikov on Gorny Street in order to identify our car. ... Maybe the KGB is trying to lure me out? ...

Tonight I finished an eight-page letter to my children that I've been working on for three days. Getting my thoughts down on paper has made me feel better.

December 1.

They've broken the chain on the front door. They probably did it last night. They're letting us know they can enter at will. We hate the thought that they may abduct us and separate us, but we're prepared even for that.

... Ten minutes before our one o'clock exercise session, something landed on our terrace. We thought it was a snowball. Two policemen ran out from behind the corner of our building, chasing a fleeing man. ... It wasn't ... however ... a snowball: it was a package containing three apples, white bread and three pieces of meat, still warm. . . . Lusia threw the package away; it's painful for her to see or smell food. . . .

Surrendering would destroy me morally. We are prepared to die. It will be murder, not suicide—the KGB began this test of wills two years ago.

December 3.

The twelfth day.

We're becoming weaker, but overall, we're in pretty good shape. Lusia said, and she's right, that we're doing better because we're *together*. She's disturbed by thoughts of her mother and children, what they're going through now and what they might suffer afterward. But once again we agree that we've made the right decision. The only alternative is complete capitulation.

"They've come to kill us," said Bonner to Sakharov on December 4 during their midday balcony stroll as she looked back into the apartment and saw it filling with KGB agents and medics in white.

"We have to hospitalize you," said a man who identified himself as representing Municipal Health.

"Will we be together?" asked Bonner.

"Yes," he lied.

They kissed. Tears came to Sakharov's eyes; Bonner was bitter that it was happening on her daughter's anniversary, to which they would have clinked glasses that evening.

As they were pushed toward separate ambulances, Bonner shouted out one last piece of practical advice: if they force-feed you, breathe deeply.

At Semashko Hospital, apart from pulse and blood pressure, Sakharov allowed no medical procedures, demanding to be reunited with his wife. "We each had confidence that the other would not abandon the hunger strike, but separation was difficult to bear —the KGB was apparently counting on that fact to break us."

Izvestiya ran a column called "Another Provocation" on December 4, the same day as the forcible hospitalization but published a few hours before, which managed both to attack Sakharov and Bonner and to allay world concern about their health with the information that they were receiving medical care. Shown it, Bonner tore the paper to pieces and threw them at a doctor, shouting: "You and your *Izvestiya* can go to hell!"

Sakharov too was active, as writer, reader, conspirator. He wrote a statement that his hunger strike would not end until Liza could emigrate. He read a little of Nabokov's autobiographical *Speak, Memory,* then wrote to Bonner on the margins of pages 114 and 115: "Lusia, I'm refusing to talk to doctors or to allow any procedures or tests while we're separated. I send kisses. You are always with me. I'm infinitely grateful to you."

He asked a nurse to bring the book to Bonner, who was not in another ward, as he believed, but in another hospital. In that surreal realm where doctors pleaded with Sakharov to allow them to keep their Hippocratic oath, the nurse kept her word.

Sakharov resisted the warnings from cardiologists and neurologists that he might already be confused, losing faculties. In fact he was quite clear of mind and determined to win the two victories at stake here: reunion with Bonner, Liza's emigration.

On December 8 Sakharov's attending physician gave him an ultimatum: he had only hours to break off his hunger strike. Were they threatening him with forced feeding? "What gives you that idea? Not at all."

Around that same time in the other hospital force-feeding apparatus, tubes and funnels, was wheeled into Bonner's room. She told the attendants she would fight it to the death.

It was now simply a question of whose will was the more tensile.

A few hours after the ultimatum, a KGB official came to Sakharov's hospital room and said: "I'm authorized to inform you that your request can now be reconsidered in a positive light, but you must first end your hunger strike."

While saying that he took the KGB's promises seriously, Sakharov replied that he had begun the hunger strike with his wife and could only end it with her.

"I'll report your answer. You'll be seeing me again."

By seven o'clock that evening he and Bonner were in each other's arms, reunited in victory.

A few days later Liza traveled to Gorky with Natalya Gesse.

He and Bonner were still in the hospital. In Liza's words: "Andrei Dmitrievich was sitting in an armchair and the nurse was taking his blood pressure. He rushed toward us with a cry: 'Liza! Natalya!' The blood-pressure machine dragged after him." The beauty of Liza's happiness was more than sufficient recompense for Sakharov. Though still shaky, Bonner could not be dissuaded from accompanying Liza back to Moscow. Four days later, on December 19, 1981 Liza telegrammed Sakharov from the airport: "I'm leaving, smiling through tears."

Sakharov had decided to remain in the hospital until Bonner returned and spent his time listening to radio reports on the situation of Solidarity in Poland. On December 13, in what seemed a very Polish act of heroic masochism, General Jaruzelski, in order to save Warsaw from Warsaw Pact troops, declared martial law.

After Liza's departure, Bonner took to her bed in Moscow for six days. Sakharov had a heart spasm on December 22, which so concerned the medical staff that they discharged him the next day —all they needed was Sakharov suddenly dying in their hospital. Three days later he had a heart attack, from which it would take a month to recover. Victory over the KGB never comes cheap.

II

Now there was an unspoken truce. Sakharov and Bonner had won the victory they so badly needed, had no other immediate cause, and peace and recuperation was their primary concern. The KGB was reexamining options. It was neither practical nor somehow seemly to attack again at once. And Andropov, still head of the KGB, was more succession minded than ever; on January 25, 1982, Mikhail Suslov—who had served Stalin, set the ideological tone for decades, and was instrumental in crushing every rebellion from Prague to Poland—died at the age of eighty. Brezhnev was only seventy-six but clearly not long for this world. His favorite was the party functionary Konstantin Chernenko but, as Andropov knew, Chernenko had no power base of his own, and Brezhnev's favor would not outlive him. Still, the question of succession required a great deal of attention, and it was just as well that the Sakharov front remained quiet for the time.

Sakharov was again at work on his memoirs—to please Elena,

who had wished the book into existence in the first place, to retrieve memory from theft and oblivion—and the book itself had become part of the overall struggle, which is measured only in victories and defeats, as Bonner reminded him when he was initially reluctant to start over. Still it was excruciating for him. Not because he lacked the will and force to begin again, but because so deeply a private man had been fated to live such a public life. He had the high-intelligentsia horror of public linen and the high-intelligentsia imperative to be honest about everything. He had to make scrupulous accounts, determining the balance between achievement, love, renown, the freedom won by courage, and all that was "false, cowardly, shameful, foolish, ill-advised, or inspired by subconscious impulses it's better not to dwell on."

But the writing had its pleasures too. He could go back to the courtyard and spin fantasies with Grisha Umansky, be on the train to Ashkhabad, run with Klava through a field, watch his Idea become explosion against the horizon of Kazakhstan. Or be his father's ardent student and speak with him of science and of life's disappointments, drink tea and apple jam with him at the dacha in October. As he often told Bonner, he was certain that he would live to be as old as his father, who died at seventy-two when the heart condition they both obviously shared took his life in the night. Sakharov, nearing sixty-one, was giving himself eleven years.

By April Sakharov had finished another draft of the memoirs, nine hundred handwritten pages, which Bonner brought in batches to Moscow for typing and which were then, by people invisible to history except for their daring, ferried out of the country. In addition to their responsibilities toward Sakharov and Bonner, Efrem and Tatiana now assumed that of the memoirs, with whose author they could have no exchanges about the text. Meanwhile, Sakharov was engaged in serious revision, devoting nearly all his time to it. After the debilitations of the theft and the hunger strike, he had little time for science though, as Bonner observed, not a day went by without him thinking about physics. The year 1982 did see the publication of one article, "Multisheet Cosmological Models of the Universe," published in the *Journal of Experimental and Theoretical Physics*. Still, his "only regret" was "to have accomplished less than I would have liked or than might have been expected of me." He paused from the memoirs in April to issue a statement protesting the imprisonment of scientists such as Sergei Kovalyov, then later made a small number of statements and appeals including one for Anatoly Shcharansky to President Mitterrand of France. And in September he received a visit from FIAN colleagues, to whom he offered his own favorite breakfast, slices of boiled beets, and whom he impressed as being quite up to date, somehow missing nothing even when he dozed off.

In early September Bonner traveled to Moscow while Sakharov put the finishing touches on the revision of the memoirs. In Moscow the KGB monitored her very closely, trying to determine how she maintained "illegal contact with sympathizers abroad by using the diplomatic mail of the American Embassy in Moscow as well as foreigners who come to the Soviet Union on scientific and cultural exchanges as tourists." The KGB was particularly angered by Bonner's "Appeal to World Public Opinion" on behalf of Sofiya Kalistratova, a Helsinki Watch Group member and one of the few lawyers in the country who would mount a genuine defense for dissidents. At this same time Bonner also announced the dissolution of the Helsinki Group, "allegedly because of the continual persecution of the Group's members in the Soviet Union." She incurred further ire when applying to OVIR in late September for a visa to travel to Italy for eye surgery, a "contrived pretext . . . to stir up interest in Sakharov (which has periodically flagged in the West)." In its own internal propaganda, the KGB's focus was now increasingly on Bonner, both as instigator and lifeline. Sakharov's operational code name had been changed from "Ascetic" to "Askold," hers remained "Vixen."

But it was at Sakharov they struck on October 11. It was four o'clock in the afternoon, still light. Sakharov and Bonner's long-suffering car—the ignition wire had recently been hooked up to the fuel pump, causing a fire—was parked by the river. Bonner left to buy railroad tickets, Sakharov remaining in the car. The manuscript, nine hundred handwritten pages, five hundred typed and revised, was in a satchel on the floor behind the driver's seat. His window was half open. A dark-skinned man with curly black hair in his midthirties approached, looking for a ride: "You've got Moscow plates—are you headed for Moscow?"

The last things Sakharov remembered were his reply—no, he wasn't going to Moscow—and a "strange odor, like that of rotting fruit." Even though the flying shards of glass cut him, he did not hear the left rear window being smashed. Then he was stumbling out of the car, speaking with three women who were standing nearby, no doubt KGB medical personnel there in case of an adverse reaction to the aerosol narcotic.

Returning from the railroad station, Bonner found Sakharov bleeding, staggering toward her "with the expression of a man who has just learned of the death of someone close to him."

Nine hundred handwritten, five hundred typed.

For days he was so suicidally depressed, seized by what he called "black anguish," that Bonner was afraid to leave him alone. But Sakharov hated defeat. Several days later he was hard at work again, restoring the memoirs from memory. "Andrei has a talent," said Bonner. "I call it his 'main talent,' to finish what he starts."

On November 3, 1982, in bad weather Brezhnev waved to the masses and missiles from on top of Lenin's tomb at the celebration of the sixty-fifth anniversary of the revolution. The next day Sakharov was summoned to the Gorky prosecutor's office and issued, for his signature, a stern warning about his unsubstantiated and slanderous accusations of KGB involvement in the theft of his manuscripts. Insisting that the only crime was the KGB's, Sakharov refused to sign and went back to work on the memoirs, "writing bits and pieces in the hope that they would fit organically into the manuscript" that Efrem and Tatiana were safeguarding and for whose publication in the West they had already begun discreet discussions.

Within a week Brezhnev died and was immediately succeeded by Andropov. If Sakharov was a prisoner of conscience, Andropov was a prisoner of consciousness, his own. Now with full power and fully aware of the problems facing the country, he could, as Volkogonov observes, "think of nothing better than yet again saddling up the worn-out nag of Bolshevik discipline." Cauterizing corruption with execution, rounding up loafers from movie theaters and bathhouses during working hours, arresting drunks—all had a bracing, tonic effect but did not treat the roots of cynical apathy at home, which now included the costly war in Afghanistan.

For Sakharov and Bonner, in practical terms, Andropov's assumption of leadership meant that the person chiefly responsible for their fate would now have less time for them. Not that there was any improvement. It was under the new KGB chief, Fedorchuk, that Bonner, who been suffering severe heart pain, was detained as she arrived in Moscow by train and strip-searched, even though she had been willing to hand over the satchel containing an eight-millimeter camera, English-language lesson tapes, and the latest 250 pages of the memoirs. It took three hours for the detective to go through all her bags and make a scrupulous list of everything confiscated, by which time the train had been shunted a distance from the station. Even after the confiscation her bags were heavy and, as a doctor, she knew that the heart pains she experienced on the long way back to the station were serious and demanded that she sit down and rest. She did rest, but only until she could drive herself on again. On the stairs of a bridge over the tracks she blacked out.

The truce, such as it was, was over.

* * * *

Sakharov, stimulated by an article by an American colleague, Sidney Drell, was again in much of a thermonuclear mind as the year 1983 opened. Not that he was in any great disagreement with Drell—the universe of Mutual Assured Destruction had an infernal logic of its own, requiring that apocalypse be discussed as dryly as possible. The perfect solution to the problem was, obviously, the elimination of all nuclear weapons. But a "world poisoned with fear and mistrust" was a difficult place to reach agreement.

Sakharov's article, "The Danger of Thermonuclear War," arose in particular circumstances and addressed them. The United States and the Soviet Union were at a particularly tricky point in defining parity, balancing silos, warheads, throw weight. Though not by name, Sakharov refers to the Soviet Union as a possible inadvertent cause of nuclear war: A country equal to the enemy in nuclear weapons, with an overwhelming superiority in the conventional, risks war in the belief that the enemy will not risk annihilation. There would be "little cause for joy" when this assumption was proved false. Sakharov supports America's spending "a few billion" on the new MX missile system but not if the Soviets, moved by their survival instinct, "take significant verifiable measures for reducing the number of land-based missiles (more precisely, for destroying them)." Completed on February 2, 1983, the article, by whatever exact mechanism, was dispatched to the editorial offices of *Foreign Affairs*.

The Szilard Award in April diverted him from working on the memoirs. His acceptance speech contained one new idea, one perennial. New was the dialectical belief that "nuclear deterrence is gradually turning into its own antithesis and becoming a dangerous remnant of the past. The equilibrium provided by nuclear deterrence is becoming increasingly unstable." Perennial was that "in the long run, convergence is the only alternative to global destruction," because the best solution removes both means and motivation. It was an idea that was rejected by Soviets as dangerous heresy and by Americans such as Secretary of State Alexander Haig, addressing the American Bar Association in New Orleans, as dangerous fantasy: "Soviet antagonism toward Western ideals is deeply rooted. We cannot count upon a convergence of Soviet and western political principles or strategic doctrines. Convergence should not be, and cannot be, a goal in negotiations."

Whatever exact mechanism brought Sakharov's manuscripts to the West also worked in reverse, and in April, in addition to the honor of the Szilard Award, he had the pleasure of receiving the first part of his memoirs in a photocopy of the manuscript made by Efrem. But April ended very badly when Bonner had what she knew was a second heart attack, confirmed by a cardiogram at the Academy Clinic on May 14. She needed hospitalization but would not risk it without him, and on May 20, at a press conference in Moscow, she made public her demand that she and Sakharov be hospitalized together.

That demand, following on Sakharov's article, incensed the KGB. An *Izvestiya* article of July 1983, titled "When Honor and Conscience Are Lost," was signed by four members of the Academy and bristled with indignation: the government and people have "been more than tolerant toward this man who is living peacefully in their midst," especially since he is essentially a traitor, identifying the enemy as "the Soviet Union, the country where Sakharov lives."

Though nearly none of his regular mail reached him, Sakharov now received more than 2,400 letters of denunciation.

The book *CIA Target—the USSR,* by Nikolai Yakovlev, had gone into its third revised edition in early 1983. Yakovlev was a serious specialist in American history who had written a biography of Franklin Roosevelt. He was as well a KGB literary hatchet man who had come to that position through his own arrest, which led him first to betray his comrades and later to cooperate in the camps. Privately, he was said to profess fairly progressive views, but those were mere accessories, cultural cuff links. His book vilified dissidents in general as wanting to cash in after leaving the country and Sakharov and Bonner in particular. He summarized his book in an article, "The Downward Path," for a popular magazine. "It looks like Academician Sakharov has become the hostage of the Zionists, who are dictating terms to him through Bonner, a cantankerous and unstable woman." All the characterizations of her—a "wanton" woman involved in two murders, who had driven out Sakharov's children and taken control of all his money, "persuading her husband to do this or that by hitting him with anything at hand"—were demonstrably false. A libel suit against Yakovlev could afford them the platform of a trial to denounce the illegality of Sakharov's exile and to reiterate their demand to be hospitalized together, which by that summer did not seem forthcoming at all.

In an act of unmitigated gall or idiocy, Yakovlev himself arrived at Sakharov's apartment at two o'clock on the day after the article was published, requesting an interview and offering to autograph a copy of his book for Sakharov. "In the nineteenth century, I would have challenged you to a duel," replied Sakharov, who was able to restrain his temper for twenty minutes of sharp exchanges until, as he knew he would from the start, he struck Yakovlev, a "left-handed slap on his flabby cheek."

Bonner observed that, after he calmed down, Sakharov was

"very pleased with himself. . . . As a physician, I felt this was a necessary release of stress for him; as a wife, I was delighted. . . ."

But the press had done its job, resulting in spontaneous attacks. On the day of Yakovlev's ill-fated visit, a woman came up to Sakharov on the street "brandishing" the *Izvestiya* article signed by the four Academicians: "Those of us who were at the front are going to show you and your Jew-wife Bonner what war means. She's the one behind all this, couldn't you find yourself a Russian woman? If there's a war, we'll all die—there won't be any survivors. At the front, your kind—traitors—were executed, and we'll execute you, *scum*. We'll tear you apart."

The message was the same all the way from the street to the Kremlin, where that summer Andropov alluded to Sakharov's mental instability when meeting with a group of American senators. But it was Andropov who was unstable, physically. His kidneys unable to cleanse toxins, he had been put on dialysis in the spring. The September 1 session of the Politburo, which dealt with the downing of a Korean passenger liner by a Soviet fighter, was the last at which Andropov presided. After that he secretly ran the country from the hospital, "the sick leader of a sick country," in Volkogonov's words.

Because of her frequent train trips to Moscow, Bonner was more frequently exposed than Sakharov to spontaneous attacks by irate citizens. In the confines of a rail car compartment, not free to flee them, she could "physically feel their hatred." As she telegrammed Sakharov after one such incident in early September: "It was very frightening and therefore I was completely calm." Later she told him that although she knew he had to write the article on nuclear war, he "should understand what it cost her."

Yakovlev's article reappeared in a new and even cruder version in the magazine *Man and the Law,* which had a circulation of almost nine million. What bothered Bonner most was the aspersions he cast on her young love for the poet Seva Bagritsky, killed in the war. When she read those passages, she felt "a heart attack coming on." She meant that quite literally, writing her complaint for suit against Yakovlev with nitroglycerine tablets always at hand. After two weeks of work, on September 26, 1983, she filed a suit against the writer and the magazine at a Moscow court.

When her health took a sharp downturn in October, she and Sakharov had the sort of conversations that take the worst into account. He said he would commit suicide if she died. She made him promise to wait six months.

Sakharov had two visitors in December. Natalya Gesse came to say good-bye. Her apartment had been searched. The other two women of Pushkin Street had died. Her son now lived in America. Bonner took Natalya's emigration hard: "They did not want me to have a friend whom I could trust completely."

To his other guest, Vitaly Ginzburg, head of FIAN's Department of Theoretical Physics, who arrived in late December 1983, Sakharov announced his intention of calling a hunger strike in the year to come if Elena were not allowed abroad for medical treatment of her eyes and heart. He was putting the word out early and often so that he could be spared the ordeal and the KGB spared another defeat. Ginzburg tried to dissuade him, but Sakharov, "cheerful and excited, was certain that, no matter what, 'they' would yield."

* * * *

Nineteen eighty-four was a year of double significance in Russia, for it would test the prophecies of both Orwell and Amalrik. The USSR still survived, but it seemed a dying country, exhausting its resources to buy food and machines from abroad, industry and agriculture both failing in the land of the hammer and sickle. So many of the country's best minds and talents were exiled abroad or internally that the Soviet Union seemed a nervous system responding only by Pavlovian conditioned response. And, after fifteen months in power, Yuri Andropov was, in January 1984, experiencing in the flesh what he had already expressed in verse: "Men live and vanish. . . ."

To the amazement of the country, and the world, a third sick man was hoisted into office. Konstantin Chernenko suffered from such advanced emphysema that it took him twenty minutes to walk through an auditorium to the podium. His speeches were composed with his infirmities in mind—large letters, short phrases. The grand clerk of the Soviet Empire, Chernenko had been in charge of all paperwork, which also afforded him highly useful information. Paper gave him power; paper was all he knew. One of his first directives had to do with proper margin size.

It was clear to Sakharov that under Chernenko nothing would change—not the war in Afghanistan, the suppression of dissent, the neglect of the economy. For all practical purposes his rule would be a vigil.

The new regime would need a little time to consolidate itself; it would be impolitic, and probably counterproductive, to present it with a crisis at once. And it would be easier and safer to go on a hunger strike in spring than in winter. Sakharov and Bonner now clashed furiously. She was opposed to a hunger strike. But he was resolved and would go it alone this time, hers now the weaker health. And if Bonner could not dissuade him, no one could.

In the meantime he was visited by colleagues from FIAN, who also experienced no success in talking him out of the hunger strike. At the moment, as one of the visitors observed, Sakharov was more interested in demonstrating "his skill in the use of a small computer sent to him from abroad. In particular, he programmed and solved the problem of the motion of Mercury around the sun, taking into account the effects of general relativity, and thus numerically determined the rate of Mercury's perihelion shift. Working with the 'clever' computer brought him real pleasure."

In April Sakharov formally announced the hunger strike, appealing to "friends the world over. . . . Save us!"

The regime reacted more swiftly, and more severely, than could have been anticipated. On May 2 Bonner was to fly from Gorky to Moscow. She said good-bye to Andrei at the airport terminal. On the tarmac she was surrounded by five men and hustled to a small Black Maria van. She had no doubts she was under arrest. Driven to the other end of the airport, she was brought to an office where she was awaited by a man in plain clothes and two women in police uniform. The man informed her that she was being charged with Article 1901, slander of the Soviet state or social system, which carried a maximum punishment of three years' labor camp. The two women took Bonner to another room, searching her body and her bag, which contained documents, some to be delivered to the American Embassy, which was itself incriminating. Then, after agreeing not to leave Gorky, she was driven home, where Sakharov told her that he had seen them arrest her from the terminal and, not wasting a second, had already telegrammed the Supreme Soviet and the KGB that his hunger strike had begun.

When she had realized that sooner or later Andrei would go on a hunger strike, Bonner had written two letters, one to her mother and children, the other to Andrei. The letter to her mother and children sums up her life, what she loved—her work as nurse and doctor, her "woman's work," writing, dancing, her friends, their kitchen—and the loves of her life, the poet killed young, the Russian medical student who had vowed to marry "those legs," and now "that incredible, unimaginable human closeness that fate has bestowed upon Andrei and me." Though not wanting her letter to be a farewell, Bonner asked forgiveness from her mother and children, for warmth ungiven and temper unbridled.

To Andrei she wrote: "I am tired of being homeless, of feeling the hatred of your children, your distrust of them and your expectation that one of them will betray you." And she says, in the harsh love that grants the other's freedom: "Neither pity, nor anxiety for your health, nor fear for your life restrains me now. I know that this is your decision and that any action now is easier than inaction."

Bonner was now called in for questioning every day, accompanied by Sakharov, who would wait for her outside the office, sipping hot water from a thermos. May 6 was a fine spring day, and Bonner was busy with her flower boxes on the balcony while Sakharov dug a flower bed for her in the ground in front of their balcony. A woman in a beige raincoat carrying a bouquet of flowers came running up beside Sakharov, who almost never recognized anyone unless he was "prepared" for their arrival. But Bonner had no such problem; she immediately informed their friend, Irina Kristi, that she had been arrested and that Andrei was on a hunger strike. The two essential points were communicated before the KGB dragged off Kristi still clutching her undelivered bouquet.

Bonner went in for questioning that day as well, the interrogation proving oddly "pallid and not too long." At the end the interrogator asked if he might have a word with Andrei Dmitrievich. The word was: "doctors have come for you, you must go to the hospital." When Sakharov protested, half a dozen men in white coats entered the room. Recognizing the futility of resistance, he switched tactics immediately, requesting that his wife be allowed to accompany him.

The request was granted but only until Sakharov was hospitalized. Then Bonner was dragged out of his arms and down the corridor, his shouts fading. She was put in a car and sent home. Again they had been forcibly separated, and who knew what this forcible separation would mean?

The next morning at nine o'clock, after presentation of a warrant, Bonner's investigator and his squad searched the apartment, the two witnesses required by law also present. Long and thorough with much attention to potential secret hiding places, the search lasted until ten o'clock that evening, a good deal of those thirteen hours spent on meticulously listing the 319 different documents confiscated as well as books, typewriters, tape recorder, cameras, and their radio, their last connection to the outside world.

The next day, buying flowers to bring Andrei, she was approached by two KGB agents, who informed her: "you can't go to the hospital. . . . You won't be let inside, you'll only make more trouble for yourself." She brought the flowers home and arranged them around the apartment. It was a bright and sunny day and, needing work to distract her, she began planting the flower bed Andrei had dug for her. The next several days were divided between gardening and interrogation, and waiting for word, some exchange of brief notes to be permitted.

After his forcible hospitalization on May 6, no further action was taken against Sakharov until the eleventh, when he was thrown to his bed by a squad of women orderlies and tied hand and foot. An intravenous feeding needle was inserted into a vein, and he was also given an injection that caused him to pass out, urinate, and immediately develop symptoms of a cerebral spasm or stroke. That procedure lasted five days. He was force-fed through a tube through the nose for the next nine days. Then on May 25 they switched to the "most excruciating, degrading and barbarous method"—hands and feet tied, a clamp over the nose so tight he could breathe only through his mouth. When he opened his mouth, it was filled with food and clamped shut so that his only choice was to suffocate or swallow. After three days, unable to bear it any longer, Sakharov promised to swallow voluntarily, which afforded him enormous relief and caused him enormous depression, for it meant that his hunger strike was over. But he intended to resume it as soon as he was certain that he could "bring it to a victorious conclusion." He remained in the hospital, but the moment did not seem to come.

Bonner could tell from certain characteristic repetitions of letters in a note she received from him that he had suffered some sort of cerebral spasm. On the day that Sakharov's force-feeding went from IV to nose tube, Bonner received a disturbing telegram from Sakharov's three children, Tanya, Lyuba, and Dima: "we, the children of Andrei Dmitrievich, ask and implore you to do everything you can to save our father from this mad undertaking, which could lead to his death. We know that only one person can save him from death—that is you. You are the mother of your children and you must understand us. Otherwise we will be forced to turn to the prosecutor's office, because you are inciting our father to commit suicide. We see no other way out. Understand us correctly. . . ."

Between interrogations Bonner tried to keep as busy as she possibly could—putting the apartment back in order after the search, shopping for winter clothes for Andrei, working on her garden (inveterate smoker, she found the scent of nicotiana "intoxicating"), and making "jam, lots of jam." On July 25, 1984, she was officially presented with the charges against her, several counts of slandering the Soviet state or social system, which, with supporting documentation, ran to six volumes and took her three days to read.

Bonner was summoned to court on August 8 for a trial that lasted two days in the same courthouse, she noted, where Gorky's hero had been tried in the novel *Mother*. Counting some eighty-five people in the courtroom, none of whom she knew, Bonner categorically refused to accept any guilt in a trial that was closed, not even her husband being allowed to attend. It took only about an hour for the verdict to be rendered—five years' exile, the place not indicated, itself a threat.

It was a perfectly 1984ish situation for both her and him, both law and medicine perverted into their exact opposites. The doctors were attempting to convince him that the trembling of his hands indicated Parkinson's disease: "We won't allow you to die. . . . However, you will become a hopeless invalid." Sakharov was keenly aware of the Orwellian nature of his predicament: "In the novel and in real life, the torturers sought to make a man betray the woman he loves. The part played by the cage full of rats in Orwell's book was played for me in real life by Parkinson's disease." But over the summer Sakharov began preparing himself inwardly to resume the hunger strike, by remembering days of his life in exact detail and occasionally by driving himself into such hysterics that his doctors, the "Mengeles of today," would leave him alone for a while. He was tormented by the thought that he was failing Elena. By the time August was coming to an end, they had not seen each other for more than a hundred days. It was unbearable to face the needle, clamp, and tube again; it was unbearable not to see her; it was unbearable not to win her a visa.

Weighing all the unbearables, he announced yet another hunger strike to commence on September 7 but was abruptly discharged from the hospital the following day.

"He was wearing the same light coat in which he had been taken away in early May from the procurator's office, and his beret. It didn't seem as if he had lost weight; on the contrary, he looked almost bloated. We embraced, both of us in tears. We got into the car. I couldn't move. We just sat and wept with our arms around each other."

After twenty minutes she broke the silence to tell briefly of her trial and sentence, of which he had known nothing. Driving to a hill overlooking the Volga, he described his 123-day ordeal. Then they went home.

His mood veered from elation to despair. Nothing could have been more natural than the happiness of being with her again. They followed each other everywhere, even into the bathroom, talking, talking, or not talking, but never "apart for a minute's time."

But the more he felt her love, the more he felt he had failed her.

And he had failed himself as well. She was, apart from everything else, his connection to the world. They were trying to turn him into a "living corpse."

He had failed to defend his love and his life; what failure could run deeper than that?

III

On September 12, 1984, as a convicted criminal, Bonner was fingerprinted, her mug shots taken. That indignity reignited Sakharov's fighting spirit. By the end of the month Sakharov was seriously discussing the next hunger strike, the one that had to prove victorious, the one whose defeat he could not survive.

In mid-October he wrote a long letter to the president of the Academy, Alexandrov. Passionate in its opening—"I appeal to you at the most tragic moment of my life"—it was dispassionate in describing the experience of force-feeding, and concluded with ultimatum and afterthought:

> If you and the Academy's presidium do not find it possible to support me in this tragic matter, which is so vital for me, or if your intervention and other efforts do not lead to resolution of the problem before March 1, 1985, I ask that this letter be regarded as my resignation from the U.S.S.R. Academy of Sciences. I will renounce my title of full member of the Academy—a proud title for me in other circumstances. I will renounce all my rights and privileges connected with that title, including my salary as an academician—a significant step since I have no savings.
>
> If my wife is not allowed to travel abroad, I cannot remain a member of the Academy of Sciences. I will not and should not participate in a great international deceit in which my academy membership would play a part.
>
> I repeat: I am counting on your help.
>
> Oct. 15, 1984 Respectfully,
> Gorky A. Sakharov
>
> P.S. if this letter is intercepted by the KGB, I will still resign from the Academy and the KGB will be responsible. I should mention that I sent you four telegrams and a letter during my hunger strike.

P.P.S. This letter is handwritten because my typewriter (together with books, diaries, manuscripts, cameras, a tape recorder, and a radio) was seized during a search.

P.P.P.S. I ask you to confirm receipt of this letter.

Sakharov was also furious about the "cruel and unfair telegram" from his own children to Elena that caused her "additional suffering and anxiety in her already horrible and almost unbearable situation. The telegram gave the KGB the 'green light' for any action against us." And as a sign of displeasure he ceased writing to them for a year and a half.

But he was in good spirits when visitors from FIAN arrived in mid-November laden with frightening quantities of heart and eye medicine, in part sent by the German writer Heinrich Böll, with whom Sakharov was acquainted, along with kilos of food, all of which was coordinated by the painter Boris Berger, whose dual portrait had so moved Sakharov. When she had been free to travel to Moscow, Bonner had been a literal lifeline of food, which was intermittently abundant in the capital and usually scarce in Gorky. The latest quip was that Sakharov had broken off his hunger strike but the people of Gorky continued to go hungry.

Sakharov had challenged the president of the Academy and his colleagues to help him now or forever part company. But there were infinite shades of cowardice. Some scientists would not even send him reprints of their articles. Some would but without dedication. Some would write dedications but only the most common and formal; some of the more audacious would even risk a phrase of their own.

This was his first visit from FIANers since April. They thought Bonner looked much worse than Sakharov did—because of her sheer presence she had always seemed the same size as he, but now she appeared much smaller, as she in fact was. But even as they followed the usual pattern of those visits—breakfast, science, lunch, rest, science, dinner, then a rush for the night train back to Moscow—the visitors could not help but feel the atmosphere of their home, "one of doubtless happiness . . . a calm and peaceful happiness."

Still, there were recurrences of his "black anguish." Describing how the doctors had threatened him with Parkinson's during the hunger strike, he called them "bastards," startling his guests, who'd never heard the word from him before. And he was "inconsolable" when he spoke of breaking off the hunger strike: "I betrayed Lusia."

Yet more confident now that come spring he would again risk death for all the old reasons and a new one now, to redeem his own honor, Sakharov was glad to see the old year out and to celebrate their thirteenth wedding anniversary on January 7, 1985, which he described for the family in "faraway Newton" in a letter some days later:

> How do we live? Tragically. Buried alive. And at the same time strange as it may seem, happy. On the seventh we celebrated our thirteenth official wedding anniversary; we did everything right, a party for two—Lusia went all out (cake and cheese pie, a "goose" [i.e., chicken] with apples, fruit liqueur), thirteen candles in a pretty design. . . . We kiss you. Wish you good health. Kiss the younger children for us. Every night at bedtime. Lusia knocks on wood eleven times, with thoughts of you and of us, naming everyone and wishing them well. Kisses, Andrei.

On February 25 Sakharov was again visited by two FIANers—they always came in pairs so that at least in principle one could keep an eye on the other. One of them had been there before and observed the "ritual" of doing the dishes according to Bonner's instructions in her absence: "Sakharov said: 'When Lusia was leaving she left a note about what to do and where to put things; so that's what we'll do. . . . This dish has to be wiped (glance at the note) with this cloth. And this saucepan has to be put . . . (another glance) right in this place.' "

But now Bonner was there to critique: "Andrei, you put the saucepan in the wrong place again!"

"Don't get excited, I got carried away talking."

The conversation was particularly far-ranging that day. They were asking the big questions—why was the universe such as it was and not some other way, why was life encoded in biological information the way it was and not some other way?

Unfortunately, there were more mundane matters to be dealt

with. In violation of protocol Sakharov had asked his colleagues to bring some materials not strictly scientific in nature back with them. He gave them time to think it over. By the time the two FIANers decided they could not be complicit, Sakharov had withdrawn the request, asking only that they bring regular materials to his colleagues, though quite quickly one of those materials proved to be Sakharov's open letter to the president of the Academy, imploring help, threatening resignation.

* * * *

When a Soviet leader died, he was revered if he was Lenin, reviled if he was Stalin, or consigned to oblivion if he was Brezhnev or Andropov. There was no question to which category Chernenko, the "spectre of the chancery," belonged when, on March 10, 1985, he succumbed to a variety of ailments. In his thirteen months of rule the bureaucracy was computerized and an underground pneumatic post installed between the Politburo in the Kremlin and Central Committee headquarters on Old Square. Documents moved with greater efficiency, but the country did not. Hard-currency reserves were spent to import meat, grain, and machinery and equipment, much of which idled and rusted. The Soviet Union could export only natural resources; no one would buy the products of their industry. Balance of payments was radically out of line. The numbers were bad.

But there were also some good numbers, of a different sort. The new leader, Mikhail Sergeyevich Gorbachev, was a vigorous fifty-four. The party had finally brought itself to hand over power to the next generation. There were other good numbers. Gorbachev had joined the party in 1952, the year before Stalin died, which meant that he had been shaped by the reformist spirit of the Khrushchev years. The first party congress Gorbachev had attended was the twenty-second, at which Khrushchev had demanded that Stalin's body be removed from Lenin's mausoleum, a speech lent extra thunder by Sakharov's Big Bomb.

Sakharov and Bonner had bought another radio after the search and tuned in to the excited Western speculation about the new leader, who had a law degree, had traveled abroad several times, and with whom Maggie Thatcher, the Conservative prime minister of Great Britain, had declared she could "do business." The hard-line Suslov had liked the earnest, energetic party man from Russia's deep south, and Andropov had made him his protégé. Suslov and Andropov were dead by the time Gorbachev's slim chance at full power came, but he was supported by fortuitous circumstance—Chernenko died suddenly, some key votes could not make it back to Moscow in time—and by the last of the Stalin-era officials, Foreign Minister Andrei Gromyko, who said of Gorbachev that he had a nice smile but iron teeth.

Guilty because she knew she did not have the strength to join Sakharov on his next hunger strike, Bonner kept finding reasons for him to delay. Not in March on Tatiana's birthday, and the weather was still too cold. And she would hate not to bake the traditional cakes for Russian Easter in April.

On April 13, 1985, Gorbachev received one of his first briefings on Sakharov and Bonner. After a comprehensive background report, Chebrikov, head of the KGB, having inherited an image of Bonner that was ideologically useful, informed or, actually, disinformed Gorbachev that Sakharov was, "at his wife's instigation," preparing to announce a new hunger strike in three days. In that event he would be hospitalized and isolated from Bonner who, to "make it impossible for Sakharov to take solid food . . . has secretly hidden his dentures."

But the final sentence of the report contained plain truth: "One must take into account the fact that, given his age and the general condition of his health, it will be more difficult for Sakharov to survive his third lengthy hunger strike in the last few years."

"Bastards! Murderers! . . . They're giving me an injection!" shouted Sakharov as he was taken by force to the hospital on April 21, five days after beginning the strike. Bonner, under guard in another room, did not see him go. And this time he disappeared as he had never disappeared before. Bonner was not allowed to communicate with him by any exchange of notes. Prior to this forced hospitalization, the KGB had already begun efforts at disinformation, changing dates and tenses—"the snow is melting" becoming "the snow has melted"—on the postcards that Sakharov sent to the family in Newton and which he numbered as pedantically as he had when separated from Klava at the end of the war. Now, in the letters and telegrams she sent, Bonner never used the pronoun "we," so as to alert recipients that she and Andrei were not together. But as far as she could tell, from the foreign radio broadcasts she picked up in the cemetery, the world had no real idea of what was going on.

In late April, watching Gorbachev speak on television, Sakharov remarked to a fellow patient: "It looks like the country's gotten lucky this time, the new leader's an intelligent man."

On May 21, 1985, Andrei Sakharov turned sixty-four in Semashko Hospital in Gorky, surrounded by KGB—two in the next room, two in the corridor, one on the stairs, and one by the exit, not to mention the other patient in his double room. But he had a few advantages of his own: high confidence, resolve, experience—he knew what they could do to him, would do to him. Within the narrow compass allowed him, he could make certain modifications in his behavior, resist less, except for fits thrown at will. Still, it was hideous. Especially when they used the squad of women orderlies to tie him down, the clamp on his nose immediately cutting off his air supply.

Everyone knew this hunger strike could be the last—the KGB, Gorbachev, Sakharov himself, and of course Bonner. And the lack of even the slightest real news made the days a slow torment for her. From the moment Sakharov was taken away, she developed an "aversion" for food and, even though she made herself eat three times a day, her weight dropped from 147 pounds to 109. But she resisted despair by tending her flowers, putting their mass of books and magazines in some order, making shelves that she sawed and planed on the balcony, causing malicious neighbors to remark: "That's Sakharov's wife making a coffin for herself."

When she was summoned to Gorky party headquarters in early June, Bonner was afraid that Andrei had died. "I did not cry; I was in a stupor." When she did weep, it was for joy, able to tell by the KGB officer's demeanor that Andrei was still alive. What the officer wanted from her, using both wheedling and threats, was not to pass on any information about the hunger strike, which officially did not exist since Sakharov was eating, the fact that he was being force-fed beside the point. She wasn't buying.

"How many heart attacks will it take?" asked the KGB official with feigned solicitude.

"For what—for me to change? No amount will do that!"

Sakharov was alive, and so again was she.

* * * *

It was a rainy summer. She worked mostly in the house and on the balcony. It had been almost three months since she'd seen Andrei when on July 11 two doctors arrived to inform her that her husband was being discharged and would be brought home in two hours. Bonner was being informed in advance so that she could come out to meet him. None of this was normal hospital procedure, which made her suspicious, but then again there was nothing normal about the entire business.

Very thin, very calm, Sakharov emerged from the car. Unseen, a KGB cameraman filmed their reunion. Later the footage would be sold to the West in order to deceive the West, an ideal combination. The official line was that Sakharov, after hospitalization for cardiac arrhythmia, was back with his wife and would be receiving visits from his colleagues at FIAN as soon as his health permitted. Hidden cameras had also recorded Sakharov being fed at the hospital. Burdened with all the problems of the country and those of any new administration, Gorbachev did not want his opening year entangled with a hunger strike by the supremely inconvenient Academician. The new media-savvy administration's response was to produce a faux documentary with all the graniness of reality.

Sakharov explained to Bonner why he was home. He had written to Gorbachev on June 30 requesting she be allowed abroad and thought it tactically better to make the initial gesture of goodwill and break off the strike. If, however, he did not hear back in a "reasonable time"—two weeks—he would resume the strike.

That first night, dreaming he was back in the hospital, Sakharov cried in his sleep, and Bonner had to wake him twice. The next day was "dreary—gray, rainy, and windy"—and it was not until July 13 that they could take their radio to the cemetery, where he was encouraged to hear his name and cause still alive.

First and foremost, those two weeks would be "a time for living." After long breakfasts, Sakharov again able to enjoy warmed cottage cheese, they went hunting for mushrooms, or to the movies or the market. They had not allowed him outside the hospital for exercise, and he had not breathed fresh air for three months. One day Sakharov and Bonner had a light outdoor lunch of buns and fruit by the Volga. KGB cameras filmed their picnic.

Sakharov was discharged on July 11, and two weeks later on the twenty-fifth he still had received no reply. He came out to the balcony where Bonner was trying to snag an errant frequency on their radio.

Though they loved to talk to each other, they did not need many words for the most important things. Seeing that he was resolved to keep to his timetable, Bonner said: "I guess you're right."

He kissed and thanked her, having already begun. Later that day they drove into town so he could send a telegram to Gorbachev. When, on the twenty-seventh, eight men in white came to hospitalize him forcibly, the procedure had an almost ritual-like familiarity, a jocular here-we-go-again quality on the part of the draggers and only passive resistance on Sakharov's.

Food was left in his room, the sight so tormenting he had to cover the dish with a napkin. Sometimes he would thrash and battle against the force-feeding, sometimes just for a symbolic length of time, a sufficient expression of will and continued intent; then he would take a little food voluntarily. Hidden cameras filmed those moments. But no cameras were running when the doctors, for reasons unknown, would suddenly switch from clamp and spoon to subcutaneous and intravenous drips that caused his legs to swell so painfully he could not walk for a day or two. He suspected that psychotropic drugs were also being fed into his system by those same tubes, as if the situation itself were not maddening enough.

Bonner was alone again: "another string of empty days, fast and slow. Reading, mending things no one needed, washing walls, sometimes necessary and sometimes not, and fussing with the flowers. All despite my weakness, holding myself as tight as a fist. . . . In the evenings, like a pendulum, pacing about the balcony, I read poetry aloud, to keep from forgetting how to speak. And to answer the question: 'Who needs poetry and why?' "

Now her task was to convince the world not to believe its eyes and ears—and somehow communicate that Sakharov had been on a hunger strike since April except for the two-week interlude in July. Before, she had attempted to communicate their separation by using "I" when she would always have said "we." Now she tried absolute silence, not even sending her mother a telegram of congratulation on her eighty-fifth birthday.

Sakharov, however, was writing. On July 29 he wrote letters to Gorbachev and Gromyko, making a quite explicit offer: if his wife was allowed to travel abroad for medical care and to see her family, he would break off his hunger strike and "discontinue my public activities apart from exceptional circumstances." Sakharov was at ease with his conscience (and with later criticism) on this point, feeling that he had "earned the right" to a private life and some time for his science. It was a small concession, since he could not, in any case, speak out on any issue from Gorky and he was reserving the right to decide what constituted an exceptional circumstance. And no cause mattered more to him than Elena, who was in dire health and had not seen her children for almost seven years.

Her children were also proving media-savvy. On August 30 Alexei declared a hunger strike in front of the Soviet Embassy in Washington, galvanizing a congressional resolution of protest against the treatment of Sakharov; the State Department also promised to ratchet up the pressure. The campaign of falsification and denial was failing. The first summit between Gorbachev and Reagan would take place in November in Geneva, the outside date for resolving the Sakharov problem, which otherwise would be on the table.

It took not two weeks but a little more than a month for Gorbachev to respond. On September 5, 1985, a KGB official, "polite, almost deferential," informed Sakharov that Gorbachev had received his letter, but there were certain questions: What chance was there of Bonner staying with her family in America? None, said Sakharov. Would she agree not to hold press conferences? "Discuss that with her," Sakharov replied.

The KGB preferred that he discuss it with her and granted him a short reprieve for that purpose.

Arriving at the apartment, he saw the key in the front door—she had gotten tired of replacing the locks after each break-in—and entered without knocking.

"Don't be overjoyed; I've only got three hours," he said to her, explaining that that was the amount of time the KGB had given him to write up certain papers.

"The KGB can go fuck itself!" said Bonner.

"Just listen to me," said Sakharov, quieting her fury with his own calm.

When he was done explaining, Bonner was glad to draw up a statement: "In case I am allowed to travel abroad to see my mother, children, and grandchildren, and also for treatment, I will not hold press conferences or give interviews."

Typing her statement and his, Bonner wondered: "but what does it all mean?"

"I'm afraid to believe it," said Sakharov. "I won't let myself believe it, but maybe they've decided to let you go to America."

"I don't dare believe it either."

It could just be a probe to see what concessions they would make, more a measurement of Sakharov's resolve than a signal of any definite intent. Typically, once something had been decided at the top, events moved quickly. But days passed and nothing happened. On September 10 the U.S. Congress passed a resolution urging the president "to protest in the strongest possible terms and at the highest levels, the blatant and repeated violations of the Sakharovs' rights by the Soviet authorities." By late September autumn had come to Gorky, the leaves turning, the days getting colder and shorter, and still not a word.

On October 5, before meeting with President Mitterrand, Mikhail Gorbachev and his wife, Raisa, went shopping in Paris, eliciting catty remarks about her taste in clothes and her reported use of an American Express card. It snowed the next day, and Bonner took a handful into her mouth, making the traditional wish during the first snowfall. Later that day, feeling no more could be done, she sent Sakharov a postcard in the hospital with the agreed-upon signal to end the hunger strike, a line from Pushkin.

In keeping with some final calculations of their own, the KGB delayed delivery of the postcard for nearly two weeks. On October

21 Bonner was summoned to OVIR and informed that her visa had been granted. Assuming the visa would take the usual three months to process, Bonner indicated she would be willing to travel as soon as it was received. But the following day, to OVIR's irritation and astonishment, she refused their offer of an immediate visa—she would not be willing to leave until she had at least one month at home with her husband.

Sakharov appeared in the doorway of the apartment the next day at five in the afternoon, his thin, gray face dwarfed by his fur hat.

Not even kissing her, he asked what was going on.

"You don't know? I've been called in to OVIR."

His face was transformed, "in fact his face disappeared; all that was left was his eyes, alive and glowing. . . . And he wiggled his rear end, as if he were dancing; I'd never seen Andrusha make a movement like that."

"We won!"

* * * *

In the month before her departure, Bonner nursed Sakharov back to health and prepared him and the house for a long interval without her. Gradually he gained back half of the thirty-seven pounds he had lost over the months of the hunger strike. They closely followed the summit between Gorbachev and Reagan on November 21 in Geneva. As Dusko Doder observed in *Gorbachev: Heretic in the Kremlin,* the Soviet leader's advisers "suspected that Reagan was trying to have a Kennedy-Khrushchev summit in reverse, with Gorbachev in the role of the untested young adversary." There were few concrete results. Predictably, Gorbachev protested against "Star Wars," worried that keeping pace with U.S. plans for such an expensive antimissile defense system could doom any efforts at reform when the Soviet military was already consuming more than half the state budget. But there also were unpredictably long, private meetings between the two leaders.

On November 25 Bonner departed for Moscow, another separation whose meaning would be revealed only in the fullness of time. Two weeks after Bonner flew to Italy on December 2, Sakharov was visited by two colleagues from FIAN—one, Evgeny

Feinberg, at whose home Sakharov had first met Solzhenitsyn. Feinberg thought Sakharov looked thin, gray, aged. He had the flu and made his guests wear gauze face masks. After greeting them, he returned at once to bed, but he was so passionately interested in the latest developments in string theory that the initial conversation lasted four hours, until lunchtime, at which point Sakharov requested some cottage cheese be warmed up for him. But there were some nonscientific interludes, and they were rancorous: Sakharov was bitter that Feinberg and others had chosen not to pass on word about his hunger strike in a letter he had sent back with an earlier visitor. But his "disappointment and bitterness" passed, and on December 20, 1985, he sent Feinberg a postcard with New Year's greetings and the forgiving largesse of the victor: "All's well that ends well."

It was a solitary New Year's for Sakharov, though the occasion was brightened by the exchange of televised greetings between Gorbachev and Reagan—and he marked their wedding anniversary on January 7, 1986, with anxiety for Elena, who on the thirteenth would undergo a sextuple bypass, "miraculous and monstrous work," as she called the open-heart surgery, performed by Dr. Cary Atkins at Boston's Massachusetts General Hospital. Once again she experienced the unsettling sensation of returning to life as she had during the war, when she was dug out of a mound of dirt, as if that "had been nothing but a rehearsal" for the operation. Still without a telephone, Sakharov went to the post office on the fourteenth, where Tatiana called from Newton to inform him that the bypass on Elena had been a success.

Without her cause to champion, Sakharov dedicated himself to science. The previous year of hunger strikes and hospitalization had hardly been conducive to intellectual labor, and the conversations on string theory with the FIANers in December had both excited him and reminded him that during his exile in Gorky "high-energy physics did not stand still." Sakharov went to work so intensely on mastering "string theory, the theories associated with it, and the new theoretical developments on the frontier between cosmology and high-energy physics" that he even gave up listening to foreign broadcasts in the cemetery or at the racetrack, which also afforded good reception. The only breaks were a surprise visit from two FIANers at the end of January and drafting a letter to Gorbachev, which he mailed on February 19. The scientific exchanges with his colleagues had a political addendum; they had been briefed to ask him about his "intentions" if he "were allowed to return to Moscow." Specifically, would he work on magnetically controlled fusion reactors (MTRs) and continue to refrain from making public statements? MTRs were no longer of interest to him. "As for public statements," Sakharov replied, "the assurances were given with respect to my life in Gorky and in anticipation of my wife's trip. My return to Moscow would create an entirely different situation; it would bring new civic responsibilities, and I'd have to rethink a whole series of questions."

The letter to Gorbachev was occasioned by an interview the Soviet leader gave to the French communist magazine *L'Humanité* on February 8, 1986. Gorbachev denied the existence of political prisoners in the USSR: "we don't have any. . . . We don't try people for their opinions." His remarks on Sakharov were entirely standard: "It is common knowledge that he committed actions punishable by law. . . . Measures were taken with regard to him according to our legislation. The actual state of affairs is as follows. Sakharov resides in Gorky in normal conditions, is doing scientific work, and remains a member of the USSR Academy of Sciences. He is in normal health as far as I know. His wife has recently left the country for medical treatment abroad. As for Sakharov himself, he is still a bearer of secrets of special importance to the state and for this reason cannot go abroad."

Sakharov's letter to Gorbachev of February 19 opens with an expression of gratitude for allowing Bonner to travel, which he assumes did not occur without Gorbachev's "personal intervention." Then Sakharov immediately takes issue with Gorbachev, arguing on strictly legal grounds that his own exile and the sentences of other political prisoners contravene Soviet law. Sakharov's reasoning, close and detailed, is designed to appeal to Gorbachev both as someone who has studied law and as a leader projecting a reformist image. Calling for the "unconditional release of all prisoners of conscience," Sakharov cites fourteen cases with which he is personally acquainted, that of his friend Anatoly

Marchenko, now serving his sixth term, at the top of the list. Sakharov concludes the letter "with respect and with hope."

Gorbachev had now been in power almost a year, but it was still too soon to assess him. His policy of perestroika and glasnost was appealing but still vague. Perestroika—restructuring, an engineering metaphor—had the right Soviet ring, but exactly what would be restructured, and exactly how, was yet to be spelled out. Sakharov could only be heartened by a leader calling for glasnost—openness, disclosure, telling the truth; he'd been calling for it for years. But little truth was being told—about the economy, the past, the war in Afghanistan, or his own exile. Still, Sakharov, who was horrified by the damage caused by alcoholism in the USSR—the rate had almost quadrupled between 1964 and 1982—and what it signified about Soviet life, welcomed Gorbachev's crackdown on alcohol sales, though the effort smacked of Andropov-style discipline, treating symptoms, not causes. The estimated annual losses to the state, which had a monopoly on alcohol, would exceed the cost of prosecuting the war in Afghanistan. Gorbachev's initial attempts at economic reform—the speedup campaign—had increased production but at the cost of quality, to rectify which new bureaucratic layers of inspectors were created. Gorbachev's own thinking was undergoing a speedup, but political resistance within the party still prevented him from being fully explicit when he presided at the Twenty-seventh Party Congress, which opened on February 25, 1986, just as Sakharov's letter was arriving. Gorbachev had not yet achieved any breakthrough accomplishment or been tested by the sort of crisis that shows a leader's true colors.

For the next two months Sakharov immersed himself in science, string theory in particular: "In contrast to quantum field theory, which treats particles as points, the new theory's fundamental units are 'strings,' minute, one-dimensional structures, which can be 'open' (resembling a worm) or 'closed' (resembling a loop). They can change from one form to the other. And they can repel or combine with other strings. Strings can create fields. Space, on the other hand, is considered to lack inherent dynamic qualities and acquires them only through interaction with strings. In other words, string theory is the realization on a new level of my old concept of induced gravity. I can't help feeling proud of myself!"

Some scientists fervently hoped that string theory would yield a Theory of Everything, but Sakharov was skeptical: "I doubt that we will ever know all the laws governing nature." But he was even skeptical of his own skepticism; that we have never discovered all such laws cannot preclude their ever being known, since the past has already demonstrated that it is not an "infallible guide to the future." In the world of human affairs, that skepticism gave rise to a moral imperative: "life's causal connections appear so abstruse that pragmatic criteria are often useless; we must rely on our moral code."

But sometimes he failed to apply his skepticism with sufficient rigor, as was the case in late April 1986, when he first learned of the explosion at Chernobyl. His deeply distracting return to pure science (one of the blessings of exile), his ceasing to monitor foreign radio broadcasts, his strong belief in the peaceful application of nuclear energy had all caused him to engage in "wishful thinking"; he downplayed the significance of the explosion, calling it "an accident, not a catastrophe" in a phone conversation with Bonner more than two weeks later. The KGB recorded that conversation and put his remark into circulation. They were also secretly filming his conversations on the street with concerned citizens who, Sakharov failed to notice, had unimpeded access to him; usually anyone trying to approach him would be immediately hustled away. Bonner, still in the West, tried to warn him, but when she strayed from health and family, the line went dead.

In that respect nothing had changed. Nor had anything changed in the initial Soviet response to world concern about the explosion in the reactor. The conditioned reflex of silence and denial was still operative. The government was unconscionably slow to inform the three million people of Kiev, eighty miles to the south of Chernobyl, of the danger they were in. Upon being informed, many assumed they were doomed and, to their later regret, behaved accordingly. It was eighteen days after the explosion that Gorbachev finally addressed the nation on television.

Sakharov's sources of information did, however, improve when he was visited by two colleagues on May 21, 1986, his sixty-fifth birthday, a day that President Reagan had declared Sakharov Day. Taking a break from science, they strolled around the high bluff overlooking the river, where they were of course tailed but whose open expanses inhibited the KGB from close physical approach. Sakharov now gained a better sense of Chernobyl's proportions. Still, his colleagues found Sakharov ebullient about the "coming of deeper changes in the country."

A sort of inadvertent technological glasnost, Chernobyl blew a hole in the iron curtain. The need for international disclosure allowed Gorbachev to press harder for the free flow of information. But his forays in restructuring were being thwarted by the vested interests of the bureaucracy. In the most acute contradiction of them all, the true enemy proved not capitalism but the party.

Gorbachev was above all else a patriot of socialism. His mission was to save both the Soviet state and socialism. Lenin had created a socialist revolution and a socialist state that was subverted and perverted by Stalin. Just because that deviation was enormous did not mean in the least that it should not be corrected. Knowing that "we can't go on living like this," Gorbachev was driven by the belief that what they did now would "determine the fate of socialism."

On June 4 Elena returned to Gorky, a reverse Eurydice, with days and days of stories to tell. She also let Sakharov know in no uncertain terms that he had been duped by the KGB about Chernobyl. Then she told him of her meetings with Thatcher and Mitterrand, with American politicians—those who cared and those who only used his cause—she told him of New York, Newton, Disney World, Paris, of family and friends, and of the book she had written in America, *Alone Together,* the story of their exile and of her travels in the West, which she had ended with the lines from Samoilov's poem about Pushkin, "Thank God you are free, in Russia, / at your estate in Boldino, in quarantine," that had for a moment united Sakharov and the camp guard. She told him of the "fog" and "darkness" that settled over her soul at the Soviet border. And he who, apart from the Theory of Everything, had little to tell was more than glad to listen.

They resumed their life, their peculiar happiness. But there was always something 1984ish about Gorky, and at times Sakharov "felt like a mouse in a glass jar from which the air is gradually pumped." Both he and Bonner still thought it likely they would live out their lives in Gorky and had already picked out their spots in the local cemetery. Change was in the air, but mostly there—as Sinyavsky remarked, it was easier to allow the publication of *Doctor Zhivago* than to produce salami. The Soviet Union was still waging war in Afghanistan, and Sakharov knew that he and that war were inextricably linked. For the regime to end Sakharov's exile would be tantamount to admitting the war was at best a mistake.

The amnesty for all prisoners of conscience that Sakharov had called for in his February letter to Gorbachev had to include himself as well. He had a quite precise, almost third-person sense of his own symbolic significance, his continued exile or his free return to Moscow a standard "by which the entire human rights situation in the USSR could be measured." If Gorbachev was intelligent, as he seemed, he would realize that he could never enlist the intelligentsia as long as Andrei Sakharov was exiled to Gorky. In late October he wrote a letter to that effect to Gorbachev, concluding it with true feeling and fine distinction: "I hope that you will find it possible to end my isolation and my wife's exile." Though marginally hoping the letter might prove the "imperceptible tremor" that touches off the avalanche, Sakharov put it out of mind as soon as he mailed it, on the principle that he took a long time writing and they took a long time replying.

He had plenty else to do, other letters to write. In the wake of Chernobyl he was rethinking the peaceful, safe application of nuclear energy. Reactors, he decided, should be built underground "so that even a worst-case accident would not discharge radioactive substances into the atmosphere. . . . Existing aboveground reactors should be protected by reliable containment structures." He also gave some thought to earthquake prevention: "burying thermonuclear charges deep underground in seismologically active areas and detonating them when strains in the earth's core approach the critical level." He sent a letter to the new president of the Academy, outlining both ideas, inviting discussion.

November passed without any response to either letter. Had they been read, had they even been received?

* * * *

They settled in for winter. It was already December, the darkness coming earlier each day. The policeman at the door would doze or yack with colleagues, the voices reverberating in the walls of the apartment. They were all hostile and aggressive with one exception, Volodya, who on occasion, a finger to his lips, would give them a little help, move something heavy.

Their radio was still jammed, their apartment still bugged. Their conversations were recorded, transcribed, relayed to Moscow, analyzed, filtered, summarized for the higher-ups. Their mail was still heavily censored, still delivered by the same woman, so good-natured that Bonner would always cut her a thick slice from the meat brought by Andrei's colleagues. After nearly seven years they still had no telephone, and the nearest pay phones were kept out of order. One could, however, assume that the state was not sparing the expense of live monitoring and would react in the event of a medical emergency.

On the evening of December 9 they were listening to their shortwave with earphones to elude direct jamming. Even so, the reception was atrocious, mostly meaningless crackle. But they did catch the name of their friend Anatoly Marchenko, serving his sixth term, on a hunger strike since August. Aware that his wife had recently been instructed to apply for an exit visa, they rejoiced in the belief that he had been released. But, straining their attention on the squall of static, they gradually came to understand that such was not the case at all. Marchenko had died at forty-eight in Chistopol Prison. His wife was not allowed to take the body. After the funeral she made a wooden cross for his grave, writing his name and dates in ballpoint.

Sakharov and Bonner went straight from hope to grief. The next day, the tenth, was Human Rights Day; Bonner lit candles in their windows, for all prisoners of conscience and for the soul of Marchenko. Sakharov had lost a brother in spirit whose life, he said, had been "astonishing, tragic, and yet happy."

At the time Marchenko was dying in circumstances that remain unclear, Mikhail Sergeyevich Gorbachev was addressing the Politburo in the matter of A. D. Sakharov: "The man appears to have a good head and seems to use it for the good of the country." The letter had indeed arrived.

On the evening of December 15 they were watching television. Glasnost may not have yet saved the Soviet Union, but it had already saved Soviet television, Bulgarian folk dances now replaced by serious discussion, KGB thrillers by investigative reporting. Bonner was sewing, Sakharov was thinking of his father, who had died twenty-five years ago to the day. What would he have said about all those turns of events? When Dmitri died, Andrei was still the thermonuclear hero, well before "Reflections," dismissal, Elena, the Nobel, Gorky. Would all that have sufficed to assuage his final disappointments?

It was after ten when the doorbell rang. It could only be approved visitors, but none ever came that late. With some misgivings, Sakharov let them in—two technicians and a KGB agent. The technicians went right to work installing a phone. All the KGB agent said was: "You'll get a call around ten tomorrow morning."

In that season of quickening glasnost, Sakharov had already received two requests for interviews, which he refused, not wanting to speak publicly as long as he had a "noose" around his neck. The installation of the phone probably indicated that the next request for an interview had the backing of high officials.

No one called at ten the next day. And the telephone remained silent during lunch. In Russia any foul-up was possible, any delay. By three o'clock Sakharov was tired of waiting and was just about to go out for bread when the phone rang.

A woman's voice said: "Mikhail Sergeyevich will speak with you."

Waiting for him to come on the line, Sakharov told Bonner: "It's Gorbachev." She went to the door where policemen were talking loudly and silenced them with the words: "Quiet, Gorbachev's on the phone."

"I received your letter," said Gorbachev. "We've reviewed it and discussed it. . . . You can return to Moscow."

Sakharov did not like the tone of voice Gorbachev used when

saying that Bonner had been pardoned, seeming to contemptuously mispronounce her last name, and interrupted Gorbachev with a reproach.

"You can return to Moscow together," continued Gorbachev. "Go back to your patriotic work."

"Thank you," said Sakharov, then voiced his second objection. "But I must tell you that a few days ago, my friend Marchenko was killed in prison. He was the first person I mentioned in my letter to you, requesting the release of prisoners of conscience—people prosecuted for their beliefs."

"We've released many, and improved the situation of others," countered Gorbachev. "But there are all sorts of people on your list."

"Everyone sentenced under those articles has been sentenced illegally, unjustly. They ought to be freed."

"I don't agree with you," retorted Gorbachev.

Sakharov was polite, but political: "I urge you to look one more time at the question of releasing persons convicted for their beliefs. It's a matter of justice. It's vitally important for our country, for international trust, for peace, and for you and the success of your program."

Gorbachev was noncommittal and, in violation of protocol, it was Sakharov who ended the conversation: "Thank you again. Good-bye."

Some two thousand five hundred days of exile were now, suddenly, over. But, grieving for Marchenko, they "had no sense of joy or victory." And Sakharov was troubled by his promise not to speak out on issues except in "exceptional cases." Had he demeaned himself? Limited his own free voice? Some of the other terms and conditions weren't entirely clear either—no mention had been made of returning his state awards.

Sakharov's incipient return to Moscow was made public and official at a December 19 televised press conference at the Ministry of Foreign Affairs. Sakharov was amused and disgusted by the formulation: "Academician Andrei Sakharov, currently living in Gorky, addressed the Soviet leadership with a request to move to Moscow. . . ." As if it were all an issue of residence.

The weather was too cold for Bonner to travel that weekend,

but Monday, December 22, was somewhat milder, and they took the night train to Moscow. Arriving in the early morning of the twenty-third, wearing a heavy jacket and fur hat for the weather but dressed for triumphal return in a suit and striped tie, Sakharov stepped out onto the platform and into yet another new life.

CHAPTER FOURTEEN

ASTONISHING TIMES

AND into a new Russia. Palpably freer, as proved by the platform packed with journalists, but disorienting, with only seven hours of night train between isolation and Moscow, the world, fame.

It was a positive sign that so many reporters were there, and not just foreign correspondents but Soviet ones as well, from all over the country. In Gorky he had been secretly filmed; in Moscow the ceaseless flashbulbs made an acrid wash of white light in the morning gloom. From every side, mikes, recorders, questions: Afghanistan? Star Wars? Gorbachev?

Against the first two and for the third, but public politeness and intellectual responsibility required some details and reasoning be given. And why rush savoring triumphal return? It took him forty minutes to bisect the throng of reporters, itself an accurate measure of the density of glasnost.

Yet it was also clear that the reporters were there for the same reason that he was—they'd been allowed to come. He was a move in Gorbachev's game, a bishop brought in at a sudden diagonal. But as Sakharov had already demonstrated in his phone conversation with Gorbachev, he was a master at the quick countermove.

He and Elena went home to the apartment on Chkalov Street, which also turned out to have been sealed in isolation—a few years of weather had blown in broken windows. It would take time to put the place in order; it would take time to retrieve all their possessions from Gorky. But the immediate question was what to do with that first day at liberty? It was a Tuesday, which, as Sakharov had recently been reminded, was the day that FIAN's theoretical seminar met at three o'clock. Of course he'd attend.

He was welcomed with love and ovation. But some of his colleagues could not help wincing a bit at "his exhausted face, his sunken affectionate eyes, his stooping shoulders, and his emaciated body." The atmosphere at the seminar was, however, as a participant put it, "one of high excitement and joyful celebration." By brilliant coincidence, the paper scheduled that day was on the baryon asymmetry of the universe, a subject that Sakharov had pioneered. The report began: "As Andrei Dmitrievich has demonstrated. . . ."

FIAN was his second home, and it was good to be back there too, not least because the moral atmosphere was so comfortable. No FIANer had ever come out against him; there was nothing to forgive. He could attend a seminar without much ado, the normal and the miraculous now one and the same. And afterward he could go back to his office, which had been kept for him, the paper and ink of the name tag on the door faded in the seven years of his exile, though A. D. Sakharov was still clearly legible. And there, to colleagues avid for every detail, he recounted the phone call from Gorbachev, the exact words that had freed him from Gorky and returned him to FIAN, his office, his desk, which had once been Tamm's.

After the roughly two thousand five hundred days of exile, in which he had been permitted seventeen one-day visits from colleagues, it was such a pleasure to be amidst scientists that Sakharov spent six hours of his first day of liberty at FIAN. But it quickly became apparent that he'd do little science in his new life, a "madhouse" from the very start. The demands for interviews and intercession were incessant, escalating. "On January 1, 1987, while normal people were resting up after New Year's celebrations," Sakharov was "slaving" over a written interview for the influential newspaper *The Literary Gazette,* wanting his "debut in the Soviet press to be as cogent as possible." But he quickly saw that glasnost was a labyrinth, some of whose corridors led to blank walls. In the end the interview would not be printed, his views on Afghanistan, prisoners of conscience, Star Wars, disarmament, still too sharp for the sanction of high-level publication.

The decision to release him had been a shrewd symbolic move but would be hollow if not followed by the release of all political prisoners. Glasnost and the existence of a single prisoner of conscience were absolutely irreconcilable concepts, as Sakharov wrote to Gorbachev in mid-January. And amnesty, he reminded him, should not be delegated to the agencies that had committed the original crime of arrest. There was no direct response to the letter, but a mean-spirited amnesty soon went into effect, freeing some prisoners of conscience, but only if they signed a humiliating promise not to engage in further "illegal" activity. Sakharov was a realist on the subject—the system was changing for the better faster than anyone could have expected, but the forces of change still encountered obstacles, as proved by his blocked interview, this grudging amnesty. Still, better to accept the level of change as it was, not as might be hoped, and to use each new development as leverage for the next. That position cost him politically in his own camp. Larisa Bogoraz, who, pregnant, had protested Czechoslovakia in 1968 and was now the widow of Anatoly Marchenko, was among those who now accused Sakharov of "having advocated shameful concessions, of having urged capitulations that could scar prisoners for the rest of their lives." A Russian-language newspaper published in New York put it more crudely, to the effect that "The Pardoned Slave Helps His Master."

The apartment on Chkalov Street became a sort of embassy, and in fact most Western ambassadors and high officials paid their respects. On February 5, 1987, Sakharov and Bonner hosted a ten-person U.S. Council on Foreign Relations delegation, including Henry Kissinger, Jeane Kirkpatrick, and Cyrus Vance. Kirkpatrick seemed an "intelligent and tough-minded woman" and took notes when Sakharov spoke about prisoners of conscience and emigration. But Kissinger posed a hard question: "Is there a danger that the USSR will first effect a democratic transformation, accelerating its scientific and technological progress and improving its economy, and then revert to expansionist policies and pose an even greater threat to peace?" Allowing it to be a fair question, Sakharov argued that an "open, stable" Soviet Union was in the interests of America and should be supported but with "eyes wide open." The evening ended with coffee and sweet nostalgia, Bonner's cheesecake reminding Kissinger of the kind his mother made when he was a boy.

Nine days later Sakharov received other distinguished visitors,

Jeremy Stone and Frank von Hippel from the Federation of American Scientists who, like him, were about to attend a conference, named with grandiose unobjectionableness Forum for a Nuclear-Free World and the Survival of Mankind. In their exchange of views, von Hippel found Sakharov both admirable—"how could such a frail body house such an indomitable spirit?"—and irritatingly certain, oracular.

The two Americans had also brought an invitation from Senator Edward Kennedy to visit the United States. Sakharov explained that he could accept no such invitations as long as the official Soviet position was that he possessed state secrets, a situation that could change only because of political pressure from abroad or "if this is needed by the authorities for political considerations." He himself would devote no time to his own situation, there being too many "other important tasks." Chebrikov, the head of the KGB, who was still monitoring Sakharov's Moscow apartment, reported that portion of the conversation verbatim directly to Gorbachev.

The Forum for a Nuclear-Free World and the Survival of Mankind was of course "staged primarily for propaganda purposes," but Sakharov also knew that a half-baked loaf was better than none: "after many years of isolation, this was my first public appearance, and my first opportunity to present my views before a large audience."

Glasnost had now rendered human rights and convergence both possible and imperative for its own success. The West must transcend suspicion and assist the process. His message was clear and urgent, though his argument was based, as time would show, on one shaky assumption: "On the other hand, if the West tries to use the arms race to exhaust the USSR, the future will be extremely gloomy. A cornered opponent is always dangerous. There is no chance that the arms race can exhaust Soviet material and intellectual resources and that the Soviet Union will collapse politically and economically, all historical experience indicates the opposite."

On the second day of the conference Sakharov spoke again, this time on Star Wars, officially the Strategic Defense Initiative (SDI)—that "Maginot line in space." Having always considered SDI cheaply circumventible and highly destabilizing, he was now against coupling it to arms reduction talks, that is, against the official Soviet position of the "package principle." Von Hippel may have been irritated by the tone Sakharov sometimes took but still saw this talk as probably "the most important event at the forum."

That view was shared by Anatoly Dobrynin, for twenty-four years the well-respected ambassador to the United States, who entered the auditorium when Sakharov began speaking and left as soon as Sakharov was done. Dobrynin reported Sakharov's position personally and immediately to Gorbachev, who was glad to learn that he now had Sakharov's scientific and moral sanction to make a political move that would allow him to recapture the initiative in arms negotiations.

Gorbachev spoke on the third day of the forum at a banquet for 1,500 in the Kremlin's Palace of Congresses. Soviet television featured shots of Sakharov applauding Gorbachev's call for the abolition of nuclear weapons by the year 2000. But the area of the banquet hall where Gorbachev and Raisa dined with Stone and von Hippel was cordoned off by bodyguards. The placement of Sakharov's table did not allow him access. Gorbachev did not want them linked any more than they were already, and it would be a year before they met and spoke. The lack of access to Gorbachev aggrieved Sakharov, who was carrying a list of additional prisoners of conscience to be liberated. He did finally manage to press it on the American industrialist Armand Hammer, tracking him down after their conversation had been interrupted by the ballerina Maya Plisetskaya, who had simply whisked the billionaire away. Sakharov had chosen an odd messenger, the only American businessman who had dealt with every Soviet leader since Lenin. Hammer, who had been excited by Sakharov's rejection of the package principle as the possible basis for a summit, accepted the list with considerably less enthusiasm.

But Sakharov would soon prove effective in both endeavors. Within two weeks the Soviet government adopted Sakharov's position—SDI was decoupled from arms negotiations—and by April nearly all the political prisoners in the country had been released. A letter that he wrote at that time to Foreign Minister Eduard Shevardnadze concerning a Georgian prisoner produced immediate action, a call from the ministry informing him that the man had "been pardoned and is at liberty." If this was being exploited, the more the better.

It was more than exploitation; it was vindication for the humiliations he had suffered from Khrushchev, from Slavsky, from Marshal Nedelin's obscene joke at the banquet celebrating the successful test of the Third Idea; as he would write in his *Memoirs,* the "emotions kindled at that moment have not diminished to this day." At last, his advice, which he had been offering his government for decades, was being sought and heeded. In fact his positions and Gorbachev's were achieving a convergence of their own. But their goals and motivations differed. Sakharov's idea of convergence would free the country from the worst of socialism and increase stability in a nuclear world, the one goal that, unachieved, made all others potentially meaningless. Gorbachev wanted to salvage the best of socialist ideas from the worst of Stalinist practice. Still, Sakharov could not have wished for a better Soviet leader to play against. Gorbachev had electrified the atmosphere with a qualitatively new sense of the possible. Before, one acted because it would be shameful not to; now because it could actually do some good. That first spring after exile had a scent of Prague.

* * * *

Sakharov, who for seven years had seen no one, was seeing everyone. From Margaret Thatcher to Daniel Ellsberg, from the French prime minister to the German Greens, from establishment to opposition. Lunching with the prime minister of England, he urged Thatcher to support perestroika while "maintaining a firm line on human rights." After a meeting with the French prime minister he reminded the world that, though perestroika was changing the Soviet Union, the country remained at war with Afghanistan. Atrocities were occurring on both sides, itself a good reason to stop the war, which in any case was not going well for the Soviets, the Afghanis having brought down their first Soviet helicopters with American Stinger missiles during Sakharov's last autumn in Gorky.

The pace accelerated in May. He met with French Prime Minister Jacques Chirac at the Academy of Sciences (which, unlike FIAN, did have something to repent) and with a delegation of

Crimean Tartars, once again taking up their cause. In a letter he reminded Gorbachev how far they had to go to achieve justice for the Tartars exiled en masse as a traitor nation in 1944, forty-three years ago. Justice demanded that they all be allowed to return to the Crimea under the most favorable conditions and granted an autonomous region in the steppe lands; residence there should be purely voluntary. Reality demanded the Crimean coastline remain all-Union, national. Granting the Crimean coastline to the Tartars would only magnify the resistance of the bureaucrats; it was one thing to allow citizens a few liberties, quite another to deprive the power elite of their privileges. Once the only choice had been to be uncompromising or be compromised, but real politics was nothing but compromise. Yet the miracle was that even this semblance of real politics existed. Sakharov thrived.

At the end of May, having turned sixty-six, he attended the Fourth Moscow Seminar on Quantum Gravity, an international event that allowed him to renew his acquaintance with John Wheeler, whom he had last seen nearly twenty years before in Tbilisi, memorable hours of feasting, science, and Zeldovich's enforced naps. Speaking of his latest work, the point at which physics blends into philosophy, Wheeler was as impressed with Sakharov's openness to ideas as he had been at their first meeting. "Never before have I met anyone so senior who communicated more strongly the aura of a humble searcher for truth, one wanting to learn about the great mysteries—learn from nature, learn from the scientific literature, learn from discussion."

Sakharov too was afforded an opportunity for admiration at the seminar. He met Stephen Hawking, paralyzed by Lou Gehrig's disease, able to communicate only with a computer-driven voice synthesizer: "His morale is amazing: he has retained his good nature, his sense of humor and his thirst for knowledge." Sakharov was particularly delighted by Hawking's witty overturn of Einstein's famous phrase that God does not play dice with the universe: "God not only plays dice, but he throws them so far they're beyond our reach."

He and Hawking connected, launching into a discussion of a lovely and arcane subject—the conditions under which the arrow of time would reverse its course. During their exchange Sakharov experienced what could be called scientific emotions—he was "glad" that Hawking had rejected the "erroneous assumption that the arrow reverses at the moment of maximum expansion of the universe and *maximum* entropy." Though Sakharov did mention this could of course occur only with *minimum* entropy, he was "too shy "to "bring up the simplest example, a closed universe in a state of false vacuum with positive energy and zero entropy." But after their exchange Sakharov was left with a different sort of emotion: "Hawking's face and eyes haunted me for a long time."

June was a month of happiness, honor, and hostility. In another sign of favor, Ruth, now almost eighty-seven, was allowed to return home to Moscow and came accompanied by her granddaughter Tatiana and two great-grandchildren, Anya and Matvei, adored by Sakharov and Bonner. A birdlike chain-smoker with a racking cough and the withering black humor of the Gulag, Ruth was at first apprehensive about readapting to Russia after her most recent and strangest exile to a place where people went where they wanted and said what they thought. But soon enough Ruth was content just to be with family, which amazed Bonner: "Could I have ever imagined that my mother, a Party member, antibourgeois and maximalist, who never allowed herself to use a tender word to . . . me, would be mending tablecloths, sewing dresses for me, dressing up Tanya, could turn into a 'crazy' grandmother and great-grandmother, for whom her grandchildren and great-grandchildren would be the chief 'light in the window.' "

And there was other happiness as well: the accursed "apartment question" was finally solved. Sakharov was allotted the apartment right under Ruth's, creating a sort of duplex family commune.

The honors came from the French, in triplicate. Sakharov would be awarded diplomas of membership from the French Academy of Sciences and the Academy of Moral and Political Sciences, as well as the medal of the Institute of France. The new head of the Soviet Academy of Sciences had rejected the French suggestion of a presentation in Paris but had agreed to an international seminar in Sakharov's honor at FIAN, only to renege at the last moment, tarnishing the ceremony at the French Embassy. Accepting his honors, Sakharov criticized the Soviet Academy for its past behavior, hoping that those who had signed statements against him would at least disavow them now, and for its current lapse, going back on its word, suddenly finding that a seminar would set an "undesirable precedent." But it was clearly not the Academy alone that had decided he had "too many honors" and needed to know his "bounds." Sakharov and Bonner's poor car was once again the instrument and victim of that message. During the ceremony the wipers were stolen, and the rear window was smashed. Now he understood the phone call of a few days before in which a "confidant" revealed to Sakharov that in the early eighties a KGB plan to "liquidate" Bonner had been nixed only at the highest level, the Politburo. Sakharov understood the "confidence" for what it was, a veiled threat.

Animosity toward Elena had always angered or offended him, depending on whether it came from the KGB or from colleagues. He was glad to admit that her influence on him had been "enormous." Only she had melted his infinite reserve. Attacks by the KGB and those in his own camp were only to be expected. And there would be a new sort now that he was a public person. As Sakharov noted in his diaries, Pasternak said it best: "I come out on stage. . . . / The darkness of the night is aimed at me / along the sights of a thousand opera glasses."

* * * *

After those first hectic six months coming on seven years of exile, hideous memories of which continued to seize upon him, Sakharov needed rest and peace, a little voluntary isolation. He spent the summer with Elena and Ruth in the Estonian countryside, pleased and intrigued by the tidy, prosperous farms. The locals ascribed success to their own virtues, but Sakharov knew it was also historical—the Baltic states of Lithuania, Latvia, and Estonia had not been run over by the "steamroller of socialism" for as long as Russia had. They still had their roots in a work ethic, which had even been noted by Romans, causing a later Estonian prime minister to quip that they were Protestants even before Christianity. Like Poland, the Baltic states were showcase nations, allowed to retain certain liberties, and like Poland they would, given half a chance, opt for independence. And those countries were all linked by secret crimes of Stalin that had yet to be aired and acknowledged—Stalin got all three Baltic countries as part of his secret deal with Hitler, which also gave him half of Poland, the rest taken after the war, the Polish officer class having been executed in the meantime. But it was time to put all that aside and slip out of history and into summer.

It all came back hard with September: Afghanistan, Star Wars, forums, chairmanships. A trip to Vilnius in October did, however, provide a certain clear standard of measure. It had been there in December 1975 that, during breaks in Kovalyov's trial, he heard Elena read his Nobel speech on the shortwave. Now, in 1987, Elena was there with him, Kovalyov was free and active, and Sakharov was not attending a trial but a disarmament conference, at which the chairman of the U.S. delegation, Wolfgang Panofsky, proposed that "all work on advanced technologies that could be used to create new weapons (for instance, the development of sophisticated lasers) should be open," an idea Sakharov liked and adopted. It was also in that summer and fall that interviews with him began appearing in print, though the first one could hardly have been more tentative—Sakharov was interviewed by *Theater* magazine about a new play based on the novel *Heart of a Dog* by Mikhail Bulgakov. It was a sign of small but increased favor. The government still held the carrot, though at times the stick seemed out of its control.

All the other socialist countries that had attempted transformation had been crushed by the Soviet Union; now it was the Soviet Union's turn to transform or perish. It would be a dereliction of duty not to devote every waking hour to fight for that almost inconceivable notion, a free Russia. Which meant that Sakharov did not see his own children much. Bad feelings still rankled from Gorky—he still felt betrayed; they still felt slighted and abandoned. Even Klava's sister had written him in Gorky to say global is good, but think of your own. With his daughters it was melancholy estrangement; with his son it was fear, shame, and disappointment—Dima had dropped out in his second year of studying physics at Moscow University, then dropped out again after a semester of medical school, married, fathered a son, divorced. He couldn't hold a job and was seeking the traditional consolation of vodka.

And there were bitter residues in Sakharov's relations with the Academy. Not a single one of his fellow Academicians had stood up for him; worse, many had joined the attack. From the hacks he expected nothing, but his "one friend," Zeldovich, could have done something, if not for friendship's sake, then for that of justice or even his own self-respect. But there was no escaping the fact that Zeldovich was a coward. Still, when he came dashing over at an Academy meeting and said, as always, on the run: "A lot has happened in the past. Let's forget the bad: Life goes on!" Sakharov was willing to entertain those terms.

Sakharov, who impressed those who worked with him with his exact sense of his own time and energy, made two wrong choices in the fall of 1987, though the time and energy were inefficiently spent for different reasons. The chairmanship of the Academy's Commission on Cosmomicrophysics did not prove purely honorary as promised; the discussion of projects such as an international space observatory, while interesting and promising, was simply not the best use of his time. He also accepted a directorship position in the International Foundation for the Survival and Development of Humanity, which proved a more complicated and "sadder affair." The organization spent more time on its own survival and development than humanity's and, because of Western interest in the subject, the foundation stressed human rights, skirting military and nuclear issues. Sakharov, as the chairman of the Human Rights Committee, a title he never quite remembered acquiring, felt in a "false position." With the release of political prisoners in April, human rights in the "classical' sense was no longer the immediate issue. Now they could get on to the overwhelming real business at hand, those issues "about which we never even dared to think . . . a constitutional restructuring of the country . . . relations among the nationalities, radical economic reform, a multiparty system; ecological problems. . . . poverty, health, education." He had his doubts if the foundation was the best venue to advance those issues, but in any case he would stay aboard until their meeting with Gorbachev in January of the coming year.

December was a month of mourning. Zeldovich died suddenly of a heart attack on the second. Sakharov spoke at the funeral: "We had our periods of friendship and our periods of alienation.

We had what we had." He signed the obituary in *Pravda,* then wrote one of his own for *Nature,* and in his memoirs, which he was still revising, he would add his final word, in which science is transcendent: "The petty and superficial aspects of our relationship have faded; what remains is his enduring, truly immense contribution to science. And all those whom he helped enter the realm of science."

It was a year since Gorbachev had called him on December 15, the anniversary of his father's death. If he was correct that he'd live as long as his father, he now had something like six years.

Ruth had only days. Her health had taken such a sharp turn for the worse on the twenty-third that she could barely puff on a cigarette. They all agreed against hospitalization; she'd stay with them to the end. Calm and peaceful, Ruth took to her bed with an enigmatic smile, making Bonner wonder: "What was Mama seeing, what was she feeling? Whom was she greeting with this radiant, light smile?"

A gentle snow fell on the green fir trees at the cemetery. Tatiana read a poem, "Blessed is he who visited this world in its fateful hours." Addressing Ruth, Andrei was moved to make the simplest statement: "I loved you and you loved me."

* * * *

Sakharov and Gorbachev finally met on January 15, 1988, at a conference held by the International Foundation for the Survival and Development of Humanity, whose "grandiloquent" name had yet to be matched by acts. After shaking hands with the Soviet leader, Sakharov once again expressed his gratitude for Gorbachev's "intervention in the fate" of his wife and his own, adding: "I received freedom, but simultaneously I feel a heightened responsibility. Freedom and responsibility are indivisible."

"I'm very happy to hear you connect those two words," replied Gorbachev.

There was a subtext to the exchange, of which each was aware. Sakharov was alluding to his and Bonner's still unsatisfactory status: she had been pardoned but not cleared; he had been released, but his state awards had yet to be returned. Gorbachev was calling for realistic restraint from Sakharov, a theme he developed later when he spoke at the meeting, devoting, as Sakharov would write, "most of his presentation to a covert and occasionally overt debate with me and other proponents of more radical changes. Gorbachev stressed the danger of rushing things, of skipping necessary intermediate stages." Sakharov's initial impression of Gorbachev was quite favorable, finding him "intelligent, self-possessed, and quick-witted in discussion" and sensing his attitude toward him as "respectful, even sympathetic." But both were aware that their relationship had yet to be tested.

Among the many things no one foresaw was that the worst was also yearning to break free and, intertwined with the good, the one came up with the other. Armenia was the first case in point. In 1923 then Commissar of Nationalities Stalin ceded a large piece of purely Armenian territory, Nagorno-Karabakh, to the neighboring Soviet republic of Azerbaijan. The Armenians had endured a holocaust by the Turks only eight years before and greatly desired the protection of the Soviet Union. For that reason Karabakh was not mentioned until the Thaw but, as with the Crimean Tartars, nothing had been done, and in the interim the claims receded.

On February 18, 1988, Armenians took to the streets with the slogan One People, One Republic! And then two days later a very strange event occurred: the Karabakh soviet, the local government, voted to support annexation. As Hélène Carrère d'Encausse wrote in *The End of the Soviet Empire:* "This was a major first in the Soviet Union: deputies elected by undemocratic balloting suddenly discovered themselves representative of the poplar will and no longer the mouthpiece of the central government. That day, modern democratic politics was born in the USSR as suffrage suddenly acquired meaning." A week later, in an industrial suburb of Baku, the capital of Azerbaijan, Armenians were slaughtered in a pogrom that lasted two days and two nights. Like the pogroms of the past, this one was a murky mix of the orchestrated and the spontaneous. Aside from the bloodshed, what shocked Sakharov most was the "vacillating and unprincipled" conduct of the Soviet leadership. There was nothing the least unconstitutional about an official request for annexation and, since there was also ample historical justification, it should, Sakharov thought, be taken under advisement.

This meant yet another cause, another devourer of hours, but at least it was one dear to him because of Elena's heritage. Sakharov now came under attack for favoring her people as he had favored her children. Bonner herself said that this only proved she was the perfect mate for Sakharov, his Jew wife when need be or, like now, his Armenian one.

One emergency interrupted the other in early March. Still working on his memoirs, though at least the difficulties were luxurious compared to those of Gorky, Sakharov had invited his American translator for dinner and a discussion of fine points, but both had to be postponed; the leader of the Crimean Tartars had flown in from Central Asia to see Sakharov, fearing a "bloodbath."

As the translator described the scene:

> The word "bloodbath" pains Sakharov, who flinches and sighs: "You know what Pushkin said in *The Captain's Daughter:* God save us from a Russian revolt, merciless, senseless." Quite tall, Sakharov has something of the English country parson about him, except for the Mongol slant of his eyes, which are both shy and fearless.
>
> The leader of the Crimean Tartars, Rashid Dzhemilev, has steel-streaked black hair brushed straight back, a chevron moustache; he wears a double-breasted suit and walks with a war-wound limp in the light, airy front room where Sakharov and Bonner receive official visitors. He is accompanied by his associate, a burly, copper-skinned man with the distracted attention of a bodyguard.
>
> Having seen them to their chairs, Sakharov sits down where he has a pen and note pad ready. He wraps one ankle around the other, then wraps both legs around the leg of his chair. After placing one elbow on the back of his chair and the other elbow on the table, he seems to be leaning backwards and forwards all at the same time.
>
> But there is nothing for Sakharov to note down because Dzhemilev is passionately recounting the injustice done to his entire people. Sakharov could not be more well aware that Stalin exiled that entire people from the Crimea to Central Asia for alleged collaboration with the Nazis.

Sakharov listens with near perfect patience. Still, he can not help but touch the ballpoint pen to paper. Doodling, one geometry gyroscopes out of the other, the science that he sacrificed but would still rather do.

"We have glasnost now," says Elena Bonner with a mischievous smile, "and so I feel free to ask you just what is it you want of Andrei Dmitrievich?"

"We want him to use his authority. . . ."

"What authority?" interrupts Sakharov with an irked sigh. "I can write Gorbachev a letter supporting your demands? Do you have a list of demands?"

"We have five demands," says Dzhemilev.

"Good," says Sakharov, glad to be down to business. "What are they?"

"The demands are," says Dzhemilev, waiting with satisfaction as Sakharov takes a fresh sheet of paper to list the points, "return of the Crimean Tartars to their homeland; restoration of the Crimean republic; state financial assistance in the relocation process; and the release of Crimean Tartars imprisoned for human rights activities."

"That's four," says Sakharov. "What's the fifth?"

"The fifth?" says Dzhemilev, looking away to review what he has just said. "I can't remember the fifth one, can you?" he says to his associate, who scowls as he thinks, shaking his head.

A faint beatitude of humor plays at the corners of Sakharov's lips. Then even the Tartars laugh, but only for a second, as if every pleasure were betrayal.

"Don't worry," says Sakharov. "We can consider ourselves fortunate if one of the demands is accepted, and the first is the most important, that your people be allowed to return home."

"No!" cries Dzhemilev. "It's been almost fifty years and my people cannot wait any longer! The government must settle all our demands, otherwise there may be a bloodbath!"

Still, the Tartars leave satisfied.

"My meat pie is ruined!" says Bonner when they sit down to eat after ten.

"No, it's fine," says Sakharov, rising from the table to heat his salad in a frying pan, stirring it gravely with a long wooden spoon and an occasional conspiratorial wink at Bonner.

Clearly, it was time for him to put the memoirs aside and do some writing that could influence events. Especially because glasnost had just revealed its negative side, and how influential that side could be. A teacher of chemistry in Leningrad, Nina Andreyevna, published an article, "I Cannot Forsake Principles," that extolled the virtues of iron order and denounced vulgar glasnost, which reduced the past to "nothing but mistakes and crimes." Blood was in the air, and so was the spirit of Stalin. Gorbachev was shocked to discover that much of the Politburo was in favor of the article, which he could view only as a vote of no confidence in him that had to be squelched at once with a threat of resignation.

In March Sakharov wrote a letter to Gorbachev about the Crimean Tartars and Karabakh, two long-standing problems whose intelligent resolution could set the tone and the example for dealing with future fissures, from Kazakhstan to Estonia. But he also sprang to Gorbachev's defense in an essay, "The Inevitability of Perestroika," that he wrote in March for the daring collection *No Other Way,* which was rushed into print two months later like a trainload of fresh troops. "Above all, I want to emphasize that I am convinced of the absolute historical necessity of perestroika. It's like war. You must win."

* * * *

"Do you know what I love most of all in life?" Sakharov asked Bonner, who later confided in a friend: "I expected he would say something about a poem or a sonata or even about me." She who knew Sakharov better than anyone did not know him well enough. "The thing I love most in life is cosmic background radiation." Perhaps he was half teasing his wife by saying that his deepest and most abiding passion was reserved for the faint echoes of the Big Bang, the lingering hiss of creation, but only half.

"What do you mean science?" replied Sakharov to a colleague. "In Gorky I had time. Now I don't have a free minute. . . . I recently received a letter from two old women who announced a hunger strike to protest the destruction of the church in their village. I have to help them but I don't know how."

In April 1988 he did publish a postscript to Zeldovich's article "Is It Possible to Create the Universe from Nothing?" but that would represent the sum total he published in the years after Gorky. There were, however, some scientific interludes. In April he and Bonner took three weeks to themselves on the Black Sea, "marvelous days, free, productive, and happy." He worked on a talk on the baryon asymmetry of the universe that he was to deliver at the Quarks-88 Conference in Tbilisi in May, she on her second book, *Mothers and Daughters,* the story of her life before the arrest of her parents in 1937.

At the conference in Tbilisi, though displaying his perennial avidity for science and picking up a conversation with a colleague, Mark Perelman, as if twelve years had not passed since they had skipped stones and talked science on a Black Sea beach, Sakharov still seemed changed, aged: "The years of exile had taken their toll on A. D.," thought Perelman. "He was stooped, seemed shorter, complained of how quickly he tired."

He had other complaints as well. His analysis of the causes and consequences of Chernobyl was being ignored, as were his proposals for the peaceful use of atomic energy. And, as a scientist, he was disgruntled that much of his work was still classified; even his dissertation had yet to be published, and in it he had been the first to propose the theory of multiphoton processes, the microscopic basis for nonlinear optics. He was aggrieved that his scientific work was not better known and had been overshadowed by the peace prize. Tamm, when he won the Nobel, had thought he'd been awarded it for the wrong work. Sakharov had done him one better and won the wrong Nobel.

He also told Perelman that he still intended to finish revising his father's physics textbook, supplemented with puzzles and problems of his own devising, some of which came to him while staring at the clock on the wall of his hospital room in Gorky in the nightly respites between the force-feedings.

But as soon as he returned to Moscow, there was no time for doing new science or revising the old. In June Sakharov was giving his favorite idea a new twist at a roundtable, "XX Century and

Peace": the more the two systems converged, the more pluralist each would become. Then Sakharov was in the Leningrad studios of "Fifth Wheel," one of the wide-open, hard-hitting programs that made everybody keep their television on all the time, though some were already beginning to weary of the endless crimes and endless victims, the KGB killing fields, the skulls held up for the camera. But his remarks on Armenia, which he feared could turn into a new Kashmir or Ulster, never reached the air. The many enemies of glasnost were gathering their forces and striking hard in that summer of 1988, with press restrictions and onerous taxes on fledging businesses.

Sakharov countered the censorship of his remarks with a telegram campaign advocating that Karabakh be directly administrated by Moscow. And he countered Gorbachev's lurch to the right by joining the Moscow Tribune, a club formed with the aim of critiquing perestroika but which was, he knew, "in effect an embryonic legal opposition." The club, hoping to accelerate change, came together at Protvino, home of the USSR's largest accelerator, fifty miles south of Moscow, where Sakharov was paying a protracted visit. They all agreed—the summer campaign must be met with a fall counteroffensive.

Another group Sakharov could not help but join was called "Memorial." Its initial purpose, to create a monument for the "victims of illegal repressions," expanded to include a projected museum, library, and archives, as Memorial quickly became a mass movement with chapters in two hundred cities by the end of the year. Sakharov agreed that one cannot even speak of justice if the victims of past injustice cannot, at the very least, be commemorated, but to others such commemoration implies indictment.

He gave the first public speech of his life at a Memorial meeting that contentious fall. He spoke extemporaneously and well, happy that the gift of speech had graced him on that important occasion. Memorial was one of the "healthy forces" and, like Tribune, it was part of a "legal opposition," an assertion of some citizenly independence, inspired in part by the Baltic states, which that spring had declared not their independence but their sovereignty, acting, as David Remnick observed, with "Sakharov's calm confidence on a mass scale."

Solzhenitsyn refused honorary membership in Memorial because, among other reasons, the movement confined itself to Stalin's crimes when in fact Lenin had been the source of all the evils, Stalin included. But Sakharov, involved in real politics, could not afford the consolations of any such maximalism and, for now, found the very existence and growing strength of Memorial sufficient victory in itself. Any direct attack on Lenin would only convince the enemies of perestroika that this was not reform but counterrevolution, to be suppressed by any means. The second law of thermodynamics must not be tempted into play.

Gorbachev was already suspicious of Memorial, seeing it as a rival, not just another force spinning out of control in a landscape of whirlwinds: "we have to somehow de-energize 'Memorial,' really give it a local character," he told the Politburo. "What this is about is not 'Memorial.' It's a cover for something else."

Memorial could be impeded by snags in registration and banking procedures, tolerated but not allowed to become national. There still had to be some limits.

In a joke of the time a dog explains glasnost: "The chain is longer, the food is still far away, but you can bark all you want." As Gorbachev put it, the USSR had become a "debating society." And that was to the good, no perestroika without glasnost. Still, there was something uncanny about it all, a strange disassociated duality, a nation drunk on a spree of free speech but one where cabdrivers no longer accepted rubles. Sakharov had that impression at times in that October of roundtables, forums, public meetings. After attending his first Pugwash Conference on peace and disarmament, Sakharov could not have been clearer in his evaluation: "let Pugwash do its work. But without me!"

But it was not all rhetoric, not all politics. Some conferences were dynamic and productive; some afforded him the chance to speak of the problems that would still await when the current crises were resolved, such as "the danger to the gene pool caused by the chemicalization of life on earth."

Science and politics merged that October when Sakharov was nominated to the Presidium of the Academy of Sciences. The initiative had originated with Roald Sagdeev who, Sakharov noticed, was "unwilling to approach me himself." Sagdeev, who had made a brilliant and honorable career in science, becoming the head of the Soviet space agency, had lapsed once in relation to Sakharov, his teacher, not signing the collective letter denouncing Sakharov after he won the Nobel prize but offering instead to write something of his own, squeezing, as he later admitted, "a few sentences of a compromise denunciation out of my conscience." But to have Sakharov on the ruling body of the Academy would, for certain others, elicit the shamed coward's hatred. Eighty people voted against him, but he still won a majority.

That very same day the Politburo removed the travel ban on Sakharov; he could go where he wished. This was a signal to Sakharov that Gorbachev was winning in the struggle behind closed doors, and to the world that the changes in the Soviet Union were real and deep, though the question of irreversibility still hung in the international air.

The decision had been reached a week earlier in the Politburo where, despite the objection of the Ministry of Medium Machine Building ("He knows of one design that is unknown to the Americans") and despite those who sought linkage and postponement (he can have his awards back "if he goes to the USA and returns"), the "healthy forces" prevailed, even the KGB chief siding for restoration. Khariton, Sakharov's old boss from the Installation, had assured Gorbachev and others that Sakharov was zero risk to state security.

Sakharov knew that the state's relation to him still gave an exact measure of freedom in the country. Less than two years ago he had been confined to the city of Gorky and under curfew; now he was off to New York.

He and Bonner had made it clear that they would not seek to travel together so as not to "complicate" things for the authorities, and so on November 6, 1988, Andrei Sakharov flew from Moscow to New York, then to Boston to be with the children and grandchildren, including Liza's baby, which he and Bonner nicknamed "Hungerstrike." Mobbed by reporters in both citys' airports, Sakharov spoke to them of the clash of change in his country, Tartars, Karabakh; but, a good ambassador, he stressed that perestroika must be supported, with "open eyes," but supported.

Then, slightly uncomfortable in the odd role of moral celebrity and never fond of the tuxedo-and-banquet circuit, Sakharov was in a whirlwind of his own—one minute in the vast marble interior of the Metropolitan Museum of Art delighting in Degas before a meeting of the International Foundation for the Survival and Development of Humanity, at which he raised hackles by comparing the organization to a centipede with so many legs "it didn't know which foot to start off with," then to the White House, where Reagan was "charming" but unbudgeable on Star Wars. And, at Sakharov's insistence, he had half an hour alone with Edward Teller before a formal banquet honoring Teller on his birthday. He told Teller that he had identified with him at certain points in his own career and, as he repeated in his speech, of the respect he had "for the principled and determined manner in which he defended his views, regardless of whether I agreed with them or not." But all the same he was startled by Teller's intransigent mind-set—Teller supported SDI out of a "profound and uncompromising distrust of the Soviet Union."

Some of the West's lingering suspicion was dispelled in early December when a resurgent Gorbachev addressed the United Nations, renouncing class struggle in favor of the "primacy of universal human values" and calling for an end to the Cold War. Earlier that year he had demonstrated his good intentions by allowing the Russian Orthodox Church to celebrate its one-thousandth anniversary with the reopening of churches and the ringing of bells in the Kremlin's cathedrals. At the United Nations, the acclaim was enormous, a high point of "Gorbymania," to use a newsism of the time. Sakharov, still in the United States, could only have been gratified; Gorbachev was sounding more like him all the time. There were, however, measurable differences: Gorbachev announced a 10 percent reduction in troops; Sakharov wanted them cut in half.

Gorbachev's UN glory lasted only a day. On December 7 Armenia was struck by an earthquake whose magnitude compelled the leader's immediate return. It was a natural catastrophe that mixed suffering and symbolism. The Azerbaijanis saw it as Allah's punishment for Armenian arrogance, while the Armenians took it as their latest martyrdom. What shocked the world were the televised images of shoddy concrete tenements and medics carrying bottles of blood stopped with rags like Molotov cocktails.

Before leaving the United States, Sakharov had some other last business of his own to attend to. The former exile from Gorky made a phone call from Newton, Massachusetts, to the still exiled Solzhenitsyn in Cavendish, Vermont. Solzhenitsyn's wife said she would, as an "exception," call him to the phone. Devoted to his work, Solzhenitsyn scheduled his life with military precision, so outrageously satirized in Voinovich's novel *Moscow 2042* that some Russian émigrés were afraid to read it, not sure of which side they'd come down on.

Solzhenitsyn congratulated Sakharov on his success in the political arena but would not himself return to any Russia that would not publish the *Gulag*. When they were done with politics, Sakharov came to the point: "there should be nothing left unsaid between us. In your book *The Oak and the Calf* you hurt me deeply, insulted me. I'm speaking of your pronouncements about my wife, sometimes explicit and sometimes without naming her, but it's perfectly clear whom you mean. My wife is absolutely not the person you describe. She's an infinitely loyal, self-sacrificing, and heroic person, who's never betrayed anyone. She keeps her distance from all salons, dissident and otherwise, and she's never imposed her opinions on me."

After a long silence, Solzhenitsyn—unused, thought Sakharov, to "direct rebukes"—replied: "I would like to be believe that it is so."

As with everything, Sakharov analyzed and evaluated the data, then synthesized it with humor: "By ordinary standards, that wasn't much of an apology, but for Solzhenitsyn it was apparently a major concession."

* * * *

After a stop in Paris where he conferred with Lech Walesa, then later found it "hard to maintain a straight face" while marching down an allée of drawn sabers at an official French reception, Sakharov returned to Moscow on December 13, having been reunited en route with Bonner, who was traveling separately, an encounter described by one person present: "when they saw each other, their faces lit up like young newlyweds. Such clear young faces. They saw nothing except each other."

That Sakharov felt his best with her can be seen in the photographs of him at the time. Alone, he seems frail; with her, robust. They were soon in operation as a team, flying to both Azerbaijan and Armenia in late December to attempt to mediate a conflict that seemed headed for war. Unfortunately, it proved more of a fact-finding mission. The Azerbaijani capital of Baku was seized by anti-Armenian rallies of a half million people, some of whom slaughtered lambs and cooked pilaf on the vast central square. Sakharov and Bonner called on the leadership to perform a "noble" act, the giving of Karabakh to Armenia, but were told: "Land isn't given. It's conquered."

In the Armenian capital of Yerevan, the prevailing atmosphere was one of "shock and panic, almost mass psychosis." Now, to violence, refugees, and earthquake had been added fear for the safety of a nuclear power plant built near a fault line by the same agency that had designed the reactors at Chernobyl. Sakharov and Bonner were helicoptered into the worst-hit areas where, in the rubble of a school that had collapsed on the children, a homework assignment marked with an "A" brought them to tears.

Sakharov was already displeased by Gorbachev's on-the-spot handling of the Armenian earthquake, which he found "childishly peevish" when, in early January 1989, Gorbachev convoked the cream of the intelligentsia for a discussion of perestroika. Sakharov now joined those who found that Gorbachev spoke too much and listened too little, calling Gorbachev's opening remarks "rather long-winded," intelligentsia understatement for interminable. But Gorbachev's message was important. There were three forces battling it out in the country now: the advocates of perestroika, its enemies in the party, and those who endangered progress by insisting on unwise acceleration. Command-and-control communism was collapsing, giving way to a democratic, modern, economically healthy and peace-loving society, the true socialism progressives had always dreamed of. It could be achieved, but only if all the steps were taken at the proper speed. History was time and timing. When he'd gone to Czechoslovakia, Gorbachev had been asked what was the difference between him and the Prague Spring and had answered: "Nineteen years."

Sakharov was for perestroika and, since Gorbachev and perestroika were still indivisible, he was still for Gorbachev. Himself an avowed evolutionist, Sakharov was in favor of avoiding the national temptation of reckless speed but also knew that evolution was not a uniformly gradual process—sometimes spurts were required. He pressed Gorbachev hard on the mismanagement of the Armenian situation, resulting in unnecessary suffering and death. Publicly stung, Gorbachev was publicly angered.

The meltdown at Chernobyl and the earthquake in Armenia were taken as bad omens, but there were good signs where they counted most, in the world of practical politics. A December 1988 law provided for something like free elections, the first since November 1917, in which Lenin's Bolsheviks came in third and then, true to the logic of their ideology and ambition, they enforced the will of history and seized the state. Gorbachev was now calling on the people to break the party in the "vise" of perestroika: "If there's pressure from only one side, it won't work."

But Gorbachev would, or could, grant only semifree elections. About one-third of the deputies to the Congress of People's Deputies would be freely nominated, the rest remaining under party or some other control. New, complicated, and unfamiliar to say the least, the nomination campaign produced a variety of oddities as far as Sakharov was concerned. A natural candidate for the office of people's deputy, he was nominated by a great number of cities, districts, and organizations from Leningrad to Siberia including the Installation and the auto plant in Gorky, each in its own way seeking to redeem its honor. But in the end Sakharov decided that the body that most needed redemption was his own Academy of Sciences; he would stand as a candidate for it and no other. Though on the practical level he welcomed the elections, Sakharov was all too aware of what was "phony" and "quasi rigged" about them and for a time even hesitated as to whether to stand at all. Bonner was dead set against any participation in any Soviet government. But, speaking to an overflow crowd of "the impoverished proletariat of mental labor . . . the kind of people you see waiting to get into a Chagall exhibition or a film festival . . . honest and intelligent, understanding everything," Sakharov was energized by contact with his core constituency, and for the first time felt he had received a "moral mandate."

He was free now to travel abroad with Elena, another small positive increment. They used the break between the nominations and the elections to travel to Rome in February, where they met with both the head of the Italian Socialist Party and the pope. And, in an event that captures the flavor of that year of wonders, Sakharov and the pope discussed perestroika in the Vatican.

But there was also time for art and honor, degrees granted him in Bologna and Siena: "solemn processions of faculties dressed in medieval robes, with heralds and maces, with ancient music and lofty rhetoric." He enjoyed a pleasant sense of belonging in Italy. Its history was his; Russians were, "after all is said and done, Europeans."

Sakharov was also much taken with Canada, where he traveled next, he and Bonner both receiving honorary doctorates from Ottawa University. He responded to the nation's health—"not smug in any way . . . a model country . . . if it weren't so difficult to follow the examples of others."

But it was also there, in Ottawa, that Sakharov ignited a time-delay explosion back home. To a Soviet correspondent's question about Soviet POWs in Afghanistan, Sakharov replied: "Our country waged a cruel and horrible war in Afghanistan. We call our enemies bandits and don't recognize them as the armed forces of a belligerent. But bandits don't have prisoners of war, they have hostages. There have been reports that our helicopters have shot surrounded Soviet soldiers to prevent them from being taken prisoner." That last sentence would be quoted in *Red Star,* the official Ministry of Defense newspaper, to show that Sakharov not only condemned the war but impugned the honor of the military.

The remainder of their trip was private. They visited Elena's children and grandchildren in Massachusetts, where Sakharov worked on the memoirs, that would ultimately be divided into a larger volume that ended with his liberation from Gorky and a shorter one that described his life since returning to Moscow. In Gorky the enemy of the memoirs had been single and evil, but now the enemies were multiple and good—family, friends, travel, causes, honors. But he finished what he started, and that included the memoirs. In the midst of everything, they would get their due hours.

* * * *

Sakharov was back in Moscow a good week before the elections of March 26, 1989. There were meetings, posters, mimeographed fliers, but nothing that came close to a Western-style campaign with its spin, hype, and buzz. Yet, to the surprise of the intelligentsia and the dismay of the establishment, the electorate displayed both festive enthusiasm and hard-nosed savvy. Glasnost was working. The new ideas in the air had entered people's minds and changed them. Whenever there was a real choice, party candidates were almost always roundly trounced, and Andrei Sakharov was elected the Academy of Sciences representative to the First Congress of People's Deputies, which would convene in two months, on May 25.

All that did not mean in the least that those in power would act like good communists and bow to the will of history. When the last Soviet troops crossed the Friendship Bridge from Afghanistan back into Soviet territory in February 1989, they were not met by any official Soviet delegation. Some in the military were insulted and felt ill used, but tended to blame the laxities of glasnost rather than the Politburo that had sent them there in the first place. Many people in the party and the army were now seriously troubled about the fate of the country and about their own.

Electoral euphoria died violently in April. Soviet forces used poison gas, truncheons, and sharpened shovels to break up a peaceful demonstration in Tbilisi, Georgia. Twenty-one were killed and hundreds hospitalized.

Sakharov and Bonner flew to Georgia to assess the situation. It was she who had the best practical ideas—to call U.S. ambassador Jack Matlock for information about poison gases, to use her contacts with Médecins sans Frontières to help treat the injured, and to arrange for American toxicologists to perform on-site analysis. Sakharov visited the young hunger strikers in the hospital in Tbilisi and promised to do everything he could; then he, the champion of hunger strikers, persuaded them to end theirs.

On May 3, at a meeting of Moscow deputies with party and state leaders, Sakharov angered Gorbachev by his insistence, made vehement by the events in Tbilisi, on the unencumbered right to free speech and assembly. On the fourteenth Sakharov flew to Milan to attend the annual convention of the Italian Socialist Party and thank them for their support: "Others parties and leaders did a great deal, but you did the most." Five days later he was in the north of Russia, supporting a candidate in a runoff election. The candidate was Revolt Pimenov, who nineteen years ago had been the defendant at the first dissident trial Sakharov had ever attended. Many circles were coming to a close.

While there in Syktyvkar, the capital of the Komi Autonomous Republic, a place not often visited by luminaries, Sakharov was of course asked for his opinion of Gorbachev and his emergent rival, Boris Yeltsin. He liked what he had seen of Yeltsin, though he was aware that he hadn't yet seen much. "A person of a different calibre" from Gorbachev, Yeltsin won Sakharov's admiration when he demanded live television coverage of the congress. Part of Yeltsin's appeal was that he had risen up early against the party and had suffered for it, part was his hardy brashness, his primitive democratic instincts. Yeltsin seemed less the temporizer, the conciliator, the vacillator that Gorbachev sometimes did; caution was never his strong suit; as a boy during the war he had lost two fingers when pilfering grenades to fight the Germans.

But speaking in Syktyvkar, Sakharov said: "I just can't see an alternative to Gorbachev at this critical juncture. Even though his actions may have been prompted by historical circumstances, it has been Gorbachev's initiatives that have completely altered the country and the psychology of its people in just four years. All the same, I don't idealize him, and I don't believe he's doing all that's needed. Furthermore, I think it's extremely dangerous to concentrate unlimited power in the hands of a single man. But none of this changes the fact that there is no alternative to Gorbachev. I have repeated these words many times in many places—Gorbachev's face lit up with joy and triumph when I said them again at the pre-congress planning session."

Two days later Gorbachev traveled to Beijing, where his mere presence elicited democratic enthusiasms. By May 17, the day of his arrival, thousands of hunger strikers had massed on Tiananmen Square. Millions poured into the streets of Beijing and other cities, creating an atmosphere of "exuberance and infectious gaiety," in the words of correspondent Michael Dobbs.

The Chinese leadership was opposed to any infection of Soviet-style reform. The Soviet Union had neither order nor abundance. The Communist Party of China would guarantee both and thereby maintain its legitimacy. No sooner did Gorbachev leave than on May 20 martial law was declared.

The next day, May 21, Sakharov turned sixty-eight, now only four years younger than his father had been when he died. In that most political of years it proved the most political of birthdays. To the immense chagrin of the French physicists who had arranged a banquet in his honor, Sakharov, at the very last minute, decided it was more important to speak at a huge rally to protest the violence in Georgia and to counter the reactionaries' attempt "to impose their own 'rules of the game.'" But somehow the event turned into a rally in support of Yeltsin, which caused Sakharov to recast his speech in haste, impeding both thought and delivery. Bonner, who had been against attending the rally in the first place, was not reluctant to remind him of that: "You should have listened to me. We could have flown to Paris for the day. The physicists had a special cake baked for you, and you've hurt their feelings."

Though he had to admit "the mood had been spoiled," there was hardly a moment for regret with the congress opening in four days.

* * * *

The grounds of the Kremlin are spacious and grassy with an almost rural tranquillity. Yet the cathedrals where tsars repented murder were at best a blur of green and gold for the deputies streaming into the Great Kremlin Palace, a Soviet modern building with a glass exterior, its vast conference hall done in the ruby reds of communism and overseen by an immense white plaster Lenin. And no sooner were the thousands seated than one deputy, a bearded actor from Latvia, rose and proposed a moment of silence for the dead in Tbilisi, a silence he won then broke with a question that would hang answerless over the congress: on whose orders had the atrocity been committed?

An honor, a signal, a test, Sakharov was the first to be given the floor. He strode to the podium at his own pace, "ambled" it seemed to one. His first point was simple and could have been communicated quickly if his very presence had not provoked such hostile shouts and jeers, countered, but insufficiently, by cheers and applause. Outwaiting the onslaught, he made his point: One of their tasks as deputies was electing the head of state, the chairman of the Supreme Soviet, a deliberative body that met in session, unlike the congress, which would convene twice a year. If Gorbachev, or anyone, were head of both the state and the party, tremendous power would be in his hands. Despite that, the situation was so critical that Sakharov still supported Gorbachev, but only conditionally. First, let Gorbachev "speak of both accomplishments and mistakes, and not be afraid of self-criticism. . . . He and the other candidates should explain what they are planning to do in the immediate future to overcome the crisis that has developed in our country." Again the great hall erupted into a pandemonium of applause and abuse. A short while later Sakharov, feeling thousands of eyes on him, walked out of the hall to protest the shameful lack of any real debate about electing Gorbachev head of state.

But that spectacle, the example of a man freely calling on his leader to give an account of himself, was watched live by millions. Nothing was produced those days except history. The stores ran out of salt and matches.

When Sakharov returned later he saw that he was not alone; he had allies among the outspoken. A former Olympic heavyweight weight-lifting champion, Yuri Vlasov, called the KGB an "underground empire" and "a threat to democracy." Others rose with other demands: a new constitution with the KGB under civilian control, a multiparty system, Lenin to be taken from his tomb and given the ordinary burial he would have wanted.

But the wounds of the nation were also exposed for all to see—the decline in everything from sheaves of wheat to the years a person lives. Chernobyl, contaminated water, dead seas, cancer, poisonous chemicals in every fifth sausage. Gorbachev was even attacked for being too much under the influence of his wife, Raisa, a formulation almost guaranteed to elicit a sympathetic smile from Sakharov, who was now in high elation as the congress "demolished . . . illusions" and "burned all bridges."

The violence in Georgia was soon back in the debate, itself becoming more violent. "If the guilty aren't punished," said one deputy, "public opinion will see it as proof of the omnipotence of the Party elite and the military." Yet the party elite and military were starting to feel anything but omnipotent; it was time to reassert their dominance while they still had the strength. In that attempt, one of them, the general who commanded the forces in Tbilisi, provided some inadvertent comedy by grousing that glasnost was worse than the Terror: "We're always harping on 1937, but things are harder now than they were in 1937. People can say whatever they like about you and you can't defend yourself."

But the moments of comedy, inadvertent or otherwise, were few and far between. This was deadly serious business, for both sides. Still, not every moment was stormy; there were well-reasoned speeches, some real debate. Sakharov twice spoke on such legal issues as having defense counsel present as soon as charges are brought. There was also a little political horse-trading off to the side. The idea of a commission headed by Gorbachev to draft a new constitution had already been approved when someone noticed that all its members were communists, prompting one delegate to ask: "Are they going to draft a new constitution or new bylaws for the Party?" Gorbachev suddenly needed Sakharov to lend that committee legitimacy and took audience applause at the idea of his inclusion to be sufficient voice vote, aware that Sakharov would probably not win an actual count. Understanding what Gorbachev had done and why, Sakharov went to the podium and expressed his conditional acceptance: "It's evident from the composition of the commission that on all issues of principle I'll be in the minority. Therefore, I can serve on the commission only if I have the right to propose alternative formulations and principles and to take issue with the recommendations that I oppose."

For a moment Gorbachev was worried that Sakharov had not

quite accepted and was greatly relieved to realize that he in fact had.

Sakharov's support of Gorbachev and his acceptance of the constitutional commission post were both conditional because he had yet to see any evidence of acts that could win his unconditional allegiance. Gorbachev seemed more interested in amassing power than in exercising it. Meanwhile nothing of substance had changed; all power was still centralized and in party control. Perhaps a private conversation with Gorbachev would be most effective at this critical juncture.

He discussed this idea with Bonner, who met him for lunch every day and otherwise watched him on television like everyone else. On June 1, as the congress entered its second week, Sakharov requested that private conversation and was "on pins and needles" for the rest of the day. They spoke late, both tired. It was a blunt exchange, Sakharov noticing that Gorbachev's usual smile for him —"half kindly, half condescending—never once appeared on his face."

Sakharov spoke first, presenting Gorbachev with a conundrum that was also an ultimatum: "Mikhail Sergeyevich, there's no need for me to tell you how serious things are in the country, how dissatisfied people are, and how everyone expects things to get worse. There's a crisis of trust in the leadership and the Party. Your personal authority has dropped almost to zero. People can't wait any longer with nothing but promises to sustain them. A middle course in a situation like this is almost impossible. The country and you personally are at a crossroads—either accelerate the process of change to the maximum or try to retain the control and command system in all of its aspects. In the first case, you will have to rely on the left and you'll be able to count on the support of many brave and energetic people. In the second case, you know yourself whose support you'll have, but they will never forgive you for backing perestroika."

Unfazed, Gorbachev shot back: "I stand firmly for the ideas of perestroika. I'm tied to them forever. But I'm against running around like a chicken with its head cut off. We've seen many 'big leaps' and the results have always been tragedy and backtracking.

I know everything that's being said about me. But I'm convinced that the people will understand my policies."

The next day it was Sakharov's turn to be attacked and to defend himself. A legless Afghan war veteran denounced him for the remarks he had made in Ottawa about Soviet troops killing their own to prevent them from being taken prisoner, then brought most the auditorium to its feet and the most thunderous applause of the congress with his battle call: "State, Motherland, Communism."

The outcry against Sakharov was murderous as he took to the rostrum. To one reporter he looked "beaten, dejected," but then, roused, "he faced the angry chorus alone, and he did not flinch or retreat," as another observed. "The last thing I wanted to do was to insult the Soviet army," said Sakharov. "The real issue is that the war in Afghanistan was itself a crime, an illegal adventure, and we don't know who was responsible for it."

His words were jammed by shouts of hatred: "Shame! Away with Sakharov!"

He may have lost in the hall but he won in the nation, as he was soon to learn. Meeting him at the end of the day, Bonner said with her usual frankness: "You spoke badly, of course, but you're a real hero." Not only to her but to the millions watching, tens of thousands of them sending him letters and telegrams of praise, an opinion later ratified by polls for the best deputy, which showed Sakharov number one, Yeltsin two, and Gorbachev seventeenth.

Some of the deputies were proposing to extend the congress for a week; there was too much to say, do, and decide. Especially since they had lost a day, June 4, set aside for national mourning when, in yet another ominous misfortune, an exploding gas pipeline derailed two passenger trains, both just happening to pass it at the wrong moment. That meant the deputies were not distracted by their urgent business from two other great events of that day, the butchery on Tiananmen Square in Beijing and the elections in Poland, where Solidarity won even more seats than it wanted, fearing a backlash from the party and military, who still "had all the guns."

It was the same choice facing Sakharov and the other deputies: rapid change or sudden violence. But the process was dialectic;

violence could beget change. Coinciding with the events in Beijing and Warsaw was a pogrom in Soviet Uzbekistan against an exceedingly small people, the Meskhi Turks, who had lived in Georgia near the Turkish border until they too became another nation exiled in its entirety by Stalin in 1944. The Uzbeks murdering them in June 1989 used the occasion to settle other scores—with Russians, Ukrainians, Tartars, Armenians, and Jews. A delegation of Meskhi Turks traveled to Moscow to beseech Sakharov for help, one of the women falling to her knees. Sakharov phoned Gorbachev to urge him to meet with the Meskhi Turks but was told that Gorbachev was about to travel to West Germany to meet with the chancellor, Helmut Kohl. "I blew up, probably as angry as I've ever been, and shouted: 'Tell Mikhail Sergeyevich that he's not going anywhere. I'll appeal to Kohl to put off Gorbachev's visit. It's impossible to receive the head of a state that permits genocide!' "

Sakharov initiated a process that quickly resulted in a reasonable compromise: to ensure their survival, the Meskhi Turks would be evacuated to Russia. They'd rather die. "That's a second exile," said one of their leaders. "We've been trying to return to our homeland for decades, and we're ready to give up our lives for that. If we agree to move to Russia now, we'll never get back to Georgia, we understand that all too clearly!"

Sakharov reminded them of the obvious: the dead cannot fight for their homeland. Accept evacuation, continue the struggle. The Meskhi Turks may not have been as angry with Sakharov as Sakharov himself had been on their behalf, but nevertheless they stormed out of his apartment.

Sakharov had to sympathize; he too was impatient with compromise. The congress had both transformed the nation and changed nothing. Even the congress's appeal for peace in China was so vaguely worded as to apply both to hunger strikers and machine gunners. Gorbachev, the head of the party, had been elected head of state. The new constitution was, with the exception of Sakharov, being written by communists. The party still monopolized power.

Discussing these problems, a group of like-minded deputies, Sakharov and Yeltsin among them, formalized their relations into the Interregional Group of Deputies, an embryo opposition, calling for a new constitutional amendment eliminating the party's monopoly, free elections of leaders, and a devolving of power from the center to the local governing agencies known as soviets. Only faster change could outpace backlash.

Gorbachev felt so betrayed by what he saw as Sakharov's shift to the left that he cut off Sakharov's speech on the last day of the congress by simply pulling the plug on his mike. An ironic moment in the annals of glasnost, it had no practical effect. The live broadcast continued, and not only the deputies close enough to hear Sakharov but the whole hall knew what he was calling for: repeal of Article 6 of the Constitution guaranteeing the "leading role of the Party," that is, its monopoly on power. In Sakharov's scheme, everyone would have to stand the trial of election. The new freely elected congress "shall have the exclusive right to elect and recall the top officials of the USSR," including the head of state and of the KGB. Freedom of speech and assembly were essential to this process, just as it was essential to realize that "any danger of armed attack on the Soviet Union vanished long ago." The real threat of violence was at home. The displaced nations could no longer be ignored; they were the fissures preceding earthquake. Sakharov wanted a "new constitutional system based on horizontal federalism" that would grant all minorities equal rights while a Soviet of Nationalities sorted out their claims.

He delivered the remainder of his final speech with a paradoxical feeling of "tragic optimism": "The Congress does not have the power instantaneously to feed the country, instantaneously to solve our nationality problems, instantaneously to eliminate the budget deficit, instantaneously to make the air and water and woods clean again, but what we are obliged to do is to establish political guarantees that these problems will be solved. *That* is what the country expects from us."

And then in a paradox possible only in that time of Marxism gone haywire, Sakharov, with Gorbachev and Lenin behind him, concluded his speech with the words: "*All Power to the Soviets!*"

* * * *

Battle-weary, Sakharov and Bonner retired to Massachusetts for the summer with stops en route to CERN (European Center for

Nuclear Research) in Switzerland and Oxford University in England for honors. They stayed with either Tatiana's family in Newton or Alexei's in nearby Westwood, where at a small desk wedged into the corner of a guest bedroom Sakharov finally had the peace and time to get back to the memoirs. Though containing small mistakes of fact, the earlier two versions, stolen by the KGB, were fresher, livelier, more vivid and impassioned. The errors could be fixed and some opinions modified—perhaps he'd been too harsh on Zeldovich—but there was no recapturing that initial ardor. Yet the events of the congress were still fresh in mind, and he gave them pride of place in the second volume of his memoirs, *Moscow and Beyond*. The writing was also influenced by current events such as the Russian miners' strike that summer for the one thing they needed most and which the economy could not supply them, soap. The working class was rising against the Marxists! These were, as he was fond of remarking in that golden July, "astonishing times."

The most pained and plaintive note is struck in the second volume's second to last paragraph when Sakharov writes of his children and grandchildren: "There is much I have failed to do, sometimes because of my natural disposition to procrastinate, sometimes because of sheer physical impossibility, sometimes because of the resistance of my daughters and son which I could not overcome. But I have never stopped thinking about this."

The last words were of course for Elena Bonner, "the only person who shares my inner thoughts and feelings. . . . We are together. This gives life meaning."

All he had left to write was the epilogue to the first volume of the memoirs and the foreword to the second, but he postponed that work: "completing a book gives one a sense of crossing a frontier, of finality. As Pushkin put it, 'Why is this strange sadness troubling me?' At the same time, there is an awareness of the powerful flow of life, which began before us and will continue after us."

Those last two pieces of writing were still undone when he returned to Moscow in late August 1989 just as the KGB was burning the last of 584 volumes of operational materials on Sakharov and Bonner. It would be a fall of all-out battle—perestroika had to accelerate to crash through the barriers. This would demand every minute of his time but, perhaps out of a sense that he had behaved somewhat badly with the French physicists who had wished to celebrate his birthday that May, he did accept an invitation to receive an honorary degree from Claude Bernard University in Lyons, France, on September 27, 1989. He did not deliver a written speech but gave a free-ranging talk, the desire for which, as Bonner observed, may have been "a direct psychological consequence of the aggressive reaction and obstruction that Sakharov had encountered at the First Congress of People's Deputies."

In any event, what emerged from the talk was a summa and a credo. Sakharov opened by posing the question of how to characterize the twentieth century, which had little more than a decade yet to run, how to reconcile its chief features—war and science. Though war and science were obviously connected in some ways, their essential nature was different: war set nation against nation; science joined all scientists in every country with identical knowledge. As an international community of reason, science had a tropism for convergence and was thus both the model and the instrument for a better world. Convergence was the only answer to international tensions and to Russia's problems: "Our country now faces the historic challenge of building a society that combines social justice with an effective economy. At the moment, we have neither one nor the other." China had one, the economy, but not the other, as the "savage" reprisals on Tiananmen Square had made clear.

But Sakharov was also moved to speak on grander scales that day in France. There could yet be a new convergence between the objective universe and "our individual fates, which are such small points in time and space." Science and religion, necessarily opposed at an earlier stage of history, will merge: "in the next stage in the development of human consciousness, there will be some deep, versatile resolution of the perceived discord between religion and science." Science will restore the awe that is at the root of all religion: "We are looking into the fantastic possibility that regions of space separated from each other by billions of light-years are, at the same time, connected to each other with the help of additional parallel entrances, often called 'wormholes.' In other words, we do not exclude the possibility of a miracle: the instantaneous crossing from one region of space to another. The elapsed time would be so short that we would appear in the new place quite unexpectedly or, vice versa, someone would suddenly appear next to us."

The objective universe is not only awe-inspiring but has meaning, best defined with the humility of vagueness: "My deep sense (not even conviction—the word 'conviction' wouldn't be right here) is that some kind of inner meaning exists in nature, in nature as a whole."

He was coming down on the side of Kant, except that the starry skies above were now known to contain billions of galaxies and black holes, where light met its death. Still, as he put it a touch more explicitly in his memoirs: "I am unable to imagine the universe and human life without some guiding principle, without a source of spiritual 'warmth' that is non-material and not bound by physical laws."

* * * *

In the kitchens of the Moscow intelligentsia the talk was of civil war, the national taste for dark operatic rumor honed by events sharp as the shovels in Georgia. The Soviet Union was disintegrating triply—it was losing the eastern bloc, its own republics were moving toward secession, and inside Russia were dozens of small nations yearning for land and independence. Unlike his hero Lenin, with his two steps forward, one step back, Gorbachev performed a political tap dance whose steps were too rapid to follow. He was now back with the right, calling Sakharov's and Yeltsin's Interregional Deputies organization "a gangster group." The atmosphere of dread and tension infected Gorbachev too: "Reading the press, you get the feeling that you are standing knee-deep in gasoline. The only thing lacking is a spark."

And sparks were already flying between Sakharov and Gorbachev. During the Supreme Soviet's sessions Gorbachev rejected the Interregional Deputies' proposal for a popularly elected president. No, said Gorbachev with historic distrust of the electorate, democracy was still too "feeble"; a president would have too much power. The legislature should be built up instead—a position on which, within months, he would do a complete reverse.

With anarchy sometimes seeming only hours away, Sakharov was frenetically busy. He attended the sessions of the Supreme Soviet every day; every evening and weekend was given to meetings of Memorial and other organizations. Unselfconsciously dozing off during the interminable harangues, he would wake and make a suggestion that had all the daylight of common sense. And when not attending meetings, he was helping striking miners find a lawyer in Moscow, not just making calls, but going around with them personally. Or flying to Chelyabinsk to speak at a ceremony commemorating Stalin's victims; the dialectic of progress was energized by a judicious assessment of the past.

He had only the odd moment to put the finishing touches on his assessment of his own past, and still had not sat down and written the epilogue and foreword. Once again the "merely personal" was outweighed by a greater task—writing a draft of the constitution that would blueprint the transformation of his country from a party police state to a democracy, the Union of Soviet Socialist Republics becoming the Union of Soviet Republics of Europe and Asia, a voluntary organization unlike its predecessor. But even this greater task was broken off by events of greater grandeur. On November 9, 1989, the Berlin Wall came down to sledgehammers and champagne, the outer walls of the empire breached by the barbarians of democracy. As Tamm had said in 1953—have we lived to see the day! And on November 24 Alexander Dubcek, leader of the Prague Spring, addressed three hundred thousand in Prague's Wenceslas Square. The crowd's roar of "Freedom, freedom!" was heard live in Moscow, exhilarating or terrifying, depending on the increasingly either-or politics of the day. It could still come down to bayonets.

The Supreme Soviet ended its session on November 29, 1989, and on that same day Sakharov was chairing the first session of the Academy of Sciences' Scientific Council on the Complex Problem of Cosmoparticle Physics, which brought scientists from various disciplines together to discuss the intersection of the largest and the smallest, cosmos and particle. The project appealed to Sakharov because of its subject and because its work had been launched by Zeldovich, whom he still found himself addressing mentally. But that all took time, that first session lasting over four hours, and the Second Congress of People's Deputies, in which Sakharov intended to play an active role, as he had in the spring, was to open in only two weeks on December 12.

Sakharov returned to work on the new constitution. Every day only made it clearer that the country would need an instrument of political change to resist the centrifugal forces of disintegration. It could hardly be a finished and ideal document. Only new government bodies could resolve some of the country's fundamental and near intractable problems such as the conflicting claims of nationalities. But the goals and means could be indicated with the clarities allowed by the present moment.

His ten-page draft guaranteed all human rights in a "flexible, pluralist, tolerant society." A president would be popularly elected for a five-year term of office. His power would be balanced by a bicameral legislature and a supreme court. The economy would be mixed, property could be owned, private and corporate businesses free to compete with the state, which would responsible for maintaining a basic level of social welfare. Sakharov modeled his constitution on its great predecessors. A liberal constitution, of which his Kadet grandfather or Uncle Ivan would have approved, Sakharov's outdid the American Constitution in one respect, positing the right not just to the pursuit but to happiness itself.

But a new constitution was impossible because of a clause in the old Constitution, Article 6, that granted a monopoly of power to the Communist Party. That was the wedge that blocked the flow of history. The best instrument for its dislodgment was a two-hour national work strike that Sakharov and four other deputies called to take place on December 11; it would send a strong signal to the Second Congress of People's Deputies that would open the next day—the people were for perestroika whether or not Gorbachev any longer was.

The appeal for the strike ended with four slogans that had a whiff of the October Revolution about them:

Property to the people!
Land to the peasants!
Factories to the workers!
All power to the soviets!

But in the meantime history was superseded by the death of individuals. On December 3 Sakharov attended the funeral of Academician Obukhov, who had been one of the forty who signed the detestable letter condemning him. Sakharov's presence was a sign that he had forgiven and, as it turned out, even forgotten. Later, in a discussion about betrayal, Sakharov was asked how he felt about Obukhov. He replied that by coincidence his physician in Gorky had also been named Obukhov, and now all his negative associations were with that name: "When they tied me down and force-fed me there, I came to understand for the first time what the slaves in ancient Rome went through when they were crucified."

Another funeral he attended, on December 8, that of Sofia Kalistratova, occasioned a different sort of forgiveness entirely. Kalistratova, the lawyer who had mounted a spirited defense of dissidents and had herself suffered for her convictions, had been one of the old crowd. December 8 was also the third anniversary of Anatoly Marchenko's death, both of their names joined during the church service. Sakharov and Bonner were given a ride to the service by his old university chum Misha Levin, with whom he had strolled through Moscow discussing Pushkin and physics. And as they drove reminiscing of Sofia and Anatoly, Pushkin interwove himself again into their conversation, Sakharov pointing out the church where Pushkin married, as had Sakharov's parents, and where he, Andrei, was taken for communion as a boy. And he laughed as they passed Public School 110, where he had been briefly enrolled and had failed to construct a simple stool but had succeeded in standing up to the bullies who jammed his fingers in a vise. He had made no friends there, and now everyone was telling the most fantastical stories about their classmate, the young genius. It was absurd, irksome, but probably inevitable, and certainly as forgivable as signing the letter of the forty.

While Sakharov was attending the funeral, the latest head of the KGB, Vladimir Kriuchkov, a die-hard reactionary who would become one of the leaders of the failed coup against Gorbachev, was reporting on Sakharov's activities both to Gorbachev personally and to the Central Committee. Some of the old notes were struck: Sakharov is receiving "propaganda support from abroad," but he is no longer portrayed as a dupe or a gadfly, rather as the dangerous leader of a movement that wants to seize power from the Communist Party. "A. D. Sakharov claims that the driving force of perestroika is an oppositionist intelligentsia and that it is capable of leading the working class." The two-hour warning strike Sakharov had called for the eleventh was a first step in that direction, as was the subject of the strike—the repeal of Article 6. It had come down, as it always does, to a struggle for power. "The Committee for State Security continues to monitor the situation."

The strike had been called for the hours of ten to twelve on December 11, a Monday, the first day of the workweek and the day before the Second Congress opened. Sakharov arrived at FIAN before ten o'clock that morning, striking one colleague as youthful and energized as he bounded up the front steps. In the lobby he ran into Boris Altshuler, a physicist and the son of Lev Altshuler from the Installation, who had suffered from nightmares about security and was one of the first people Andrei Sakharov had stood up for, out of sheer collegial decency. Sakharov told Altshuler the younger that it was a "good sign" that the "partocrats," as he now called them, were angry: "If we hadn't used the word 'strike,' no one would have heard about our appeal."

The auditorium on the third floor of FIAN held eight hundred and was more than full, journalists as well as scientists allowed entry. Addressing them, Sakharov said: "The country's fate is at a crossroads. Or, as one could say in this hall, at a point of bifurcation." Those afraid of losing power, those who had been frightened by the "example" of Eastern Europe, saw that bifurcation as keeping power or losing it, Berlin or Beijing. The "partocracy"—the bureaucrats, the KGB, the military-industrial complex—was on one side and the left opposition on the other. The winner would be the side that won the people, who once again now counted. But the people did not have full confidence in perestroika and feared it could all yet be reversed, as such things already had many times in the past.

And as if to prove Sakharov's point, his own speech was interrupted by the sudden and tumultuous arrival of foreign correspondents who had broken through a line of security, one not summoned by FIAN and whose demeanor left no doubts as to who had sent them. And it now turned out that there were also, suddenly, problems with the recording systems, but ABC, on the spot, volunteered to share its tapes. Only then could Sakharov resume speaking and state his conclusion: the people's confidence could be restored only by the repeal of Article 6—as long as the party monopolized power legally and officially, people would have good reason to fear. The strike was the perfect solution because it both activated the people and restored their confidence while at the same time tipping the political balance in favor of the liberal minority. In the end Sakharov agreed with his enemies—the strike *was* a "test of forces."

Sakharov left soon after he spoke, due at noon for a preelectoral meeting of the Academy of Sciences, through some of which he could safely doze. After the session he met one-on-one with Sergei Kovalyov, who had come to their kitchen so many years ago for one last taste of Bonner's cabbage soup before facing prison. Sakharov overcame Kovalyov's objections and convinced him to run for election to the Supreme Soviet. By six that evening he was addressing Memorial and accepting a petition with fifty thousand signatures. FIAN was already inundated with telegrams of support: "We the fisherman of the North Ice Ocean. . . ." Some preferred to avoid the inflammatory word of "strike" and instead held "meetings" or "rallies," which was good enough. He would enter the congress the next day with a position of some strength.

On the opening day of the congress, Sakharov took the podium with a sack of mail in his hand and told his fellow delegates that over sixty thousand telegrams and signatures had already arrived in support of abolishing Article 6. Although the "aggressive-obedient" majority drowned him out at points, Gorbachev did not let him speak a second longer than the time allotted and cut him off with an angry bell and a warning to stop "manipulating" the people. And to prove that he, not Sakharov, controlled the critical forces, Gorbachev had the move to repeal Article 6 squelched on the spot by voice vote. The party was in control, both constitutionally and in fact, making any repeal of its power both unnecessary and impossible.

It was a defeat, but not necessarily permanent or decisive. Gorbachev was retreating for tactical reasons of his own, having decided, correctly or mistakenly, that it was too soon to risk alienating the party. But sooner or later he would have to either defend his original vision or cease to matter. The lingering question was whether Gorbachev himself was ultimately a prisoner of the mentality that had formed him. Would he be able to take every step but the needful one?

The shout-down and backlash had more of a dispiriting effect on Sakharov's Interregional Group of Deputies than on him. On the next day of the congress, December 13, his natural allies told him they now doubted their political validity as an organized opposition and questioned the political efficacy of Sakharov's call for a strike.

He spent that evening working out a response to those doubts and questions. While at his desk he also wrote the long-postponed foreword to the second volume of his memoirs and the epilogue to the first:

> The main thing is that my dear, beloved Lusia and I are united—I have dedicated this book to her.
>
> Life goes on. We are together.

On the morning of the fourteenth he gave Bonner those pages, both as a gift, since she had wished the memoirs into existence in the first place, and as a task, since she was in charge of typing and editing, always his first reader. The thought of finishing the story of his life had always troubled him with a "strange sadness" and, though not superstitious, it was better he give her the final pages that day than the next, which was the anniversary of his father's death.

He spoke that day to the Interregional Group of Deputies like a general to troops who have lost heart: "What is the meaning of political opposition? We simply cannot share responsibility for the actions of a government that is leading the nation to disaster and postponing the realization of perestroika for years to come. During that time the country will fall apart, collapse." And he spoke to the "accusation that the call for a two-hour political strike was a gift to the right wing, and the gift would be augmented by our declaration of opposition to the government. I categorically reject this notion. Our appeal initiated a nation-wide discussion. . . . What matters is that the people have finally found a means to express their will and are ready to support us politically. . . . For us to fail to take action now would be a *real* gift to the right wing; that is all they need from us in order to triumph."

How many had he rallied, how many were lost? He would know tomorrow. That evening he ate a quick dinner, then told Bonner he was going for a nap, but to wake him at nine: "There'll be a battle tomorrow."

But when she came to wake him he was dead.

EPILOGUE

The Life After Death

THE days of the funeral were cold and bitter with grief and politics. The government proposed to Bonner that Sakharov's body lie in state in the Hall of Columns, but she wouldn't hear of it; where Lenin and Stalin had lain was no place for Sakharov. The Palace of Youth was more appropriate. The following day the cortege would stop at the Academy of Sciences so his fellow Academicians would have one last chance to define themselves in relation to Sakharov, then to FIAN. From there to Luzhniki Stadium for a memorial and finally to the same cemetery of snow and pine where Ruth was buried.

It was as if Gorbachev begrudged Sakharov the fame of his death. He would not postpone the congress except for a moment of silence and refused to declare a day of mourning because "it was not the tradition." The hostile majority of the deputies took Sakharov's death with unconcealed pleasure as a windfall of good fortune. They hissed when, after some delay, it was announced that the congress would in fact pause for a few hours during the funeral. And history does not ring with Gorbachev's voice silencing their scorn.

Sakharov's death occasioned other posturings. Now by natural inheritance the leader of the opposition, Yeltsin made sure his fervent words about Sakharov were heard by correspondents. The ink on Yevtushenko's poem on the death of Sakharov was no sooner dry than the poet made it available to the world press. Videos of Sakharov's last days were on sale for fifteen hundred dollars, no rubles accepted.

The KGB was on high alert. The funeral would occasion a mass and potentially seditious assembly that could not be prohibited but could, and must be, controlled by a strong police presence. Not only did the KGB continue to report on Sakharov after death, it was once again portraying Bonner in the time-honored tradition. Discussing the cause of Sakharov's death in a report to the Central Committee, the KGB chief states: "To some degree, the academician's widow, E. Bonner contributed to this by fanning her husband's political ambitions and by attempting to play on his pride." This is the ominous retrograde tone of those who would strike against Gorbachev, as Sakharov had predicted. Gorbachev ignored Sakharov to his peril and would thus go down in history as both Liberator and Fool.

* * * *

On a frigid day tens of thousands came to the lying in state at the Palace of Youth—Russians, Tartars, Armenians, Azerbaijanis, Lithuanians, Estonians, Ukrainians, Poles, in specific gratitude or simple homage.

Sitting by the open coffin, Bonner wept and kissed Sakharov's face, or else lapsed into silence, cigarettes the only consolation.

The next day was not as cold: drizzly, slush and mud underfoot. When the rain fell, some in the crowds repeated the saying, Nature weeps for the good. There were two stops for the cortege, which was a police Mercedes and a few old yellow buses, Sakharov's coffin in one, Bonner on a bench beside it. The first stop was at the Academy of Sciences, where Bonner told Gorbachev: "You've lost your best opponent." The second was at FIAN, where no one needed to be told what they had lost.

Fifty thousand people streamed into Luzhniki Stadium. The Albinoni concerto, the music at which Sakharov had wept for the happiness of love, was played. In the orations Sakharov was called a "man of the twenty-first century," but one whose life had deep parallels to the Russian past. Deputy Anatoly Sobchak, future mayor of St. Petersburg, pointed out that Sakharov had died on December 14, the day in 1825 when the Decembrists, a "handful of Russia's best people rose up against slavery." Another deputy was reminded of Tolstoy, whose funeral had also come on December 18, eighty years before. "There's been a sea of blood and grief between the two events. Dawn still hasn't come, but it's near."

Boris Bolotovsky, communist, physicist, visitor of Sakharov in

Gorky, recalled a short story about Tolstoy's funeral by Leonid Andreyev, "The Death of Gulliver." Miniature creatures scurry to retail their connections to the great man—the time they fell into his pocket—but some good Lilliputians had grown used to falling asleep to the rhythms of that enormous heart and now tossed in their beds.

Below the podium was the crowd of fifty thousand—leather jackets, fur caps, puffs of cigarette smoke and breath visible in the cold, flags of all the nations seeking freedom, placards with a "6" struck out, referring to Article 6 of the Soviet Constitution. Some argued about what killed Sakharov: was it Gorky that cut years off his life, or was it the way Gorbachev cut him off with that bell that gave him the heart attack? People inclined to the latter. Cries of anger: "You didn't die, you were killed!" "We know who tortured you and we'll avenge you." And, as in a medieval painting, off to one side, missing the great moment entirely, a few were busy complaining that the flower sellers jacked up their prices, and hockey fans relived yesterday's game.

But that only made it feel more like the funeral of a saint. On that point the speakers, the correspondents, and the masses were in agreement. And it was a point on which Sakharov, to use the postmortem subjunctive, would have vehemently dissented. Sainthood was an alien concept, an elevation that alienated him from people while also effectively removing him from the political debate.

In the orations Sakharov was compared to Gandhi, but he would have preferred someone to have struck Orwell's tone: "Saints should always be judged guilty until they are proved innocent." In that essay, written in 1949 as Sakharov was first visiting the Installation, Orwell had asked: "Is there a Gandhi in Russia at this moment?"

And was there? The two masters of the hunger strike did share a similar mix of the moral and the political, a defenselessness, a touch of sanctity, a quiet triumphancy.

But other comparisons were also made. Sakharov was like Leonardo, a cool and brilliant blend of the humane and the scientific. And, like Leonardo, he was a builder of weapons. Or a Soviet Buddha who had walked out from the palace of the Installation to discover a world of nuclear sickness and nuclear death. Or

Giordano Bruno telling his persecutors: "You are more afraid, in handing down this conviction, than I am in hearing it." Or Robert Oppenheimer, who turned against the weapon he helped create, though Sakharov himself preferred to think of himself as Teller in his Installation phase and Oppenheimer-like thereafter, though he would "go even further than Oppenheimer had."

But comparisons only illuminate aspect, not essence. Sakharov eludes them all because he was so individuated and paradoxical, the two perhaps going together. Gallant, old-fashioned, yet a visitor from the future, a nineteenth-century "Martian." Shy defier of tyrants. The friendless friend of humankind. More of a success as the "father of the H-bomb" than as the father of his children. Ill at ease with women and unable to dance, he was a man of two great loves, the second, with Bonner, a grand waltz through the ballroom of history.

To sickly sweet canonization, Sakharov would have much preferred a just assessment of his victories and failures as a political man and as a scientist. Inevitably, his scientific work is judged against a standard of what he might have otherwise achieved, but that does not mean that what he did in fact achieve cannot be fully and fairly adjudicated. But scientific opinions can be issued only by scientists. The two cultures are still a one-way street. It would have been interesting to read Sakharov's stolen essays on Pushkin and Faulkner, but there are very few nonscientists capable of rendering detailed, accurate judgments on scientific work. Sakharov on Faulkner's work is appealing, Faulkner on Sakharov's unthinkable.

Even among those able to render such judgments, there are other factors that inhibit any full or final word. Time is one. Andrei Linde, a colleague of Sakharov's at FIAN, said: "It . . . took eleven years (from 1967 to 1978) to see the importance of Sakharov's ideas concerning the baryon asymmetry of the universe. . . . Almost an equal amount of time was required to fully understand its value in the development of cosmology. Many other ideas of Sakharov had a similar fate."

And others might yet have. According to the eminent American physicist Sidney Drell, recent experiments at the Stanford Linear Accelerator Center appear to confirm Sakharov's ideas on asymmetry. But final judgment cannot yet be rendered on his theoretical work on cosmology, gravitation, particles or on his more applied concepts, such as using atomic charges to blast asteroids close enough to Earth to mine their mineral wealth or detonating them underground to prevent earthquakes, ideas that Jules Verne and H. G. Wells would have liked.

His greatest contribution may yet prove to be in the field of controlled thermonuclear fusion as a source of energy, a field he pioneered with Tamm, who always acknowledged Sakharov's insights as the more important. On that aspect of his work two scientists from the Kurchatov Institute of Atomic Energy in Moscow took the long, and optimistic, view: "Sakharov received his highest recognition as a scientist mainly for his work on the hydrogen bomb. Compared with his role in the peaceful development of fusion power, this side of his activities may in the future seem merely incidental. He will then remain in the memory of our grateful offspring not as the designer of weapons of terrible power, but as the founder of the ecologically clean and plentiful energy source of the future—energy to serve a new human society, based on Reason and Humanity, for which he struggled and to which he offered without reserve all his strength."

None of Sakharov's colleagues, foreign or Soviet, doubted that his talent and range were of the first order, though within that category there are many gradations. Lev Landau, arguably the greatest Russian physicist, had a system by which he ranked physicists; he placed Einstein, Bohr, and a few others at the very top (he put himself in the very next rank down.) Landau does not seem to have assigned Sakharov a specific slot in that system but did say of him once: "He is an outstanding man. While I would not consider him a genuine theoretical physicist, his is rather a 'constructive genius.' "

Great achievement in science or art requires a devotion that is both selfless and ruthless. Happiness is usually not associated with the lives of the greatest achievers, and Sakharov was a happy man. Yet who, in total, outachieved him?

The question of what he might have conceived and discovered if not for his work on thermonuclear weapons and human rights is a reasonable question and one that he asked himself, but, fundamentally, it is meaningless. Like everyone else, but more than everyone else, he created himself through his choices and, like everyone else, was formed by a fate he did not get to choose. Sakharov's often quoted "Who else if not me?" was not said in the high register of a Martin Luther or a Martin Luther King, but as a matter of fact. As the creator of nuclear weapons, he was uniquely positioned to act.

The assessment of Sakharov the scientist is necessarily shifting, but one image of it is fixed, visually, as befits an icon-conscious land. Anyone entering FIAN's imposing yellow and white Soviet neoclassical edifice will pass between two sculptural reliefs, the one on the left dedicated to Tamm and the one on the right, abstract, ovaloid, futuristic, to Sakharov.

His political image is not as clear. The sudden collapse of the Soviet Union and the subsequent rush of events had the effect of shifting Sakharov into the past along with the country whose existence roughly coincided with his own. Though invoked, he was forgotten, or so it seemed until a poll on the most influential people of the Russian twentieth century placed him third after Lenin and Stalin, this time the linkage with them unavoidable.

Some thought Sakharov would have been particularly aggrieved by the disintegration of the Soviet Union, a new form of which he had envisioned in his constitution. But he also might have quickly seen that the ease and swiftness of its collapse proved its inevitability. Exactly why the USSR imploded will be argued for generations; as Chou En-lai said of the causes of the French Revolution —it's too soon to tell. But the Union of Soviet Socialist Republics had been described as "four words, four lies," and when glasnost killed those four with a thousand hair-raising truths, the USSR simply vanished. And for that reason the end of the Soviet Union was surprisingly "mild by Soviet standards."

The Russia that emerged from the collapse was hardly the best Sakharov could have envisioned, but he would have evaluated it on its merits. People were free to travel, to leave and return, some of the intelligentsia dividing their time between Moscow and some European or American city. The problem now was money, and it was foreign countries, not OVIR, making the visa problems. The economy sinking like a rusty tanker. Soldiers unpaid. The country looted by seven oligarchs, some of whom built media networks that advanced freedom of the press while furthering their own political agendas. But the oligarchs' power declined, and freedom of the press with it, under President Vladimir Putin, called "Lilliputian" by some intelligentsia who have not lost their taste for punning. Tragic pairings of events—Yeltsin standing on the tank in front of the parliamentary White House during the attempted coup against Gorbachev in 1991 (the patriot sluts of Moscow lowering themselves into the tanks to distract the boys), Yeltsin shelling that same building a short while into his own presidency in 1993 to put down a revolt of the right. Two wars against Chechnya. Yet a furious press criticized Yeltsin for the shelling and the wars. Freedom of religion and ethnic violence. The Armenians and Azerbaijanis at war over Karabakh until 1994, the twentieth century ending with no solution in sight. Civil war in Georgia. Brezhnevian stagnation and fundamentalist revolt in Central Asia. The Crimean Tartars now caught in a new snag. Khrushchev had grandly made a present of the Crimea to Ukraine in 1954 to commemorate the three-hundredth anniversary of Ukraine's voluntary association with Russia, in the certain faith that it didn't make any difference, it was all Soviet. Now the Tartars had to petition the government of Ukraine, which was also corrupt and imploding. By contrast, some of the empire's former provinces—the Baltic states, Poland, Hungary, the Czech Republic, acquitted themselves rather well politically and economically, the three former eastern bloc countries having joined NATO, and the Baltic states aspiring to. A Russia that did not produce a great outpouring of literature and art when finally set free, on the contrary, the arts seeming to wither in the stampede for riches and the struggle to survive. Science pauperized. A Russia in which neither Gorbachev nor Yeltsin are popular. A nation with neither identity nor power. And one that sooner or later will want both, but of what kind?

And yet the world was no longer in the nuclear danger it had been in during the days of the Cold War, the dread gone as a constant if intermittent feature of life. In that respect Sakharov's country and the world breathe freer, a convergence of relief. But the pessimists contend that the peace is illusory, a ruined Russia no safeguard of its nuclear arsenal and nuclear wastes.

Was it collapse or convergence? There is no question that in the twenty years between Sakharov's exile and the end of the century Russia has become more like the West. In fact the new Russia's brutal markets caused many Russians to leave for America and its welfare system—the joke was that they were fleeing capitalism for socialism.

Whatever precisely Russia underwent, collapse or transition or some dialectical combination of both, the transformations it wrought in the process were, as usual, dizzying. The police station to which Sakharov went to report the invasion of his apartment by Black September terrorists is now the Andrei Sakharov Museum and "Peace, Progress, Human Rights" Public Center; recently in financial straits, it was bailed out by Boris Berezovsky, one of the oligarchs who had looted the economy. That nicely balances Sakharov's childhood home on Granatny Lane having become Police Station No. 38. The Cathedral of Christ the Savior to which Sakharov was taken as a boy before Stalin had it dynamited has now been restored at enormous expense by Moscow's Mayor Luzhkov. And it was in that cathedral in August 2000 that the bones of Tsar Nicholas II, having first been subjected to DNA analysis, were transposed and he was, as a martyr of evil, glorified as a saint. Even the remains of Saint Serafim of Sarov, which had vanished in 1920, suddenly turned up in the cellar of the Museum of Atheism, which once had been and would yet again be Kazan Cathedral in St. Petersburg.

Past and future converged in a Russian magic spell of name changes. Having been twice renamed—as Petrograd and Leningrad, St. Petersburg is again St. Petersburg, though due to some Gogolesque bureaucratic inconsistency St. Petersburg is now the capital of Leningrad Province. The cabdrivers in Moscow carry volumes of street directories from various periods since now even Chekhov Street has gone back to its old name. Moscow's main thoroughfare, Gorky Street, is again, as before the revolution, Tverskaya. Even the city of Gorky reverted to Nizhny Novgorod, the name Maxim Gorky had known it as. The locals sometimes forget and say Gorky, and they haven't gotten around to changing it on train tickets. And Sakharov's apartment in Gorky is now a museum, one of whose original staffers, in a touch of unforeseen convergence, has now quit to seek her fortune working for Herbalife. Ashkhabad, where Sakharov finished his undergraduate work, has now become the more Turkmenian Ashgabad. Kazakhstan, where most of his thermonuclear weapons were tested, has dropped the "h," supposedly there because of the way Stalin, with his Georgian accent, pronounced the country's name. And the KGB turned into the FSB, as if that settled that.

Some of Sakharov's compatriots and colleagues themselves present the most improbable instances of convergence. Khrushchev's son, Sergei, became an American citizen and is a professor at Brown University. Roald Sagdeev, Sakharov's student and head of the Soviet Space Agency, married the granddaughter of President Eisenhower and moved to the United States in what he called a "heart drain." Agrest, the "Soviet rabbi," ended up in North Carolina trying to reconcile the Bible and cosmology.

Returning at last to Russia in a train that took the anti-Leninist route of east to west, even Solzhenitsyn became in time an absurd symbol of convergence, a prophetic bearded Russian talk-show host so carried away by his own ideas that he dispensed with guests until his ratings dispensed with the show.

Anatoly Shcharansky became Natan Sharansky, a conservative elected official in the government of the Israel, of which he had so dreamed when in prison. He keeps a portrait of Sakharov in his office to remind him of Sakharov's "straight, pure, clear moral thought. This connection of simplicity and directness with his greatness, it was from the heavens. It was a very strange feeling, which I have practically had with no one else. So it's important to always remind myself that the distance from the heavens to the earth is very close and it all depends on you how close it will be. That is what that picture on the wall reminds me of all the time."

Sakharov became a part of people's inner dialogue. Sagdeev says that, in the years of turbulent crisis after Sakharov's death, he was not alone in addressing himself: " 'Andrei would have objected' or 'Andrei would have approved' or even 'Andrei would have known.' "

In China, Fang Lizhi, the dissident astrophysicist, says Sakharov "embodied the role that intellectuals are called upon to play in the creation of civil society and inspired scientists working under other dictatorships, including myself in China, to become leaders in the struggle for democracy." Sakharov's "exhortations against totalitarianism might seem anachronistic" to the West but not to anyone who wishes to live free in China, which has foregone socialist economics but not party power. And, as Sakharov would have done, Fang Lizhi reminds the West not to be lulled by wealth and triumph: "freedom is fragile . . . if democratic societies are not protective of their liberties, even they may lose it."

Perhaps Sakharov's best legacy is in any given Russian's reflexive assumption of freedom, with no more thought of him than Rosa Parks is remembered by a black child boarding a bus in the American South.

But just how important was Sakharov in the transformations wrought on his country? The leaders of the Soviet Union from Khrushchev to Gorbachev, and all the KGB chiefs, especially Andropov, considered him significant, influential, even dangerous. Sakharov himself had no doubts about his own importance, which he gauged frequently to ensure effectiveness—accurate calibration precluding the delusions of self-importance. In an interview two months before his death he replied to a question about fate: "In fate as a destiny, I do not believe. I believe that the future is unpredictable and indeterminate; it is created by all of us, step by step, in our endless, complex interactions. But freedom of choice is our inherent right. This is why the role of an individual whom fate has singled out at some crucial moment in history is so important." He was speaking of himself.

The genius who can demonstrate the exact political consequences of courageous moral action has yet to be born. But the poll linking Sakharov with Lenin and Stalin as the three who defined the Russian twentieth century is revelatory in several ways. The conjunction of names also has an implication—on one side Lenin and Stalin, on the other, Sakharov. The first two built the house of communism; he brought it down. To ask whether it would have happened without him is as imponderable as asking whether communism would have been different without Lenin and Stalin.

As great as the question of what Sakharov might have achieved as a scientist is the question of what difference Sakharov would have made in Russia if he had been granted the few more years he expected to live. Yuri Orlov, scientific colleague and humans rights comrade-in-arms, thought the difference would have been crucial: "Sakharov had already transformed Russian history. Had he lived, he would have transformed it again. For near the end of his life, almost overnight, he developed into a brilliant, committed politician, not only leading tough battles in Congress, but meeting with industrial workers, and drafting a new constitution. . . . The only public figure acceptable to all parts of his too vast country, Sakharov had become just the person to lead a decaying nation exhausted with itself, yet still capable of responding to great honesty, great professional achievement, great suffering, and great precision of thought."

If Orlov is right and Sakharov could twice have been the savior of his country, that would make for a perfect symmetry. His scientific work had made his country a superpower. (And that the Soviet Union was a superpower only because of those weapons became abundantly clear when the empire collapsed like cheap prefab concrete.) To lead that superpower into democracy would have balanced his thermonuclear contribution, though he never had the least regret about that work; at the time patria and parity were one.

The two sides of his life, science and politics, were hardly separate and discrete. There could be no science more political than the thermonuclear, no politics more scientific than his. His politics grew out of his science. The magnitude of his scientific contribution conferred a commensurate authority and a relative impunity. It was science that woke him to conscience while calculating, to the nearest thousand, future probable deaths per megaton tested.

But he was not granted the years to achieve that symmetry. He was torn from life too suddenly to have suffered much, his face serene in the open coffin, serene in the death mask. But there had been some violent separation; Bonner found him not in bed but on the floor.

In the afterlife that exists here on earth, the people in Sakharov's life went on with theirs. His brother Yura is still alive but very ill. He's looked after by Lyuba, Sakharov's younger daughter who, with her husband and son, lives in the apartment Klava loved, the portrait of Don Quixote that Sakharov was given on his fiftieth birthday still where it always was. A librarian, Lyuba has her father's reserve and deep earnest. His elder daughter, Tanya, a research biologist who has some of her mother's Volga mix of gaiety and melancholy, has been living in Cambridge, Massachusetts, divorced from the physicist who married into the Sakharovs at the wrong time, but not before producing Marina, a tall, brilliant, and striking young woman who switched from physics to finance, a convergence that would have chagrined Sakharov even though he would have recognized it as the new template of the times (and really nothing new either—hadn't Newton become warden of the Mint?). Dima continues to cavort about Moscow, madcap, grandiose, easily offended, knowing everything and everybody, suspicious, obsessed with not being obsessed about his father, all laced with currents of secret warm tenderness, a character of Dostoevskian complexity to say the least.

Of Bonner's children it was Alexei and Liza—of whom it was caviled they'd break up after Sakharov's hunger strike reunited them—who proved to have the enduring marriage, Tatiana and Efrem having gone their ways, hers to the head of the Sakharov Archives at Brandeis University and his to Israel, some science, a life of his own. Their children, Matvei and Anna, took differing directions as well. Matvei is at the edge of the literary avant-garde in New York and Moscow, bound to both America and Russia, at home in each. Anna married an American, lives in Colorado, studies Spanish, and in 1998 made Elena Bonner a great-grandmother.

Never just Sakharov's wife, Bonner has never been just Sakharov's widow. In their love she was not the one who spoke of suicide rather than life without the other. She would not have been true to herself or to him if she had not gone on with her life, their cause.

She does what all widows do, all widows of great men especially—works so that his name will be remembered. Her energy helped create the Sakharov Museum in Moscow and the Sakharov Archives at Brandeis. She is editing Sakharov's diaries when she can summon the will. She did three years of archival work and travel to produce her third book, short but rich, *Free-hand Notes toward a Genealogy of Andrei Sakharov,* communing with his spirit after death by tracing the life that had preceded his birth.

Though concerned that Sakharov's achievement be commemorated, Bonner is not the least like that widow who, according to André Maurois, "kept a careful watch both upon the portrait of her husband and upon the attitude which she wished to see herself assume before the eyes of posterity. The results are only too familiar: 'books so well stuffed with virtues' as someone said, 'that I begin to doubt the very existence of virtue.'"

When so moved, Bonner speaks as frankly *about* Sakharov as she would sometimes speak *to* him. And whatever regard she has about her place in history, the good fight was always her passion. In Bremen, receiving the Hannah Arendt Award for 2000, Bonner proceeded at once to "dangerous words." The age of state ideology and propaganda may be over, but the "inertia of falsehood is stronger even than the inertia of fear. We lived, and continue to live, in a state of lies. The great lie calls Russia a democratic state. The barely created election procedures were violated during the elections in Chechnya, which took place during the first Chechen war, and, subsequently, in Yeltsin's 1996 election victory, which was decided largely by money and not the will of the voters. Then came the appointment of Putin as Yeltsin's heir, as if Russia were a monarchy. . . . And where there are no valid elections, then, by definition, there is no democracy. . . .

"But the greatest disaster and shame of the new Russia are the two Chechen wars and the de facto genocide of the Chechen people. The first war was needed by Yeltsin to raise his ratings in the polls and it was used by his entourage to enrich themselves with billions of rubles. It ended with the total destruction of Grozny, which had a population of half a million people. . . . This was called the 'reestablishment of constitutional order.' Western leaders (friend Bill, friend Helmut, and all the rest) took this phrase at face value (or pretended to—lying is as contagious as the plague). . . .

"The second Chechen war was needed by the heir of Russia's first president. . . . Now the Chechen war has a new, respectable-sounding name: 'the fight against international terrorism.' The Western nations accept this (or say they do); they also make a show of defending human rights by passing nonbinding resolutions deploring atrocities in Chechnya at the Council of Europe and similar forums. . . . And with these two wars Russia has lost its newborn democracy."

Though as always he would have delighted in her blistering gusto, Sakharov would also have argued every point and nuance, his vision never coinciding with anyone's, not even hers. But he could not have argued her conclusion, for it was he she quoted on humanity's capability to determine its own fate: "The future can be wonderful, but it might not be at all. That depends on us."

One more time, his words, her voice.

His last words to her had been of tomorrow, which meant he had no idea he was in his final hour. But what had moved him to finish the last of the memoirs on the day before his father's death —chance, duty, premonition?

Andrei Sakharov is as elusive in death as he was in life, a consistency that is reflected in the starry skies. In 1979 the International Astronomy Catalog named one of the roughly five thousand asteroids in the belt between Mars and Jupiter in his honor. And so even at this instant, Sakharov's asteroid streaks through space, only one tenth of one percent off perfect orbit.

NOTES

Epigraphs

ix Two great things . . . of *the* age. Ernest Gellner, *Encounters with Nationalism*, p. 92.

ix My fate was . . . keep up with it. Drell and Kapitza, eds., *Sakharov Remembered,* p. 248. Original in *Molodezh Estonii,* October 11, 1988.

Prologue. Reports from KGB Chief Andropov to the Central Committee

1 13 June 1968 . . . conversation with him. Freeze et al., eds., *KGB Files,* pp. 25–26.

1 20 April 1970 . . . *Sakharov's apartment.* Ibid., p. 37.

2 17 April 1971 . . . eavesdropping, etc. Ibid., p. 58. I changed "frondeur" to "malcontent."

Chapter 1. A Difficult Birth

3 "the Word of God, the bread of angels." Radzinsky, *Last Tsar,* p. 65.

4 "We magnify . . . our God." Timberlake, ed., *Religious and Secular Forces,* p. 211.

4 "deep silence . . . them home." Ibid., p. 210.

4 "At night . . . silver water." Radzinsky, *Last Tsar,* p. 67.

6 "constant lawyer for strikes." Bonner, *Volniye zametki,* p. 87.

6 "Since you're . . . Sakharov." Ibid., p.72.

7 "Do not hurt anyone." Drell and Kapitsa, eds., *Sakharov Remembered,* p. 238.

7 "They lived . . . same day." Bonner, *Volniye zametki,* p.78.

8 "Russia would . . . revolutionaries." Sakharov, *Memoirs,* p. 4.

9 "Mongol cast." Ibid., p. 3.

9 "obstinacy" and "awkwardness . . . people." Ibid., p. 4.

11 "What a beautiful. . . . you see one." Bonner, *Volniye zametki,* p. 122.

11 "almost physically" . . . "identified tones . . . colors." Sakharov, *Memoirs,* p. 7.

11 "optically . . . notes." Nabokov, *Speak, Memory,* p. 35.

12 "shots broke. . . . begun!" Leon Trotsky, *Stalin* (New York: Harper & Row, 1946), p. 128.

12 "Unspeakably . . . (. . . ocean)." Timberlake, ed., *Religious and Secular Forces,* p. 182.

12 "the whole cast . . . divisions." West, *H. G. Wells,* p. 4.

12 "Belgian beach. . . . fog." Sakharov, *Memoirs,* p. 8.

14 "provided . . . profession." Ibid., p. 3.
14 "Today I met . . . on them." Bonner, *Volniye zametki,* p. 45.
14 "abandoned . . . Russian Empire." Figes, *People's Tragedy,* p. 384.
15 "It was a cold. . . . people's hearts." Paustovsky, *Story of a Life,* p. 481 and p. 489.
16 "disgrace to democracy." Figes, *People's Tragedy,* p. 509.
16 "There is nothing . . . its right." Ibid., p. 510.
16 "The year 1918 . . . gray paper." Paustovsky, *Story of a Life,* p. 506.
16 "The temperature . . . Herzen." Bonner, *Volniye zametki,* p. 46.
17 "Today. . . . from us." Ibid., pp. 46–47.
18 "majestic . . . twenty-first century." Quoted in Rubenstein, *Tangled Loyalties,* p. 37.
18 "live banner." Massie, *The Romanovs,* p. 16.
18 "crowned executioner." Ibid., p. 15.
18 "The execution . . . ruin." Ibid., p. 16.
19 "maybe. . . . Russia!" Bonner, *Volniye zametki,* p. 126.
19 "night . . . steppe." Sakharov, *Memoirs,* p. 9.
20 "Something wonderful. . . . difficult birth." Bonner, *Volniye zametki,* p. 52.

Chapter 2. A Soviet Zodiac

21 "terribly . . . self." Bonner, *Volniye zametki*, p. 49.
22 "The train. . . . mistake." Quoted in Lourie, *Russia Speaks,* p. 93.
23 "All sounds have stopped." Figes, *People's Tragedy,* p. 785.
23 "time of hope . . . Soviet history." Heller, Nekrich, *Utopian Power,* p. 167.
23 "it was in the twenties . . . self-abasement." Mandelstam, *Hope Against Hope,* p. 168.
23 "Still, the Bolsheviks . . . people now." Sakharov, *Memoirs,* p. 23.
24 "Today. . . . interesting." Ibid., p. 12.
25 "stepping . . . Red Square." Quoted in Lourie, *Russia Speaks,* p. 15. The translation is slightly altered here.
25 "The house. . . . banisters." Sakharov, *Memoirs,* p. 10.
26 "crisp. . . . truly work." Ibid., p. 13.
26 "You must . . . science." Krementsov, *Stalinist Science,* p. 11.
27 "virtually . . . ladder." Sakharov, *Memoirs,* p. 18.
27 "There was a ferment . . . if at all." Ibid., p. 15.
28 "spelling out . . . steamships." Ibid., p. 27.
28 "perfect." Ibid.
29 "brilliantly . . . gestures." Ibid.
29 "eternal games" . . . "something . . . larger." Ibid., p. 18.
30 "a morose . . . shoemaker" . . . "stout and loud." Ibid.
30 "enormous pale blue eyes." Ibid.
30 "We would walk . . . fairy tales." Ibid., p. 19.
30 "fiercest animosity." Ibid., p. 18.

30 "Jewish intelligence. . . . Jewish spirit." Ibid., p. 19.
31 "tall fellow . . . his eyes." Ibid., p. 20.
31 "Leave me alone. . . . one piece." Ibid.
32 "home study . . . my life." Ibid., p. 28.
33 "slightly absent-minded . . . his manner." Ibid., p. 29.
33 "opening up . . . 'humanities.' " Ibid.
33 "who considered . . . likes of me." Ibid.
34 "I made . . . enemies." Ibid., p. 30.
34 "During one. . . . a stool." Ibid.
35 "caricature of Stalin . . . the moustache." Ibid., p. 21.
35 "I can think . . . over the grass." Ibid., p. 16.
35 "don't realize . . . are listening." Ibid., p. 20.
35 "tool compartments . . . pulled out." Ibid.
36 "I began to hear . . . more often." Ibid., p. 21.

Chapter 3. The World Aglow

37 "Papa made . . . of me." Bonner, *Volniye zametki*, p. 153.
37 "There is no national . . . multiplication table." Drell and Kapitsa, eds., *Sakharov Remembered,* p. 86.
37 "unlimited . . . improvement." Rhodes, *Atomic Bomb,* p. 107.
37 "that he had never . . . about this." Sakharov, *Memoirs,* p. 15.
38 "I was fascinated. . . . in their power." Ibid., p. 31.
38 "free and confident manner." Ibid.
38 "during the year . . . event to me." Ibid., p. 32.
38 "To our right. . . . of my interest." Ibid.
39 "often the first . . . embarrassing position." Altshuler et al., eds., *On mezhdu,* p. 18.
39 "There are two . . . completely dark." Remark by Mark Kac, quoted in Gleick, *Genius,* pp. 10–11.
39 "We are living . . . physics!" Rhodes, *Atomic Bomb,* p. 157.
40 "Before the war . . . physics." Altshuler et al., eds., *On mezhdu*, p. 780.
40 "If the university . . . any more." Ibid., p. 780.
40 "seemed perfectly . . . of proof." Ibid., p. 780.
40 "He greatly disliked. . . . Andrusha." Ibid., pp. 781, 778.
40 "In the stories. . . . hybrid field." Gleick, *Genius,* p. 52.
40 "rockets. . . . aglow." Bernstein, *Cranks,* p. 157.
40 "It was taken . . . university." Sakharov, *Memoirs,* p. 16.
40 "I entered . . . automatically." Ibid., p. 36.
40 "enjoy life . . . (. . . indeed)." Ibid., p. 15.
41 "merely personal. . . . with it." Bernstein, *Cranks,* p. 124.
41 "Somehow it entered. . . . scientist." Sagdeev, *Making Soviet Scientist,* pp. 7–8.
41 "Both Fermi's . . . fourteen years old." Rhodes, *Atomic Bomb,* p. 205.
41 "This was my first . . . of nature." Sakharov, *Memoirs,* p. 4.
42 "the chanting . . . the icons." Ibid., pp. 4–5.

42 "dazzling . . . understand." Ibid., p. 12.
42 "moved by . . . the times" . . . "subtle influence." Ibid., p. 4.
42 "One of the books. . . . in God." Lourie, *Russia Speaks,* p. 51.
42 "The electric . . . induction." Sakharov, *Memoirs,* p. 34.
43 "simply faded away." Ibid., p. 36.
43 "I have absolutely. . . . with him" . . . "no interest whatsoever." Altshuler et al., eds., *On mezhdu,* p. 779.
43 "counter-revolutionary . . . Menshevik Center." Bonner, *Volniye zametki,* p. 127.
44 "I was and I remain. . . . whatsoever." Bonner, *Volniye zametki,* p. 128.
44 "to break. . . . we are powerless." Pirozhkova, *At His Side,* p. 63.
45 "by our fatherland's . . . mild." Bonner, *Volniye zametki,* p. 129.
45 "The whole school . . . her tears." Sakharov, *Memoirs,* p. 31.
45 "caught a glimpse. . . . nothing to him." Ibid.
46 "I prefer . . . convictions change." De Jonge, *Stalin,* p. 461.
46 "I didn't make . . . the university." Sakharov, *Memoirs,* p. 37.
47 "tall, skinny. . . . of him?" Altshuler et al., eds., *On mezhdu,* p. 143.
47 "head thrown . . . gaze." Ibid., p. 771.
47 "One thing . . . didn't dance." Ibid., pp. 771–772.
47 "Einstein. . . . free of them." Ibid., p. 339.
48 "Kalashnikov's right . . . a different way?" Ibid.
48 "One must not . . . agonizing." Ibid.
48 "malice" and "enmity passing into hatred." Ibid., p. 341.
49 "But not. . . . landslide." Ibid., p. 343.
50 "deported en masse . . . state." Crankshaw, *Shadow,* p. 33.
50 "It never . . . philosophy." "attempt . . . meaning." Sakharov, *Memoirs,* p. 36.
51 "cursory manner," "inadequate grounding." Ibid., p. 37.
51 "I can't understand . . . failed to notice.)" Ibid., p. 37.
51 "contorted with grief," "the soul . . . Lane." Ibid., p. 38.
52 "For the first . . . Russian again!" Ibid., p. 40. I have altered slightly the translation, which was mine to begin with.
52 "for reasons of inertia, not ideology." Ibid., p. 41.
53 "psychological boost." Ibid., p. 42.
53 "helping . . . over Moscow." Ibid., p. 42.
53 "with the glare . . . my way." Ibid., p. 43.

Chapter 4. War and Love

55 "More than once . . . from a plane." Lourie, *Russia Speaks,* p. 227.
55 "As office. . . . 'It's . . . for himself!' " Sakharov, *Memoirs,* p. 43.
55 "What're you reading?" Altshuler et al., eds., *On mezhdu,* p. 143.
56 "To my . . . 'It's interesting!' " Ibid.
56 "goofy." Leon Bell, interview with author.
56 "new insights . . . subjects." Sakharov, *Memoirs,* p. 43.

56 "visited . . . soldiers." Ibid., p. 44.
56 "The German. . . . the German invaders!" Overy, *Russia's War,* p. 136.
56 "realized . . . on me." Sakharov, *Memoirs,* p. 44.
57 "junctions . . . tracks." Shklovsky, *Five Billion,* p. 42.
57 "During. . . . silent category." Sakharov, *Memoirs,* p. 44.
58 "To distract . . . amuse ourselves." Shklovsky, *Five Billion,* p. 44.
58 "Every evening . . . laughter." Ibid.
58 "With a vacuum . . . can never win!" Ibid., pp. 39–40.
59 "Tall, skinny. . . . to hell." Ibid., p. 41.
59 "wounded by war. . . . somehow alike." Sakharov, *Memoirs,* p. 44.
59 "One of them . . . to her charms." Ibid., p. 45.
60 "The offensive. . . . serious defeat." De Jonge, *Stalin,* p. 391.
60 "the Red Army . . . not lose." Ibid., p. 391.
61 "As spring. . . . transparent." Sakharov, *Memoirs,* p. 48.
62 "Bourgeois scientists . . . to capital." Auden and Kronenberger, *Aphorisms,* p. 261.
62 "varied from . . . at anyone." Sakharov, *Memoirs,* p. 46.
63 "Aside from . . . in me." Ibid., p. 47.
63 "I had been. . . . might be." Ibid., p. 48.
63 "talked endlessly . . . haunting them." Ibid., p. 50.
63 "they took. . . . to be quick." Wat, *My Century,* p. 132.
64 "There is. . . . what to do." Fischer, *Lenin,* p. 9.
65 "thoroughly absurd." Sakharov, *Memoirs,* p. 51.
65 "without . . . a glance." Ibid.
66 "served on . . . tin cup." Ibid., p. 53.
66 "there were . . . the door." Ibid., p. 54.
66 "giant . . . child." Ibid.
67 "Comrade Stalin. . . . plant!" Ibid.
68 "cores . . . hardened." Ibid., p. 55.
69 "costly optical . . . alloys." Ibid., p. 63.
69 "miracle of technology." Ibid.
69 "constantly . . . storeroom." Ibid.
69 "where Klava. . . . together." Ibid., p. 64.
70 "The loudmouths . . . again." Ibid., p. 60.
70 "dig up . . . peasants." Ibid., p. 57.
70 "He didn't. . . . novels!" Altshuler et al., eds., *On mezhdu,* p. 345.

Chapter 5. "A Change in Everything"

71 The letter from Sakharov to his mother, her reply, and all of Sakharov's wartime letters to Klava are unpublished and were made available to me through the kindness of Tanya Sakharov.
73 "Manstein. . . . Moscow." Overy, *Russia's War,* p. 217.
74 "largest . . . in history." Ibid., p. 220.
74 "symphony from hell." Ibid., p. 223.

74 "The air . . . scorched." Ibid., p. 228.
74 "Our detractors. . . . high summer." Nikita Khrushchev, *Khrushchev Remembers,* p. 208.
74 "recognized . . . methods." Sakharov, *Memoirs,* p. 107.
75 "not exactly . . . of design." Ibid.
75 "gaining confidence . . . beginner." Ibid., pp. 62–63.
75 "sensed . . . promise" . . . "couldn't quite figure out." Ibid., p. 63.
75 "the two of us . . . daybreak." Ibid., p. 59.
75 "intensive study . . . physics." Ibid., p. 63.
75 "warm and well lit." Ibid.
76 "was not reading . . . books." Ibid.
76 "so politely . . . reproach." Ibid.
76 "hellish and unreliable" . . . "all . . . of them!" Ibid., p. 66.
76 "I had been. . . . considerations." Ibid., p. 67.
77 "with the same . . . actions." Ibid., p. 69.
77 "truthfulness . . . work." Andreev, ed., *Kapitsa, Tamm, Leontiev,* p. 247.
77 "calmly and tactfully." Ibid.
77 "quickly . . . to the bottom." Ibid.
78 "rather modest . . . superficial." Ibid.
78 "You've got to learn. . . . requirements." Ibid.
78 ". . . (Theoretical . . . safe.") Bernstein, *Cranks,* p. 85.
79 "it's entirely . . . yourself." Andreev, ed., *Kapitsa, Tamm, Semyenov,* p. 233.
79 "Bravo, Tamm!" Sakharov, *Memoirs,* p. 120.
79 "fanatics." Andreev, ed., *Kapitsa, Tamm, Semyenov,* p. 258.
79 "Solve . . . the wall." Ibid., p. 363.
80 "I can vouch . . . brother." Ibid., p. 334.
80 "I, as an . . . corner." Ibid., p. 334.
81 "The Russian soldiers. . . . to the Gulag." Lourie, *Russia Speaks,* pp. 254–255.
82 "The spontaneous. . . . to the winds." Werth, *Russia at War,* p. 969.
82 "spent nearly. . . . my world." Sakharov, *Memoirs,* p. 72.
82 "truly . . . best book" . . . "written . . . twenty-one!" Ibid., p. 69.
83 "Extreme . . . musicians." Snow, *The Physicists,* p. 138.
83 "like an artist . . . then another." Altshuler et al., eds., *On mezhdu*, p. 470.
83 "Well . . . later on." Ibid.
83 "gentle . . . his face." Ibid., 467.
83 "So . . . talented?" "Not . . . as I am." Ibid., p. 468.
83 "He . . . a fact." Ibid.
84 "I always tended . . . least resistance." Sakharov, *Memoirs,* p. 58.
84 "of a very high spiritual level." Bonner, *Volniye zametki*, p. 49.
84 Tanya Sakharov's remarks from interview with author.
85 "There was. . . . had survived." Sakharov, *Memoirs,* p. 75.

85 "Dickensian." Altshuler et al., eds., *On mezhdu*, p. 523.
85 "ordinary link . . . whole country." Sakharov, *Memoirs,* p. 77.
86 "Each time. . . . attitude." Ibid., p. 76.
86 "When the war. . . . made easy." Rothberg, *Heirs of Stalin*, p. 106.
86 "On my way. . . . two and a half years." Sakharov, *Memoirs,* p. 92.
87 "We have been. . . . for a while." Ibid., p. 93.
87 "I hadn't left . . . everything now." Ibid.
87 "estimated . . . by Tamm." Drell and Kapitsa, eds., *Sakharov Remembered,* pp. 108–109.
88 "Even Beria . . . mesons are." Sakharov, *Memoirs,* p. 79.
88 "Every article. . . . in publication." Ibid., p. 80.
89 "I felt . . . of the gods." Ibid., p. 73.
89 "never before . . . cutting edge." Ibid., p. 85.
89 "to the point of obsession." Ibid., p. 82.
89 "neither support, nor approval." Ibid., p. 83.
89 "quite simply wrong." Ibid.
89 "If our intuition. . . . of that era." Ibid.
89 "excessively truthful." Ibid., p. 81.
90 "financial blow." Ibid.
90 "When you read. . . . for him." Drell and Kapitsa, eds., *Sakharov Remembered,* pp. 109–110.
90 "liveliest . . . joke, " a "teddy bear." Rhodes, *Dark Sun,* p. 30.
90 "first and foremost . . . in water." Sakharov, *Memoirs,* p. 209.
91 "secrecy . . . secret away." Rhodes, *Atomic Bomb,* p. 501.
91 "If a child. . . . be refused." Holloway, *Stalin and Bomb,* p. 132.
91 "larger than life. . . . state hierarchy." Sakharov, *Memoirs,* p. 93.
92 "So began . . . family life." Sakharov, *Memoirs,* p. 5.
92 "furtive" . . . "startling news." Ibid., p. 94.
92 "in 1946. . . . my consent." Ibid.

Chapter 6. Chain of Command

93 "The open . . . in its place." De Jonge, I p. 279.
94 "squeeze his eyes . . . away." Altshuler et al., eds., *On mezhdu*, p. 473.
94 "Here's the situation. . . . even nine." Ibid., pp. 521–522.
95 "ravaged eyes" . . . "What's wrong . . . ?" . . . "You don't. . . . What am I doing?" . . . "Inwardly . . . hysterical." "Get out of here!" Altshuler et al., eds., *On mezhdu*, p. 474.
95 "completely loyal . . . ideology." Ibid., p. 661.
95 "Our initial zeal . . . intellect." Sakharov, *Memoirs,* p. 97.
95 "The physics. . . . my grasp." Ibid., pp. 96–97.
95 "What was . . . *essential.*" Ibid., p. 97.
96 "We all . . . duty." Rhodes, *Dark Sun,* p. 202.
96 "a soldier . . . war." Sakharov, *Memoirs,* p. 97.
96 "If we were. . . . threaten us." Rhodes, *Dark Sun,* p. 225.
96 "possibility . . . to oneself." Sakharov, *Memoirs,* p. 427.

97 "bizarre and fantastic" . . . "striking . . . pursuits." Ibid., p. 96.
97 "From the beginning. . . . routine." Ibid., p. 103.
97 "bound . . . in 1948." Ibid., p. 101.
97 "verify and refine." Ibid., p. 94.
98 "It was clear. . . . those problems." Drell and Kapitsa, eds., *Sakharov Remembered,* p. 126.
99 "inferiority complex." Sakharov, *Memoirs,* p. 132.
99 "Zorba of cosmology." Overbye, *Lonely Hearts,* p. 154.
99 "almost childishly . . . keenly." Ibid., p. 151,
99 "Do you know . . . kilometers." Holloway, *Stalin and Bomb,* p. 54.
99 "behavior of . . . different" . . . "There is . . . publication." Holloway, *Stalin and Bomb,* p. 58.
100 "the only . . . history." Rhodes, *Dark Sun,* p. 259.
100 "basic level. . . . first try." Ibid., p. 162.
100 "One of. . . . uranium deposits." Lourie, *Russia Speaks,* p. 239.
101 "TOP SECRET . . . information." Rhodes, *Dark Sun,* p. 62.
101 "wonderful . . . lacking." Holloway, *Stalin and Bomb,* p. 95.
102 "irreproachable, dryish correctness." Altshuler et al., eds., *On mezhdu,,* p. 293.
102 "would sink . . . our blood." Rhodes, *Dark Sun,* p. 29.
102 "you can't . . . country too." Holloway, *Stalin and Bomb,* p. 106.
102 "for the dual . . . disinformation." Rhodes, *Dark Sun,* p. 270.
102 "then silently. . . . 'No. . . . come again.' " Ibid.
103 "personification . . . history," . . . "courteous . . . demanded it." Ibid., p. 182.
103 "For us. . . . very quickly." Knight, *Beria,* p. 137.
104 "it is high . . . scholars." Boag et al., eds., *Kapitza in Cambridge and Moscow,* p. 370.
104 "that greater . . . government." Ibid.
104 "I'll dismiss . . . like," . . . "but don't . . . him." Ibid., p. 368.
104 "magnificent modern . . . deceive." Allilueva, *Twenty Letters,* p. 8.
104 "in a good . . . together" . . . "laughed up his sleeve." Ibid.
105 "put the winch. . . . swindling him." Rhodes, *Dark Sun,* p. 275.
105 "We can . . . later." Holloway, *Stalin and Bomb,* p. 211.
105 "Once," . . . "as I was . . . were made." Sakharov, *Memoirs,* p. 102.
106 "alternate design . . . energy released." Ibid.
106 "light-heartedly . . . wife," . . . "rather tense." Ibid., p. 103.
107 "privilege" . . . "responsibility," . . . "great cause." Ibid., p. 104.
107 "arrest . . . campaign." Ibid.
107 "The Party . . . shotguns?" Ibid.
107 "rectified" . . . "very serious . . . suggestion." Ibid.
107 "Klava seemed . . . find a job." Ibid., p. 58.
108 "cautious, clever, and cynical." Ibid., pp. 104–105.
108 "Sakharov," . . . "should be. . . . needs him." Ibid., p. 105.
108 "Yes. . . . arguing." . . . "Yes . . . them." . . . "I have . . . Pavlovich." . . .

"He strongly . . . our offer." "Things," . . . "seem . . . a serious turn." Ibid.

Chapter 7. The Savior of Russia

109 "Fock's arrest . . . performance." Boag et al., eds., *Kapitza in Cambridge and Moscow,* p. 339.
109 "Soviet physics. . . . to the party." Krementsov, *Stalinist Science,* p. 289.
111 "was once. . . . that time." Sakharov, *Memoirs,* p. 133.
111 "distinguished . . . ideas." *Sakharov Remembered,* p. 127.
111 "We have to . . . right away." Ibid., p. 106.
112 "what kept . . . reaction." Ibid., p. 107.
112 "dense woods . . . river bank." Rhodes, *Dark Sun,* pp. 286–287.
113 "The pale light. . . . plowed land." Sakharov, *Memoirs,* p. 107.
114 "Is it here?" "Yes." Where?" "In the storehouse." Ibid., p. 108.
114 "So, the man. . . . Welcome!" Ibid.
114 "The 'chiefs' are. . . . you'll be." Ibid.
114 "professional . . . necessary." Ibid., p. 113.
115 "test model of Communism." Ibid.
115 "The building. . . . labor camp." Holloway, *Stalin and Bomb,* p. 172.
117 "Nothing . . . Igor." . . . "It worked." Rhodes, *Dark Sun,* p. 366.
117 "So I may. . . . Good!" Ibid., p. 367.
117 "It's urgent . . . up!" . . . "What . . . ? Why . . . calling?" "Everything went right," . . . "I know already." Ibid., p. 367.
117 "We promise . . . Your confidence." *Atomny Proekt CCCPTT Atomnaya Bomba, 1945–1954* (Moscow: Sarov), p. 658.
118 "those asiatics," those "pagan wolves." Rhodes, *Dark Sun*, pp. 373 and 436.
118 "very small stature" . . . "Kremlin complexion." Djilas, *Conversations with Stalin,* p. 61.
118 "Not even . . . firm" . . . "yellow eyes" . . . "mixture . . . roguishness." Ibid.
119 "A powerful . . . pow-er-ful!" Ibid., p. 153.
119 "You cannot . . . bomb." Volkogonov, *Autopsy,* p. 164.
119 "long lines . . . heels." Sakharov, *Memoirs,* p. 114.
120 "squinting . . . sun." Ibid., p. 118.
120 "valued . . . himself." Altshuler et al., eds., *On mezhdu,* p. 555.
120 "original." Ibid., p. 568.
120 "Perhaps the great . . . Tamm." Sakharov, *Memoirs,* p. 129.
121 "In those early . . . notice." Ibid., p. 122.
121 "The NKVD . . . touch me." Ibid., p. 123.
121 "Blindness. . . . fanatics?" Ibid.
122 "took this . . . meaning." Ibid., p. 126.
123 "an extraordinary. . . . place of work." Drell and Kapitsa, eds., *Sakharov Remembered,* p. 132.
123 "having satisfied . . . go." Sakharov, *Memoirs,* p. 119.

123 "If anything . . . head." Khabarov, *Etot Fatalny,* p. 102.
123 "The regime. . . . to have them." Altshuler et al., eds., *On mezhdu*, p. 116.
124 "We listened . . . mankind." Sakharov, *Memoirs,* p. 109.
124 "master . . . worker." Ibid., p. 110.
124 "The only. . . . sailors." Sagdeev, *Making Soviet Scientist,* p. 128.
124 "somehow radiant." Sakharov, *Memoirs,* p. 134.
124 "Who would . . . this heart?" Ibid., p. 134.
125 "covered . . . thick." Ibid.
125 "Life. . . . same breath." Lourie, *Russia Speaks,* pp. 182–183.
126 "Zavenyagin. . . . select group." Sakharov, *Memoirs,* p. 136.
127 "You say. . . . he behaves." Ibid., p. 16.
128 "Physics . . . over." Sagdeev, *Making Soviet Scientist,* p. 129.
128 "We took off. . . . home." Sakharov, *Memoirs,* p. 119.
128 "You're . . . kittens," . . . "What . . . without me?" Avtorkhanov, *Zagadka smerti Stalina,* p. 147.
129 "When he. . . . good at it." Nikita Khrushchev, *Khrushchev Remembers,* p. 263.
130 "So no one . . . kike." Sakharov, *Memoirs,* p. 123.
130 "Jews . . . paintings." Ibid., p. 112.
131 "valid grounds." Ibid.
131 "one foolproof. . . . dangerous." Ibid., p. 124.
131 "How . . . without you?" "Pervoye izgnanie," *Yevreiski Mir,* Feb. 7, 1997, no. 44.
131 "We don't know . . . gone." "On delal raschety sovetskoi superbomby," ibid., Jan. 31, 1997, no. 43.
131 "Don't bring . . . religious." "Pervoye izgnanie," ibid., Feb. 7, 1997, no. 44.
131 "verged . . . ingratiating." Sakharov, *Memoirs,* p. 146.
132 "Is there anything . . . me?" Ibid.
132 "Why are . . . race?" Ibid.
132 "Because," . . . "we lack . . . facilities." Ibid.
132 "plump . . . cold" . . . "their fear . . . conscious" . . . "face . . . terrifying human being." Ibid.
132 "The theoretical groups . . . results." Ibid., p. 156.
133 "the Martian." Ibid., p. 150.
133 "world apart. . . . color." Ibid., p. 157.
133 "calm . . . in the evening." Ibid.
134 "wit, spirit, and wine." Drell and Kapitsa, eds., *Sakharov Remembered,* p. 139.
134 "simple countrywoman." Mates Agrest, interview with author.
134 "Klava . . . coat!" . . . "This . . . suits me fine." Altshuler et al., eds., *On mezhdu*, p. 524.
135 "I'm finished. . . . myself." Nikita Khrushchev, *Khrushchev Remembers,* p. 307.

136 "we have plenty . . . prisons." Sakharov, *Memoirs,* p. 159.
136 "five feet . . . bird of prey." De Jonge, *Stalin,* p. 472.
137 "Passions . . . offing." Sakharov, *Memoirs,* p. 163.
138 "I am under . . . humanity." . . . "death's universal dominion." Ibid., p. 164.
139 "impressionable." Ibid.
139 "hypnotic . . . ideology." Ibid.
139 "Nothing . . . supports itself." Drell and Kapitsa, eds., *Sakharov Remembered,* p. 141.
139 "I needed . . . myself." Sakharov, *Memoirs,* p. 164.
139 "I still believed . . . future." Ibid.
139 "They've freed. . . . moment?" Ibid., p. 166.
139 "And it was . . . to light." Ibid.
140 "a man without. . . . simple." Roy Medvedev, *All Stalin's Men,* p. 140.
140 "a man of . . . ambition." Nikita Khrushchev, *Khrushchev Remembers,* note by Crankshaw, p. 546.
140 "Beria underestimated. . . . finger." Quoted in Knight, *Beria,* p. 209.
140 "We're heading. . . . knives." Nikita Khrushchev, *Khrushchev Remembers,* p. 330.
141 "new era . . . dawning." Sakharov, *Memoirs,* p. 166.
141 "We had all . . . fallout problem." Ibid., p. 171.
142 "struck . . . gray." Ibid., p. 172.
142 "Malenkov's . . . maximum." Ibid., p. 173.
142 "headlights . . . horizon." Ibid., p. 174.
142 "chiefs . . . were." Ibid.
142 "Two minutes to go." . . . "One minute." . . . "We saw a flash . . . blue-black color." Ibid.
143 "It's thanks . . . savior of Russia!" Altshuler et al., eds., *On mezhdu,* p. 596.
143 "well-aimed kick." Sakharov, *Memoirs,* p. 175.
143 "fused black . . . tower." Ibid., p. 175.
144 "nonchalantly." Ibid., p. 181.
144 "self-confidence . . . peak," . . . "euphoria." . . . "outwardly modest," . . . "actually . . . the opposite." Ibid., p. 180.

Chapter 8. Complicities

145 "Yes, perhaps . . . right." Altshuler et al., eds., *On mezhdu,* p. 606.
145 "if there. . . . correct." Ibid., p. 726.
147 "Andrei Dmitrievich. . . . come in!" Ibid., p. 578.
147 "technically . . . insubordination" . . . "reckless . . . fate." Sakharov, *Memoirs,* p. 183.
148 "feverish." Ibid., p. 185.
148 "for the resolution . . . calculations." Ibid., p. 185.
148 "foresee a result . . . calculations" and "to represent . . . to be them." Altshuler et al., eds., *On mezhdu,* pp. 37–38.

148 "in arrests." . . . "tactful . . . obliging." Sakharov, *Memoirs,* p. 188.
149 "almost . . . family." Goldansky, ed., *Yuli Borisovich Khariton, Put' dlinoyu v vek,* p. 375.
150 "himself eschewed . . . affair." Sakharov, *Memoirs,* p. 200.
151 "One night. . . . for oneself." Ibid., p. 189.
151 "The dark. . . . imitation." Ibid.
152 "One more . . . retiring." . . . "very long day." Ibid., p. 190.
152 "sinister . . . strike." Ibid., p. 191.
152 "The bomb. . . . parachute has opened." Ibid., p. 191.
152 "I stood. . . . arms around me." Ibid., p. 191.
153 "I did not . . . complicity." Ibid., p. 194.
154 "May all . . . cities." Ibid.
154 "An old man. . . . getting hard." Ibid.
154 "lashed by a whip." Ibid.
154 "We, the inventors . . . our control." Ibid.
154 "You won't . . . secret!" Nikita Khrushchev, *Khrushchev Remembers,* p. 347.
154 "Word . . . at us." Ibid., p. 347.
155 "It was as though . . . dead." Ibid., p. 343.
155 "Beria version . . . Beria." Ibid., p. 351.
156 "the year of passion." Rothberg, *Heirs of Stalin,* p. 15.
156 "The ideas . . . thinking." Sakharov, *Memoirs,* p. 195.
156 "One death . . . a statistic." De Jonge, *Stalin,* p. 329.
157 "moral dilemma" . . . "total . . . staggering." Sakharov, *Memoirs,* p. 198.
157 "What sort . . . cited?" Ibid., p. 202.
158 "just beginning . . . position." Ibid., p. 204.
158 "crushed . . . nations." Heller and Nekrich, *Utopia in Power,* p. 544.
159 "After the tanks. . . . with fraud." Ibid., p. 626.
159 "light blue. . . . her face." Akhmatova, *Selected Poems,* p. 99.
159 "I am a Khrushchevite . . . my son." Roy Medvedev, *Khrushchev,* p. 102.
161 "Your curiosity. . . . today." Altshuler et al., eds., *On mezhdu,* p. 130.
161 "pale, almost white." Roy Medvedev, *All Stalin's Men,* p. 65.
161 "During the war . . . deeds?" Sakharov, *Memoirs,* p. 206.
161 "slightly condescending air" . . . "inadmissible." Ibid.
161 "began . . . quality." Ibid.
161 "loose ends." Ibid.
162 "reckless . . . schemers." Ibid.
162 "What monsters rule us!" Altshuler et al., eds., *On mezhdu,* p. 41.
162 "relatively mild treatment." Sakharov, *Memoirs,* p. 207.
162 "So, the bomb squad's here!" Ibid., p. 214.
162 "And now," . . . "I'm involved. . . . to say." Ibid.
162 "Could devices. . . . circumstances?" Ibid., p. 207.
163 "completely unacceptable . . . morally." Ibid.

163 "only person . . . sympathetic." Ibid., p. 208.
163 "great success . . . point of view." Ibid.
164 "Immensely thankful . . . abashed." Rothberg, *Heirs of Stalin,* p. 31.
164 "internal émigré" . . . "actually become . . . paradise." Ibid., p. 34.
164 "If I were a poet," . . . "and a poet . . . help him." Mandelstam, *Hope against Hope,* p. 146.
164 "Considering . . . belong." Rothberg, *Heirs of Stalin,* p. 35.
165 " I am linked. . . . extreme." Ibid., p. 35.
166 "as usual . . . affected." Sakharov. *Memoirs,* p. 215.
166 "volunteered the opinion" . . . "little . . . at this time." Ibid.
167 "I am convinced. . . . peace?" Ibid., p. 216.
167 "Here's a note . . . Sakharov." . . . "Sakharov says. . . . dumber than we are?" . . . "But Sakharov. . . . doesn't belong." . . . "Sakharov, don't try. . . . people like Sakharov!" Ibid., pp. 216–217.
168 "test a 'clean' version. . . . five thousand years." Ibid., 218.
168 "Does Sakharov . . . wrong?" . . . "My opinion . . . orders." Ibid., p. 218.
169 "for the first . . . string-pulling." Ibid., p. 222.
169 "ionized . . . blackout." Ibid., p. 221.
169 "There's been no . . . victory!" Ibid.
170 "When you were. . . . happier." Ibid., p. 223.
171 "You don't have to call Adya yet." Ibid.
172 "giant torpedo. . . . casualties." Ibid., p. 221.
172 "idea . . . slaughter." Ibid.
173 "a crime . . . supply." Ibid., p. 225.
173 "My intervention. . . . devices." Ibid., p. 226.
174 "You can do. . . . tested too." . . . "tantamount to murder!" Ibid., p. 227.
174 "I've already agreed to that." Ibid.
174 "die. . . . criminal." "The decision is final." . . . "If you don't . . . anymore." . . . "You've double-crossed me." "You can . . . want." . . . "I don't . . . leash." Ibid., p. 228.
175 "I'm listening . . . Sakharov." . . . "I don't understand." . . . "What . . . from me?" Ibid.
175 "It was. . . . I wept." Ibid., p. 229.

Chapter 9. Critical Mass

176 "The criticism. . . . bad." Volkogonov, *Autopsy,* p. 231.
177 "Suddenly. . . . classic." Rothberg, *Heirs of Stalin,* p. 56.
177 "There's a Stalinist. . . . evil." Ibid., p. 57.
178 "No matter. . . . proposal." Sakharov, *Memoirs,* p. 231.
178 "new tolerance." Ibid., p. 232.
179 "why galaxies . . . formed." Ibid., p. 244.
179 "small . . . grow." Ibid., p. 245.
179 "indignation . . . Lysenko" had "boiled . . . again." Ibid., p. 233.

179 "one of . . . scientists." . . . "Yes, I know. . . . you?" Ibid.
179 "Why not?" Ibid., p. 234.
180 "The Academy's. . . . against Nuzhdin." Ibid., p. 234.
180 "People like . . . on trial!" Sakharov, *Memoirs,* p. 234.
180 "forger of true science." Sagdeev, *Making Soviet Scientist,* p. 61.
181 "science remains . . . impermissible." Sakharov, *Memoirs,* p. 235.
181 "First Sakharov . . . belong." Ibid., p. 237.
181 "Can we poison him?" . . . "I'm not a murderer." . . . "just won't fly." Sergei Khrushchev, *Khrushchev on Khrushchev,* p. 68.
181 "What's going on here?" Ibid., p. 209.
182 "grave mistakes." . . . "wanton . . . bodies" . . . "great deal . . . hands." . . . "Lysenko's nonsense." Roy Medvedev, *Khrushchev,* pp. 237–243.
182 "Well, that's it. . . . arrested." Ibid., p. 245.
182 "Sakharov. . . . help him." Sakharov, *Memoirs,* p. 232.
183 "could be . . . ranks." Ibid.
183 "I still believed . . . Zeldovich." Ibid.
183 "horror . . . warfare." Ibid., p. 268.
184 "It would . . . better pair." Tanya Sakharov, interview with author.
185 "Our Party . . . coexistence." Rothberg, *Heirs of Stalin,* p. 236.
185 "So that prisons. . . . communism." Andrei Sinyavsky, *On Socialist Realism,* p. 38.
185 "Stalin's successors . . . leader." Rothberg, *Heirs of Stalin,* p. 150.
186 "Now . . . speech!" Rothberg, Ibid., p. 189.
186 "I write . . . myself." Alexeyeva, *Soviet Dissent,* p. 12.
187 "For the first . . . conduct." Ginzburg, *Journey,* p. 78.
187 "Recognition . . . secede." Stalin, *Pravda,* Oct. 20, 1920.
188 "an odd thing to do." Sakharov, *Memoirs,* p. 273.
188 "And long . . . the fallen." Translation mine.
189 "hermetic." Sakharov, *Memoirs,* p. 272.
189 "first intervention . . . dissidents." Ibid., p. 275.
189 "We'll . . . measures." Ibid.
190 "more . . . desired." Ibid.
190 "collection" . . . "pretended . . . copy." Ibid., p. 277.
191 "The credo . . . pointed." . . . "Shoulder . . . co-existence." Gorelik, *Andrei Sakharov: Nauka I Svoboda,* p. 339.
191 "interesting . . . incorrectly." Ibid., p. 276.
192 "Do you. . . . Slavsky!" Ibid., p. 278.
193 "We proposed . . . project." Ibid., p. 279.
193 "good measure." Ibid.
193 "is not provided . . . illegal." Rothberg, *Heirs of Stalin,* p. 193.
193 "The person . . . Jew." D. M. Thomas, *Alexander Solzhenitsyn,* p. 344.
193 "Our reliance . . . justified." . . . "The situation. . . . events." Volkogonov, *Autopsy,* p. 285.
194 "What so many. . . . '. . . a human face.'" Sakharov, *Memoirs,* pp. 281–282.

194 "a perfectly happy person." Gorelik, *Sakharov,* p. 361.
195 "taken that into account." Sakharov, *Memoirs,* p. 285.
195 "Well, what . . . ?" "It's awful." "The style?" "No, not . . . awful!" "The contents. . . . withdraw it." Ibid.
196 "Government I . . . government." Ibid., p. 299.
196 "the most . . . satisfaction." Ibid., p. 286.
196 "universal suicide." Sakharov, "Reflections," p. 36.
196 "division . . . crime." Ibid., p. 27.
196 "Is it not . . . policy." Ibid., p. 66.
197 "We shall . . . death." Ibid., p. 51.
197 "crippling censorship. . . . force." Ibid., p. 63.
197 "sad fate . . . Baikal." Ibid., p. 49.
197 "forces . . . humanity." Ibid., p. 38.
197 "fifteen-year tax . . . nations." Ibid., pp. 46–47.
197 "white citizens . . . black citizens" . . . "serious decline . . . planet." Ibid., p. 46.
198 "American blacks . . . own." Kalugin, *First Directorate,* p. 105.
198 "profoundly socialist." Sakharov, *Reflections,* p. 54.
198 "socialism . . . labor" . . . "egotistical . . . capital." Ibid., p. 71.
199 "convergence" is a "Western" . . . "socialist . . . meaning." Ibid., p. 79.
199 "The capitalist . . . socialist." Ibid., p. 78.
199 "typical . . . bourgeoisie" . . . "especially . . . Kennedy." Ibid., p. 79.
199 "not one . . . all ills." Ibid., p. 82.
200 "collaboration . . . ownership." Ibid., p. 82.
200 "saving the . . . world." Ibid., p. 83.
200 "Gigantic fertilizer. . . . amino acids." Ibid., p. 83.
200 "thousands . . . explosion" . . . "will make . . . heredity." Ibid., p. 84.
200 "honorable . . . indifference." Ibid., p. 85.
200 "I can just . . . immaturity." Ibid., p. 74.
201 "You've got to. . . . we've got." Sakharov, *Memoirs,* p. 287.
201 "warned . . . approach" . . . "in which . . . open debate." Ibid., p. 287.
201 "would appear . . . party." Ibid., p. 120.
201 "barely . . . Jewish." Ibid., p. 138.
201 "should have . . . talk." Amalrik, *Will Soviet Union Survive?* p. 6.
202 "in third . . . Christie." Sakharov, *Memoirs,* p. 288.
202 "considered . . . idiots." Ibid., p. 289.
202 "A single flight . . . the millions." E. B. White, *Essays of E. B. White,* p. 132.
202 "deliberations . . . forestalled." Ibid., p. 133.
202 "deliberate . . . planned." Freeze et al., eds., *KGB Files,* p. 31.
203 "apolitical" . . . "susceptible . . . influence" . . . "in order . . . conversation with him." Ibid., pp. 25–26.
203 "nauseated." Remnick, *Lenin's Tomb,* p. 518.
203 "That's absurd—by who?" "Those are . . . time being." . . . "tantamount to being fired." Sakharov, *Memoirs,* p. 288.

204 "This is . . . life!" William Shawcross, *Dubcek* (New York: Simon and Schuster, 1970), p. 138.
204 "At some time. . . . our necks." Volkogonov, *Autopsy,* p. 288.
204 "I believed . . . down!" Ibid., p. 291.
204 "seven days. . . . occupation." Keane, *Václav Havel,* p. 214.
204 "faith . . . reform." Sakharov, *Memoirs,* p. 290.
205 "Throughout . . . his mind." . . . "For ten minutes . . . citizen." Rothberg, *Heirs of Stalin,* p. 249.
205 "venal . . . intelligentsia" . . . "miracle." Solzhenitsyn, *Oak and Calf,* p. 367.
205 "Merely to see . . . suit jacket." Ibid., p. 370.
205 "lively . . . beard" . . . "deliberate, precise gestures" . . . "tongue-twistingly . . . treble." Sakharov, *Memoirs,* p. 292.
206 "What is . . . reception?" Feinberg, *Epokha i lichnost',* p. 92.
206 "feeling that . . . bomb." Solzhenitsyn, *Oak and Calf,* p. 370.
206 "breaking . . . top." Sakharov, *Memoirs,* p. 293.
206 "almost word for word." Ibid.
206 "there never was . . . Lenin's teaching." Solzhenitsyn, *From under the Rubble,* p. 10.
206 "It was then . . . generous nature." Solzhenitsyn, *Oak and Calf,* p. 370.
207 "not at all . . . principle." Sakharov, *Memoirs,* p. 292.
207 "Communist Parties . . . sentenced." Ibid., p. 294.
208 "The present day . . . Stalin." . . . "In some. . . . know it." Marchenko, *My Testimony,* pp. 20–21.
208 "I remember . . . that way." . . . "The first time. . . . 'The Stare.' " Ibid., p. 141.
209 "What if suddenly . . . (when it . . . trivia)?" Rothberg, Heirs, p. 245.
209 "immaculate honesty." Sakharov's introduction to Marchenko's *To Live like Everyone Else,* p. v.
209 "enemy . . . ram." . . . "chilled." Sakharov, *Memoirs,* p. 306.
210 "an elasticity . . . meat!" Drell and Kapitsa, eds., *Sakharov Remembered,* p. 79.
210 "munched. . . . nap's end." Ibid., p. 75.
211 "should . . . insistent." Sakharov, *Memoirs,* p. 296.
213 "But . . . homeland!" Ibid., p. 297.
214 "Close. . . . cold." Ibid., p. 298.

Chapter 10. Vita Nuova

215 "In case . . . other two." Freeze et al., eds., *KGB Files,* p. 35.
216 "with . . . possible." Sakharov, *Memoirs,* p. 299.
216 "still . . . government." Ibid.
216 "A wolf . . . lair." Altshuler et al., eds., *On mezhdu,* p. 231.
217 "returning to life." Sakharov, *Memoirs,* p. 300.

217 "the introduction. . . . epoch." *Sakharov Speaks,* p. 120.
217 "The source . . . system." Ibid., p. 121.
217 "Democratization . . . disruptions." Ibid., p. 117.
218 "We are . . . leadership." Ibid., p. 118.
218 "What awaits. . . . at best 'stagnation.'" Ibid., pp. 132–133.
218 "to install . . . apartment" . . . "timely . . . acts." Freeze et al., eds., *KGB Files,* p. 37.
219 "spiritual murder . . . chamber." Rothberg, *Heirs of Stalin,* p. 300.
219 "After a breakfast. . . . accurate." Lourie, *Russia Speaks,* p. 312.
220 "cosponsor a complaint" . . . "protesting . . . hospital." Sakharov, *Memoirs,* p. 309.
220 "Stop . . . apartment." . . . "It belongs . . . apartment." Zhores and Roy Medvedev, *Madness,* p. 26.
220 "The Grigorenko. . . . between us." Sakharov, *Memoirs,* p. 310.
221 "I am collecting. . . . at home." Ibid., p. 311.
221 "such things . . . country." Zhores and Roy Medvedev, *Madness,* p. 70.
221 "psychiatric . . . repression." Ibid., p. 116.
221 "obsessive reformist delusions." Ibid., p. 131.
221 "that he had . . . bomb." Altshuler et al., eds., *on mezhdu,* p. 640.
223 "That was . . . life." Sakharov, *Memoirs,* p. 315.
223 "I know . . . fence." . . . "on the other side." Ibid., p. 316.
223 "We were all . . . arm." Lourie, *Russia Speaks,* p. 324.
223 "A short . . . world." Sakharov, *Memoirs,* p. 316.
224 "thorough and objective." "The trial . . . farce." Ibid., p. 317.
224 "Andrei . . . kefir?" "No . . . no!" . . . "What's. . . . clean." Elena Bonner, interview with author.
224 "well reasoned" . . . "Citizen judges. . . . Be just." Sakharov, *Memoirs,* p. 317.
224 "bad . . . standards." Ibid.
224 "Hide this. . . . husband." Ibid.
225 "You'll recognize. . . . mother." Ibid., p. 318.
225 "same glaring . . . Pasternak" . . . "painstaking editing" . . . "not to execute . . . embraces." Thomas, *Solzhenitsyn,* pp. 368–369.
225 "This book. . . . Universe." Rothberg, *Heirs of Stalin,* p. 350.
226 "Human Rights . . . USSR." Sakharov, *Memoirs,* p. 318.
226 "such a. . . . would be!" Ibid., p. 319.
226 "Sakharov. . . . State Security [KGB]." . . . "Moreover . . . antisocial acts." . . . "discussion with Sakharov" . . . "explain . . . Soviet state." Freeze et al., eds., *KGB Files,* p. 39.
227 "Not having . . . human contact." Sakharov, *Memoirs,* p. 320.
228 "to frighten . . . irreversible." . . . "liberalization . . . country." Rothberg, *Heirs of Stalin,* pp. 304, 310.
228 "hazardous and criminal." . . . "This affair . . . emigration." Sakharov, *Memoirs,* p. 322.

228 "Fascists!" Ibid., p. 323.
229 "politically incorrect . . . court." Freeze et al., eds., *KGB Files,* p. 48.
230 "on foot . . . lash." . . . "were all against. . . . monarchists." Lourie, *Russia Speaks,* pp. 33–34.
230 "And there I was. . . . 'good' means." Ibid., p. 35.
231 "handsome . . . color." Bonner, *Mothers and Daughters,* p. 236.
231 "It didn't worry . . . nails." Ibid., p. 109.
231 "loud, masterful steps." Ibid., p. 261.
231 "Well, that's it." . . . "She quickly . . . aside." Ibid., p. 322.
232 "A little. . . . mutinies." Lourie, *Russia Speaks,* p. 174.
232 "Listen to this one." . . . "She thinks . . . Trubetskaya." . . . "Trubetskaya . . . countess." Ibid., p. 37.
232 "Look how . . . is." Ibid., p. 192.
233 "How did I . . . view." Bonner, *Mothers and Daughters,* p. 325.
233 "When . . . rang" . . . "and I opened. . . . still hurts." Ibid., p. 328.
233 "My daughter's. . . . alienated." Lourie, *Russia Speaks,* p. 270.
233 "I'm going . . . legs." . . . "You better . . . you." Elena Bonner, interview with author.
233 "Take . . . plant it." Ibid.
234 "at least . . . thrown." Ibid.
234 "We hope . . . leadership." *Sakharov Speaks,* p. 137.
234 "The basic aim . . . citizens." Ibid., p. 142.
234 "confirmation . . . policies." Ibid., p. 149.
235 " SAKHAROV is striving . . . (and, possibly, abroad)." . . . "extremely undesirable." . . . "We therefore . . . delay" . . . "the question . . . Central Committee." Freeze et al., eds., *KGB Files,* pp. 52–53.
235 "In daily. . . . requires care." . . . "proudly . . . with him." . . . "Having made . . . war." . . . "In general . . . our adversary." Ibid., pp. 50–51.
237 "sometimes even . . . families." Sakharov, *Memoirs,* p. 335.
237 "If I lived . . . heretics." Ibid., p. 337.
237 "During the XXIV . . . activity." Freeze et al., eds., *KGB Files,* p. 58.
237 "Now the Communist . . . inevitable." Ibid., p. 63.
238 "I have been. . . . installation." Ibid., p. 65. All such transcripts should be taken with a grain of salt, if not a pood.
238 "If she's . . . who am I?" . . . "You?" . . . "You are me." Sakharov, *Memoirs,* p. 576, and Elena Bonner, interview with author.
239 "they were . . . nations." Altshuler et al., eds., *On mezhdu,* p. 496.
240 "Zeldovich." . . . "Zeldovich . . . all?" . . . "Zeldovich . . . all." Elena Bonner, interview with author.
241 "Six months . . . company." Sakharov, *Memoirs,* p. 355.
241 "Recently . . . collaborators." Freeze et al., eds., *KGB Files,* p. 71.
241 " SAKHAROV's . . . family." Ibid., p. 71.
242 "insolent . . . power." Sakharov, *Memoirs,* p. 359.

Chapter 11. Escalations

243 "skis. . . . overhead." Smith, *New Russians,* p. 445.
245 "Americans. . . . human face." Amalrik, *Will Soviet Union Survive?* p. 29.
245 "We cannot . . . beginning." Richard Lourie, "Soviet Intellectuals," *Dissent* (Winter 1974):18.
245 "Everyone . . . frontiers." Universal Declaration of Human Rights, Article 19.
245 "various . . . detention." Sakharov, *Memoirs,* p. 367.
246 "as anti-Soviet . . . are." Ibid.
246 "had been . . . me." Ibid., p. 374.
246 "You write . . . one?" "No," . . . "I never. . . . answer me." Boris Bolotosky, interview with author.
247 "emitted . . . struck." Sakharov *Memoirs,* p. 364.
247 "There are more. . . . starvation." Ibid., p. 365.
247 "known . . . provocations" . . . "who . . . flames." Freeze et al., eds., *KGB Files,* p. 79.
247 "quasi- Biblical . . . architecture." Sakharov, *Memoirs,* p. 374.
248 "the children . . . hostages." Ibid., p. 375.
248 "distinguished mathematician" . . . "displayed . . . community." . . . "less than honest" . . . "was powerless . . . decision." Ibid., pp. 378–379.
248 "To this day," . . . "I can't . . . rector." Shklovsky, *Five Billion,* p. 144.
248 "marched up . . . slapped his face." Sakharov, *Memoirs,* p.375.
249 "What are you. . . . children!" "Stay out. . . . children?" "Why are you . . . system?" . . . "It's given . . . work." "No one gave. . . . and night." . . . "It's all . . . bitter." . . . You keep on . . . isn't so." Sakharov, *Memoirs,* pp. 376–377.
249 "The greatest . . . KGB." Ibid., p. 380.
249 "soured." Ibid., p. 380.
249 "practically inevitable." Ibid.
250 "people . . . flowers" . . . "world government . . . 'John Birch Society.' " Dozmarova, ed., *Andrei Sakharov, Pro et Contra,* p. 31.
250 "ideologically . . . essay." Freeze et al., eds., *KGB Files,* p. 78.
250 "positive mark" . . . "next . . . campaign." . . . "It would . . . Soviet press." . . . "The consent . . . secured." Ibid., p. 81.
250 "just why . . . no idea." Sakharov, *Memoirs,* p. 382.
250 "actions . . . reality." Ibid., p. 633.
251 "the thought . . . face." Sakharov, *Memoirs,* p. 381.
251 "I began . . . human ones." *Sakharov Speaks,* p. 166.
251 "It seems . . . stability." Ibid., p. 171.
251 "because . . . hope." Ibid., p. 173.
251 "if a man. . . . silent." Ibid.
251 "tiredness . . . cynicism." Ibid.

251 "much more . . . need be." Ibid., p. 176.
252 "intellectual life . . . exist." Ibid., p. 177.
252 "on the whole. . . . society." Ibid., p. 172.
252 "I have not. . . . closer to us." Ibid., p. 178.
252 "Western clients." Sakharov, *Memoirs,* p. 631.
252 "This conversation . . . warning." . . . "harmful . . . character." . . . "By the nature. . . . violate them." *Sakharov Speaks,* p. 181.
253 "You seem . . . life," . . . "even though . . . freedom?" Ibid., p. 190.
253 "I am . . . life." . . . "They probably. . . . democracy." Ibid., p. 190.
253 "just a show . . . nothing." Ibid., p. 190.
253 "In particular . . . secrets." Ibid., pp. 191–192.
253 "appropriate . . . terms." Ibid., p. 188.
253 "evidently . . . overflowing." Ibid., 196.
254 "I like . . . understood." "Loyal to what?" "Loyal . . . lawful." Ibid., p. 198.
254 "I have viewed . . . minds." Ibid., p. 202.
254 "cultivation. . . . teeth." Ibid., pp. 204–5.
254 "tantamount . . . violence." Ibid., p. 214.
255 "went so far. . . . socialist nations." Sakharov, *Memoirs,* p. 632.
255 "I know. . . . Japan." Smith, *New Russians,* p. 459.
255 "unanimously . . . words." Sakharov, *Memoirs,* p. 637.
255 "*enemy of détente* . . . citizens." Ibid., p. 387.
256 "We want . . . October 11." . . . "sprang . . . blocked her." "I am not . . . duress." . . . "You'll be sorry." "What . . . kill us?" . . . "You're not . . . that!" We could. . . . grandchild. . . ." Ibid., pp. 392–393.
257 "Black September. . . . more!" Ibid., p. 393.
257 "I guess not." Interview with Naum Korzhavin.
257 "The hysteria, . . . anti-Sovietism." . . . "forces one" . . . "more radical . . . Sakharov." Freeze et al., eds., *KGB Files,* p. 92.
257 "Sakharov's . . . violence." Sakharov, *Memoirs,* p. 389.
258 "for the human . . . on Earth." Dozmarova, ed., *Pro et Contra,* p. 143.
258 "peremptory . . . others." Sakharov, *Memoirs,* p. 407.
258 "Don't give . . . shit!" . . . "You make . . . people!" Ibid., p. 403.
258 "under her heel." Ibid., p. 404.
258 "sharp tongue" . . . "explosive temper." Bonner, *Alone,* pp. 102, 67.
258 "tiny package . . . cloth." Ibid., p. 390.
259 "quite indifferent." Ibid., p. 405.
260 "devouring the masterpiece" . . . "angry, mournful, sardonic" . . . "somber world . . . Norilsk." Ibid., p. 406.
260 "the shock . . . forcefully." Ibid.
260 "international . . . *Gulag Archipelago.*" Ibid.
260 "touch . . . banks." Freeze et al., eds., *KGB Files,* p. 108.
261 "many . . . sung." Sakharov, *Memoirs,* p. 407.
261 "no public . . . six months." Freeze et al., eds., *KGB Files,* p. 112.
261 "on orders . . . organs." Ibid.

286 "total . . . violence." Sakharov, *Alarm and Hope,* p. 60.
286 "WARNING. . . . land." Ibid., pp. 70–71.
286 "Sakharov looked. . . . day." Bassow, *Moscow Correspondents,* p. 271.
286 "our duty and yours" to "defend . . . violated." Sakharov, *Memoirs,* p. 686.
287 "Human rights. . . . conscience." Ibid., p. 687.
287 "outright attempts . . . Soviet Union." Sakharov, *Alarm and Hope,* p. 54.
287 "considerable . . . arrived." . . . "an innocent . . . inexperience." Ibid., pp. 54–55.
287 "with a few . . . 'capitalism . . . human face,' " . . . "less certain . . . pluralism." Ibid., p. 102.
287 "Injustice . . . everywhere." Ibid., p. 99.
287 "major blow." Sakharov, *Memoirs,* p. 467.
288 "through the joint efforts. . . . her parents." Efrem Yankelevich, interview with author.
288 "expedient . . . propaganda." Freeze et al., eds., *KGB Files,* p. 169.
289 "Those sendoffs . . . heart." Altshuler et al., eds., *On mezhdu,* p. 788.
289 "I am sorry . . . wanted." . . . "We . . . do." Efrem Yankelevich, interview with author.
289 "minor formality." . . . "Andrei . . . differently?" Sakharov, *Memoirs,* p. 383.
289 "Psychologically . . . emigration!" Ibid., p. 381.
290 "vicious and grating" . . . "man's . . . prisoner." Ibid., p. 478.
290 "this wretched . . . dead." Ibid., p. 481.
291 "savage . . . society." . . ." Such. . . . achieved." Ibid., p. 652.
291 "I'd never been. . . .thrust." Ibid., p. 476.
292 "Whenever . . . committee." . . . "solitary . . . umbrella" with "an infallible. . . . in him." Amalrik, *Notes,* pp. 300–301.
292 "certain traits . . . country." Babyonyshev, *On Sakharov,* pp. 28–29.
292 "being . . . foolish," . . . "which one of Christ's. . . . the question." Efrem Yankelevich, interview with author.
293 "Human strength . . . power." . . . "All the more . . . firm." Sakharov, *Memoirs,* p. 483.
293 "I was right . . . regret it." Ibid., p. 484.
294 "slapped . . . policeman." Freeze et al., eds., *KGB Files,* p. 156.
295 "looked like an old man." Amalrik, *Notes,* p. 313.
295 "sweet . . . moustache." Sharansky, *Fear No Evil,* p. 92.
296 "an essential. . . . fear." Ibid., p. 23.
296 "Met him?" . . . "Who do . . . day?" Ibid., p. 332.
296 "bit less . . . usual." Altshuler et al., eds., *On mezhdu,* p. 177.
297 "hopeless sclerotic." Ibid., p. 176.
297 "To this day . . . superiors." Mates Agrest, "Pervoye Izgnanie," *Yevreiski Mir,* Feb. 14, 1997, no. 45, p. 24.
298 "The Committee. . . . milieu." Freeze et al., eds., *KGB Files,* p. 175.

299 "Why weren't . . . trial?" "For fear . . . relatives." Sakharov, *Memoirs*, p. 491.
299 "If you don't. . . . do more!" Ibid., p. 494.
300 "young . . . heart." . . . "How did . . . creep in?" Ibid., p. 361.
301 "limited . . . achieved." Volkogonov, *Autopsy*, p. 298.
301 "expansionism." Sakharov, *Memoirs*, p. 507.
301 "The USSR . . . Afghanistan." . . . "If it doesn't . . . war." Ibid., p. 509.
301 "something peculiar . . . gloating." Ibid., p. 510.
301 "A month . . . seriously," . . . "but now . . . possible." Ibid.

Chapter 13. The Blessings of Exile

303 "We were too. . . . happy." Sakharov, *Memoirs*, p. 514.
303 "Home." Ibid.
304 "I am prepared . . . trial." Ibid., p. 675.
304 "turn. . . . idiots." Ibid., p. 520.
304 "now . . . soul." Babyonyshev, *On Sakharov*, p. 35.
304 "greatly disturbed." . . . "Wouldn't it be better," . . . "simply . . . reverse?" Sakharov, *Memoirs*, p. 303.
305 "gilded cage." Ibid., p. 506.
305 "Who can know . . . in store." Bonner, *Alone*, p. 26.
306 "greedy." Sakharov, *Memoirs*, p. 540.
307 "I'll go anyway." "Sakharov . . . deeply." "When . . . H-bomb?" Lourie, *Russia Speaks*, pp. 334–335.
308 "the historically . . . mankind." Babyonyshev, *On Sakharov*, p. 239.
308 "everything is . . . Stalin." Ibid., p. 234.
308 "a closed . . . control." Ibid., p. 224.
308 "I believe. . . . goals." Ibid., p. 229.
309 "until you can't stand it." Bonner, *Alone*, p. 198.
310 "Is the leadership . . . punishment?" Babyonyshev, *On Sakharov*, p. 221.
310 "the mole . . . expected." *Chronicle* 46 (April–June 1982):14.
310 "Teeth . . . humankind." Altshuler et al., eds., *On mezhdu*, p. 137.
311 "Don't worry . . . here." Sakharov, *Memoirs*, p. 529.
311 "in a state . . . trembling." Ibid., p. 530.
312 "all honest . . . him." Babyonyshev, *On Sakharov*, p. 26.
312 "If the state . . . bars," . . . "then. . . . prisoner." *Chronicle* 43 (July–Sept. 1981):45.
312 "the only . . . today." Babyonyshev, *On Sakharov*, p. 205.
312 "to rob . . . memory" . . . "hostage . . . state." Ibid., p. 210.
312 "Western scientists. . . . responsibility." Ibid., p. 211.
312 "public opinion . . . meet it." Ibid., p. 267.
313 "one . . . time." *Chronicle* 42 (April–June 1981):35.
313 "the KGB's . . . ignoble." *Chronicle* 44 (Oct.–Dec. 1981):7.
313 "hunger strike . . . action." Sakharov, *Memoirs*, p. 558.
314 "Whenever . . . suffered." Bonner, *Alone*, p. 108.

314 "to argue . . . bed." Sakharov, *Memoirs,* p. 560.
315 "Our fortunes. . . . needed." . . . "something of a 'cult of personality' " . . . "regarded . . . human being." Ibid., p. 561.
315 "November 21. Lusia. . . . complete capitulation." Ibid., pp. 562–568.
318 "They've come . . . kill us." . . . "We have . . . you." . . . "Will we be together?" Ibid., p. 569.
318 "We each had . . . break us." Ibid., p. 570.
319 "You . . . hell!" Ibid., p. 560.
319 "Lusia, . . . you." Ibid., p. 571.
319 "What gives. . . . at all." Ibid.
319 "I'm authorized . . . hunger strike." Ibid.
319 "I'll report. . . . me again." Ibid.
320 "Andrei Dmitrievich was sitting . . . him." *Chronicle* 44 (Oct.–Dec. 1981):9.
320 "I'm leaving . . . tears." Ibid., p. 574.
321 "false, cowardly . . . dwell on." Ibid., p. 15.
321 "only regret . . . of me." Ibid., pp. 580–581.
322 "illegal contact . . . tourists." Freeze et al., eds., *KGB Files,* p. 202.
322 "allegedly . . . Soviet Union." Ibid.
322 "contrived . . . (which has . . . West)." Ibid., pp. 203–204.
323 "You've got . . . Moscow?" . . . "strange . . . fruit." Sakharov, *Memoirs,* pp. 532–533.
323 "with the expression . . . to him." Bonner, *Alone,* p. 7.
323 "black anguish." Altshuler et al., eds., *On mezhdu,* p. 367.
323 "Andrei . . . talent." . . . "I call . . . starts." Bonner, *Alone,* p. 7.
323 "writing bits . . . manuscript." Sakharov, *Memoirs,* p. 535.
323 "think of nothing . . . discipline." Volkogonov, *Autopsy,* p. 346.
324 "world poisoned . . . mistrust." Sakharov, *Memoirs,* p. 667.
325 "little . . . joy." Ibid., p. 666.
325 "a few billion" . . . "take significant . . . (more precisely . . . them)." Ibid., p. 669.
325 "nuclear deterrence. . . . unstable." Ibid., p. 661.
325 "in the long . . . destruction." Ibid., p. 662.
325 "Soviet antagonism. . . . negotiations." *Chronicle* 43 (July–Sept. 1981): 34.
326 "been . . . midst." Sakharov, *Memoirs,* p. 671.
326 "the Soviet Union . . . lives." Ibid., p. 670.
326 "It looks . . . woman." . . . "wanton" . . . "persuading . . . hand." Bonner, *Alone,* pp. 46, 31, 32.
326 "in the nineteenth . . . duel." Sakharov, *Memoirs,* p. 589.
326 "left-handed . . . cheek." Ibid., p. 591.
327 "very pleased. . . . delighted." Bonner, *Alone,* p. 10.
327 "Those of us . . . you apart." Sakharov, *Memoirs,* p. 592.
327 "the sick . . . country." Volkogonov, *Autopsy,* p. 358.
327 "physically . . . hatred." Bonner, *Alone,* p. 25.

327 "It was very . . . calm." "should . . . her." Sakharov, *Memoirs,* pp. 595, 598.
327 "a heart . . . on." Bonner, *Alone,* p. 30.
328 "They did not . . . completely." Ibid., p. 52.
328 "cheerful . . . yield." Altshuler et al., eds., *On mezhdu*, p. 239.
329 "his skill. . . . pleasure." Drell and Kapitsa, eds., *Sakharov Remembered,* p. 146.
329 "friends. . . . Save us!" Bonner, *Alone,* p. 241.
330 "woman's work" . . . "those legs" . . . "that incredible . . . me." Bonner, *Alone,* p. 65.
330 "I am tired . . . you." . . . "Neither pity . . . inaction." Ibid., pp. 66–67.
330 "pallid . . . long." . . ."doctors . . . hospital." Ibid., pp. 70–71.
331 "you can't . . . yourself." Ibid., p. 72.
331 "most excruciating . . . method." Sakharov, *Memoirs,* p. 703.
332 "bring . . . conclusion." Ibid., p. 704.
332 "we, the children. . . . correctly." Ibid., p. 73 and p. 690.
332 "intoxicating" . . ."jam . . . jam." Bonner, *Alone,* p. 97.
333 "We won't. . . . invalid." Ibid., p. 104.
333 "In the novel. . . . disease." Sakharov, *Memoirs,* p. 704.
333 "Mengeles of today." Bonner, *Alone,* p. 134.
333 "He was wearing. . . . each other." Ibid., p. 102.
333 "apart . . . time." Ibid., p. 104.
334 "living corpse." Ibid., p. 104.
334 "I appeal . . . life." Sakharov, *Memoirs,* p. 698.
334 "If you and the Academy's. . . . receipt of this letter." Ibid., pp. 705–706.
335 "cruel . . . telegram" . . . "additional. . . . against us." Ibid., p. 690.
335 "one of doubtless . . . happiness." Altshuler et al., eds., *On mezhdu*, pp. 156–157.
336 "bastards." . . . "inconsolable" . . . "I betrayed Lusia." Ibid., p. 160.
336 "How do we live? . . . Andrei." Bonner, *Alone,* pp. 117–118.
336 "Sakharov said. . . . place." . . . "Andrei . . . again!" "Don't . . . talking." Altshuler et al., eds., *On mezhdu*, p. 759.
337 "spectre . . . chancery." Volkogonov, *Autopsy,* p. 422.
338 "at his wife's instigation." . . . "make it . . . dentures." Freeze et al., eds., *KGB Files,* p. 224.
338 "One must . . . few years." Ibid.
338 "Bastards! . . . injection!" Bonner, *Alone,* p. 141.
339 "It looks . . . man." Sakharov, *Moscow and Beyond,* p. 10, and Elena Bonner interview with author.
339 "aversion." Bonner, *Alone,* p. 151.
339 "That's . . . herself." Ibid., p. 149.
339 "I did not . . . stupor." Ibid., p. 145.
340 "How many . . . will it take?" . . . "For what? . . . do that!" Ibid., p. 147.

340 "reasonable time." Ibid., p. 152.
340 "dreary . . . windy." Ibid., p. 155.
340 "a time for living." Ibid.
341 "I guess you're right." Sakharov, *Memoirs,* p. 600.
341 "another . . . 'Who needs . . . why?' " Bonner, *Alone,* pp. 157–158.
342 "discontinue . . . circumstances." . . . "earned the right." Sakharov, *Memoirs,* p. 600.
342 "Discuss . . . her." Ibid., p. 602.
343 "Don't be . . . hours." Bonner, *Alone,* pp. 158–159.
343 "The KGB . . . itself! . . . "Just . . . me." Ibid., p. 159.
343 "In case . . . interviews." Ibid., p. 159.
343 "but what . . . it mean?" "I'm afraid . . . it." . . . "I won't . . . America." "I don't . . . either." Sakharov, *Memoirs,* p. 602.
343 "to protest . . . authorities." Bonner, *Alone,* p. 161.
344 "You don't. . . . OVIR." . . . "in fact . . . like that." "We won!" Ibid., p. 167.
344 "suspected . . . adversary." Doder and Branson, *Gorbachev,* pp. 106–107.
345 "disappointment and bitterness" . . . "All's . . . well." Altshuler et al., eds., *On mezhdu*, p. 696.
345 "miraculous . . . work." Bonner, *Alone,* p. 191.
345 "had been . . . rehearsal." Ibid., p. 190.
345 "high-energy . . . still." Sakharov, *Memoirs,* p. 605.
345 "string . . . physics." Ibid., p. 606.
346 "intentions" . . . "allowed . . . Moscow." Ibid.
346 "As for . . . statements," . . . "the assurances. . . . questions." Ibid.
346 "we don't. . . . opinions." . . . "It is common. . . . abroad." Ibid., p. 607.
346 "personal intervention." Brandeis Sakharov Archives.
346 "unconditional . . . conscience." Sakharov, *Memoirs,* p. 607.
347 "with respect and hope." Brandeis Sakharov Archives.
347 "In contrast. . . . myself!" Ibid., p. 605.
348 "I doubt . . . nature." . . . "infallible . . . future." Ibid., p. 605.
348 "life's causal . . . code." Ibid., p. 561.
348 "wishful thinking" . . . "an accident . . . catastrophe." Ibid., p. 608.
349 "coming . . . country." Altshuler et al., eds., *On mezhdu*, p. 650.
349 "we can't . . . like this." Volkogonov, *Autopsy,* p. 445.
349 "determine . . . socialism." Doder and Branson, *Gorbachev,* p. 76.
349 "fog" and "darkness." Bonner, *Alone,* p. 227.
350 "felt like . . . pumped." Altshuler et al., eds., *On mezhdu,* p. 373.
350 "by which . . . measured." Sakharov, *Memoirs,* p. 611.
350 "I hope . . . exile." . . . "imperceptible tremor." Ibid., p. 612.
350 "so that even . . . structures." . . . "burying . . . level." Ibid., pp. 612–613.
351 "astonishing . . . happy." Ibid., p. 614.

352 "The man . . . country." Altshuler et al., eds., *On mezhdu*, p. 899.
352 "You'll get . . . morning." Sakharov, *Memoirs*, p. 615.
352 "noose." Ibid.
352 The telephone conversation beginning "Mikhail Sergeyevich will speak" and ending with "Good-bye." Ibid., pp. 615–616.
353 "had no . . . victory." . . . "exceptional cases." Ibid., p. 617.
353 "Academician Andrei . . . Moscow." Ibid.

Chapter 14. Astonishing Times

356 "his exhausted . . . body." Drell and Kapitsa, eds., *Sakharov Remembered*, p. 146.
356 "one of high . . . celebration." Altshuler et al., eds., *On mezhdu*, p. 755.
356 "madhouse." Sakharov, *Moscow and Beyond*, p. 4.
356 "On January 1 . . . celebrations" . . . "slaving." Ibid., p. 7.
356 "debut . . . possible." Ibid., p. 8.
357 "having advocated . . . lives." Ibid., p. 7.
357 "intelligent . . . woman." Ibid., p. 13.
357 "Is there a . . . peace?" Ibid., p. 12.
357 "open, stable" . . . "eyes wide open." Ibid., p. 13.
358 "how could such . . . spirit?" Altshuler et al., eds., *On mezhdu*, p. 742.
358 "if this . . . considerations." . . . "other . . . tasks." Freeze et al., eds., *KGB Files*, p. 242.
358 "staged . . . purposes." Sakharov, *Moscow and Beyond*, p. 15.
358 "after many . . . audience." Ibid., p. 18.
358 "On the other. . . . opposite." Ibid., p. 19.
358 "Maginot . . . space." Ibid., p. 22.
359 "the most . . . forum." Altshuler et al., eds., *On mezhdu*, p. 742.
360 "been pardoned . . . liberty." Sakharov, *Moscow and Beyond*, p. 28.
360 "emotions kindled . . . day." Sakharov, *Memoirs*, p. 195.
360 "maintaining . . . rights." Sakharov, *Moscow and Beyond*, p. 28.
361 "Never before . . . discussion." Drell and Kapitsa, eds., *Sakharov Remembered*, p. 80.
361 "His morale . . . knowledge." . . . "God not only . . . reach." Sakharov, *Moscow and Beyond*, p. 30.
362 "glad" . . . "erroneous . . . entropy." . . . "too shy" to "bring up . . . entropy." . . . "Hawking's . . . time." Ibid.
362 "Could I . . . 'light . . . window.' " Bonner, *Mothers and Daughters*, pp. 89–90.
363 "undesirable precedent." Sakharov, *Moscow and Beyond*, p. 31.
363 "too many honors" . . . "bounds." Ibid., p. 32.
363 "I come out. . . . glasses." Ibid., p. 33.
363 "steamroller of socialism." Ibid., p. 40.
364 "all work . . . open." Ibid., p. 41.
365 "A lot. . . . goes on!" Sakharov, *Memoirs*, p. 138.
365 "sadder affair." Sakharov, *Moscow and Beyond*, p. 42.

365 "false position." . . . "about which . . . education." Ibid., p. 44.
365 "We had our . . . had." Altshuler et al., eds., *On mezhdu*, p. 438.
366 "The petty. . . . science." Sakharov, *Memoirs*, p. 138.
366 "What was. . . . smile?" Bonner, *Mothers and Daughters*, p. 6.
366 "I loved . . . me." Ibid., p. 8.
366 "I received. . . . indivisible." "I'm very . . . words." Sakharov, *Moscow and Beyond*, p. 44.
367 "most. . . . stages." Ibid., p. 45.
367 "intelligent . . . discussion" . . . "respectful . . . sympathetic." Ibid., pp. 45–46.
367 "This was a. . . . meaning." Carrère d'Encausse, *End of Soviet Empire*, p. 52.
367 "vacillating and unprincipled." Sakharov, *Moscow and Beyond*, 46.
368 American translator—i.e., me.
368 "The word 'bloodbath' pains. . . . wink at Bonner." Lourie, *Russia Speaks*, pp. 367–368.
370 "nothing . . . crimes." Remnick, *Lenin's Tomb*, p. 75.
370 "Above all. . . . win." Afanaseva, *Inovo ne dano* (No other way), p. 126.
370 "Do you know . . . life? . . . "I expected . . . me." . . . "The thing . . . radiation." Remnick, *Lenin's Tomb*, p. 165.
370 "What . . . science?" . . . "In Gorky. . . . know how." Altshuler et al., eds., *On mezhdu*, p. 545.
371 "marvelous . . . happy." Sakharov, *Moscow and Beyond*, p. 52.
371 "The years . . . A. D." . . . "He was . . . tired." Altshuler et al., eds., *On mezhdu*, p. 505.
372 "in effect . . . opposition." Sakharov, *Moscow and Beyond*, p. 56.
372 "Sakharov's calm . . . scale." Remnick, *Lenin's Tomb*, p. 236.
373 "we have . . . character." . . . "What this. . . . else." Volkogonov, *Autopsy*, p. 518.
373 "debating society." Ibid., p. 312.
373 "let Pugwash . . . without me!" Sakharov, *Moscow and Beyond*, p. 64.
373 "the danger . . . earth." Ibid.
373 "unwilling . . . himself." Ibid., p. 65.
374 "a few . . . conscience." Sagdeev, *Making Soviet Scientist*, p. 246.
374 ("He knows . . . Americans") . . . (. . . "if he goes . . . returns)." Freeze et al., eds., *KGB Files*, p. 242.
375 "it didn't know . . . with" . . . "charming." Sakharov, *Moscow and Beyond*, pp. 68–69.
375 "for principled . . . not." . . . "profound . . . Soviet Union." Ibid., p. 69.
375 "primacy . . . values." Dobbs, *Down with Big Brother*, p. 215.
376 "exception." Sakharov, *Moscow and Beyond*, p. 59.
376 "there should be . . . on me." . . . "direct rebukes" . . . "I would . . . so." . . . "By ordinary . . . major concession." Ibid.
377 "when they saw. . . . each other." Remnick, *Lenin's Tomb*, p. 285.

377 "Land. . . . conquered." Sakharov, *Moscow and Beyond,* p. 84.
377 "shock . . . psychosis." Ibid., p. 84.
377 "childishly peevish." Ibid., p. 79.
377 "rather long-winded." Ibid., p. 94.
378 "Nineteen years." Doder and Branson, *Gorbachev,* p. 229. There attributed to Gorbachev's spokesman.
378 "If there's . . . work." Smith, *The New Russians,* p. 442.
378 "phony." Sakharov, *Moscow and Beyond,* p. 97.
378 "quasi rigged." Remnick, *Lenin's Tomb,* p. 220.
378 "the impoverished . . . everything," . . . "moral mandate." Sakharov, *Moscow and Beyond,* p. 98.
379 "solemn . . . rhetoric." . . . "after all . . . Europeans." Ibid., p. 101.
379 "not smug . . . others." Ibid., p. 103.
379 "Our country. . . . prisoner." Ibid.
381 "Other parties . . . most." Ibid., p. 102.
381 "A person . . . calibre." Ibid., p. 115.
381 "I just. . . . session." Ibid.
382 "exuberance . . . gaiety." Dobbs, *Down with Big Brother,* p. 243.
382 "to impose . . . 'rules of the game.'" Sakharov, *Moscow and Beyond,* p. 116.
382 "You should have. . . . feelings." Ibid., p. 116.
382 "the mood . . . spoiled." Ibid.
383 "speak. . . . country." Ibid., p. 119.
383 "underground empire" and "a threat to democracy." Smith, *New Russians,* p. 464.
384 "demolished . . . illusions" and "burned all bridges." Sakharov, *Moscow and Beyond,* p. 120.
384 "If the guilty . . . punished," . . . "public . . . military." Ibid., p. 125.
384 "We're always harping. . . . defend yourself." Ibid., p. 125.
384 "Are they . . . Party?" Ibid., p. 129.
384 "It's evident. . . . oppose." Ibid., p. 130.
385 "on pins and needles." Ibid., p. 131.
385 "half kindly . . . face," Ibid.
385 "Mikhail Sergeyevich. . . . perestroika." Ibid., pp. 131–132.
385 "I stand. . . . policies." Ibid., p. 132.
386 "beaten, dejected." Dobbs, *Down with Big Brother,* p. 260.
386 "he faced . . . retreat." Smith, *New Russians,* p. 470.
386 "The last . . . army." . . . "The real issue . . . for it." Sakharov, *Moscow and Beyond,* p. 134.
386 "Shame! . . . Sakharov!" Ibid., p. 135.
386 "You spoke . . . hero." Ibid.
386 "had all the guns." Dobbs, *Down with Big Brother,* p. 268.
387 "I blew up . . . genocide!" Sakharov, *Moscow and Beyond,* p. 139.
387 "That's . . . exile." . . . "We've been. . . . clearly." Ibid.

388 "shall have . . . USSR." . . . "Any danger . . . long ago." . . . "new . . . federalism." Ibid., pp. 152–154.
388 "tragic optimism." Ibid., p. 156.
388 "The Congress. . . . from us." Ibid., pp. 155–156.
388 "*All power to the Soviets!*" Ibid., p. 156.
389 "astonishing times." Lourie, *Russia Speaks,* p. 374.
389 "There is much. . . . this." Sakharov, *Moscow and Beyond,* p. 160.
389 "the only. . . . meaning." Ibid., p. 160.
389 "completing a book. . . . after us." Ibid., p. 159.
390 "a direct . . . People's Deputies." Bonner's introduction to Sakharov's "Lecture in Lyons," p. 22.
390 "Our country . . . other." Sakharov, "Lecture in Lyons," p. 24.
390 "our individual . . . space." Ibid.
390 "in the next . . . science." Ibid., p. 23.
390 "We are . . . us." Ibid.
391 "My deep . . . whole." Ibid.
391 "I am unable . . . laws." Sakharov, *Memoirs,* p. 4.
391 "a gangster group." . . . "Reading . . . spark." Smith, *New Russians,* p. 503.
393 "flexible . . . society." Sakharov, *Moscow and Beyond,* p. xiv.
394 "Property to . . . soviets!" Freeze et al., eds., *KGB Files,* pp. 249–250.
394 "When they . . . crucified." Drell and Kapitsa, eds., *Sakharov Remembered,* p. 73.
395 "propaganda . . . abroad." . . . "A. D. Sakharov . . . class." . . . "The Committee . . . situation." Freeze et al., eds., *KGB Files,* pp. 250–251.
395 "good sign" . . . "If we hadn't . . . appeal." Altshuler et al., eds., *On mezhdu.* p. 904.
395 "The country's fate. . . . bifurcation." . . . "partocracy." Ibid., p. 906.
396 "test of forces." Ibid., p. 908.
397 "The main thing. . . . together." Sakharov, *Memoirs,* p. 619.
397 "What is. . . . collapse." . . . "accusation. . . . triumph." Sakharov, *Moscow and Beyond,* pp. xv–xvi.
398 "There'll . . . tomorrow." Remnick, *Lenin's Tomb,* p. 282. Also interview with Elena Bonner.

Epilogue. The Life after Death

399 "it was not the tradition." Remnick, *Lenin's Tomb,* 283.
400 "To some. . . . pride." Freeze et al., eds., *KGB Files,* p. 254.
400 "You've lost . . . opponent." Altshuler et al., eds., *On mezhdu,* p. 499.
400 "handful . . . slavery." . . . "There's been . . . near." Nikishin, "Pokhorony Akademika A. D. Sakharova," p. 184.
401 "You didn't . . . killed!" "We know . . . avenge you." Ibid., p. 179.
401 "Saints . . . innocent." Orwell, *Essays,* p. 171.
401 "Is there a Gandhi . . . moment?" Ibid., p. 178.

402 "You are more . . . hearing it." Babyonyshev, *On Sakharov,* p. 50.
402 "go even . . . Oppenheimer had." Sakharov, *Memoirs,* p. 100.
402 "It . . . took. . . . similar fate." Drell and Kapitsa, eds., *Sakharov Remembered,* p. 214.
403 "Sakharov received. . . . strength." Ibid., p. 164.
403 "He is. . . . 'constructive genius.'" Sagdeev, *Making Soviet Scientist,* p. 47.
404 "Who else if not me?" Drell and Kapitsa, eds., *Sakharov Remembered,* p. 145.
407 "straight, pure. . . . all the time." Remnick, "The Afterlife," *The New Yorker*, p. 58.
407 "'Andrei . . . would have known.'" Drell and Kapitsa, eds., *Sakharov Remembered,* p. 89.
407 "embodied . . . democracy." . . . "exhortations . . . anachronistic." . . . "freedom is fragile . . . lose it." *Time* magazine, June 14, 1999, pp. 188–190.
408 "In fate . . . important." Drell and Kapitsa, eds., *Sakharov Remembered,* p. 248.
409 "Sakharov had already. . . . thought." Orlov, *Dangerous Thoughts,* pp. 319–320.
411 "kept a careful. . . . 'that I begin . . . virtue.'" Maurois, *Aspects of Biography,* p. 18.
411 "dangerous words." . . . "inertia of falsehood. . . . newborn democracy." Bonner, "Remains of Totalitarianism."
412 "The future . . . on us." Ibid.

BIBLIOGRAPHY

Afanaseva, Yu. N., ed. *Inovo ne dano*. Moscow: Progress, 1988.

Akhmatova, Anna. *Selected Poems of Akhmatov,* trans. Stanley Kunitz with Max Hayward. Boston: Little, Brown, 1973.

Alexeyeva, Ludmilla. *Soviet Dissent: Contemporary Movements for National, Religious, and Human Rights*. Middletown, Connecticut: Wesleyan University Press, 1985.

Alliluyeva, Svetlana. *Only One Year*. New York: Harper & Row, 1969.

———. *Twenty Letters to a Friend*. New York: Harper & Row, 1967.

Altshuler, B. L., et al., eds. *On mezhdu nami zhil: Vospominaniya o Sakharove*. Moscow: Praktika, 1996.

Amalrik, Andrei. *Notes of a Revolutionary*. London: Weidenfeld and Nicolson, 1982.

———. *Will the Soviet Union Survive until 1984?* New York: Harper & Row, 1970.

Andreev, A. F., ed. *Kapitsa, Tamm, Semyenov*. Moscow: Vagrius, 1998.

Auden, W. H., and Kronenberger, Louis. *The Viking Book of Aphorisms*. Viking/Dorset, 1962.

Avtorkhanov, Abdurakhman. *Zagadka smerti Stalina*. Frankfurt am Main: Possev-Verlag, 1976.

Babyonyshev, Alexander, ed. *On Sakharov*. New York: Knopf, 1982.

Bassow, Whitman. *The Moscow Correspondents: Reporting on Russia from the Revolution to Glasnost*. New York: William Morrow, 1988.

Bernstein, Jeremy. *Cranks, Quarks, and the Cosmos*. New York: Basic Books, 1993.

Billington, James H. *The Face of Russia*. New York: TV Books, 1998.

Blumberg, Stanley A., and Panos, Louis G. *Edward Teller: Giant of the Golden Age of Physics*. New York: Scribner's, 1990.

Boag, J. W.; Rubinin, P. E.; and Shoenberg, D. *Kapitza in Cambridge and Moscow: Life and Letters of a Russian Physicist*. New York: North Holland, 1990.

Bonner, Elena. *Alone Together*. New York: Knopf, 1986.

———. *Mothers and Daughters*. New York: Vintage, 1993.

———. "The Remains of Totalitarianism." *New York Review of Books*, March 8, 2001.

———. *Volniye zametki k rodoslovnoi Andreya Sakharova*. Moscow: PRAVA CHELOVEKA, 1996.

Butson, Thos. G. *Gorbachev: A Biography*. New York: Stein and Day, 1985.

Carrère d'Encausse, Hélène. *The End of the Soviet Empire: The Triumph of the Nations*. New York: Basic Books, 1993.

Cohen, Stephen F. *Failed Crusade: America and the Tragedy of Post-communist Russia.* New York: Norton, 2000.

Cohen, Stephen F., and vanden Heuvel, Katrina. *Voices of Glasnost: Interviews with Gorbachev's Reformers.* New York: Norton, 1989.

Crankshaw, Edward. *The Shadow of the Winter Palace: Russia's Drift to Revolution, 1825–1917.* New York: Viking, 1976.

De Jonge, Alex. *Stalin and the Shaping of the Soviet Union.* New York: Morrow, 1986.

Djilas, Milovan. *Conversations with Stalin.* San Diego: Harcourt Brace Jovanovich, 1962.

Dobbs, Michael. *Down with Big Brother: The Fall of the Soviet Empire.* New York: Knopf, 1997.

Doder, Dusko, and Branson, Louise. *Gorbachev: Heretic in the Kremlin.* New York: Penguin, 1990.

Dozmarova, Galina, ed., *Andrei Sakharov, Pro et Contra.* 1973 god. Dokumenty, fakty, sobytiya. Moscow: PIK, 1991.

Drell, Sidney D., and Kapitza, Sergei P., eds. *Sakharov Remembered: A Tribute by Friends and Colleagues.* New York: American Institute of Physics, 1991.

Einstein, Albert. *Ideas and Opinions.* New York: Wings Books, 1954.

Feinberg, E. L. *Epokha i lichnost': ocherki I vospominaniya.* Moscow: Nauka, 1999.

Figes, Orlando. *A People's Tragedy: The Russian Revolution, 1891–1924.* New York: Penguin, 1998.

Fischer, Louis. *The Life of Lenin.* New York: Harper & Row, 1964.

Flanagan, Denis. *Flanagan's Version: A Spectator's Guide to Science on the Eve of the 21st Century.* New York: Knopf, 1988.

Freeze, Gregory; Gribanov, Alexander; Kline, Edward; and Szulkin, Robert, eds. *The KGB Files on Andrei Sakharov.* A Publication of The Andrei Sakharov Archives, Brandeis University. Waltham, Massachusetts, n.d.

Gellner, Ernest. *Encounters with Nationalism.* New York: Blackwell, 1994.

Gessen, Masha. *Dead Again: The Russian Intelligentsia after Communism.* New York: Verso, 1997.

Ginzburg, Eugenia Semyonova. *Journey into the Whirlwind.* New York: A Helen and Kurt Wolff Book/Harcourt, Brace & World, 1967.

Ginzburg, V. L. O *Nauke, o Sebe I o Drugikh.* Moscow: Nauka, 1997.

Gleick, James. *Genius: The Life and Science of Richard Feynman.* New York: Vintage, 1993.

Goldansky, V. I., ed. *Yuli Borisovich Khariton: Put' Dlinoyu V Vek.* Moscow: Editorial URSS, 1999.

Gorbachev, Mikhail. Perestroika: *New Thinking for Our Country and the World.* New York: Harper & Row, 1987.

Gorelik, Gennady. *Andrei Sakharov: Nauka I Svoboda.* Moscow: R & C Dynamics, 2000.

Graham, Loren R. *Science in Russia and the Soviet Union: A Short History.* New York: Cambridge University Press, 1993.

Hayward, Max, ed. *On Trial: The Soviet State versus "Abram Tertz" and "Nikolai Arzhak."* New York: Harper & Row, 1966.

Heller, Mikhail, and Nekrich, Aleksandr M. *Utopia in Power: The History of the Soviet Union from 1917 to the Present.* New York: Summit Books, 1986.

Highfield, Roger, and Carter, Paul. *The Private Lives of Albert Einstein.* New York: St. Martin's Press, 1993.

Holloway, David. *Stalin and the Bomb: The Soviet Union and Atomic Energy, 1939–1956.* New Haven, Connecticut: Yale University Press, 1994.

Jones, Roger S. *Physics for the Rest of Us: Ten Basic Ideas of Twentieth-Century Physics That Everyone Should Know . . . and How They Have Shaped Our Culture and Consciousness.* Chicago: Contemporary Books, 1992.

Kalugin, Oleg. *The First Directorate: My 32 Years in Intelligence and Espionage against the West.* New York: St. Martin's Press, 1994.

Keane, John. *Václav Havel: A Political Tragedy in Six Acts.* New York: Basic Books, 2000.

Khabarov, Yu. A. *Etot Fatalny Mesyats Oktyabr.* Moscow: IzdAT, 1997.

Khrushchev, Nikita. *Khrushchev Remembers.* Boston: Little, Brown, 1970.

Khrushchev, Sergei. *Khrushchev on Khrushchev: An Inside Account of the Man and His Era.* Boston: Little, Brown, 1990.

Knight, Amy. *Beria: Stalin's First Lieutenant.* Princeton, New Jersey: Princeton University Press, 1993.

Krementsov, Nikolai. *Stalinist Science.* Princeton, New Jersey: Princeton University Press, 1997.

Levin, Mikhail Lvovich. *Zhizn', Vospominaniya, Tvorchestvo.* Nizhny Novgorod: Chuvashiya, 1995.

Ligachev, Yegor. *Inside Gorbachev's Kremlin.* New York: Pantheon, 1993.

Litvinov, Pavel. *The Demonstration in Pushkin Square.* Boston: Gambit, 1969.

Lourie, Richard. *Letters to the Future: An Approach to Sinyavsky-Tertz.* Ithaca, New York: Cornell University Press, 1975.

———. *Russia Speaks: An Oral History from the Revolution to the Present.* New York: HarperCollins, 1991.

Mandelstam, Nadezhda. *Hope against Hope: A Memoir.* New York: Atheneum, 1970.

Marchenko, Anatoly. *From Tarusa to Siberia.* Royal Oak, Michigan: Strathcona, 1980.

———. *My Testimony.* Harmondsworth, Middlesex, England: Penguin, 1971.

———. *To Live like Everyone Else,* with a foreword by Andrei Sakharov. New York: Henry Holt, 1989.

Massie, Robert K. *The Romanovs: The Final Chapter.* New York: Ballantine Books, 1995.

Maurois, André. *Aspects of Biography.* New York: Frederick Ungar, 1966.

Medvedev, Roy. *All Stalin's Men.* Garden City, New York: Anchor Press/Doubleday, 1984.

———. *Khrushchev.* Garden City, New York: Anchor Press/Doubleday, 1984.

Medvedev, Roy, and Chiesa, Giulietto. *Time of Change: An Insider's View of Russia's Transformation.* New York: Pantheon, 1989.

Medvedev, Zhores. *Gorbachev.* New York: Norton, 1987.

Medvedev, Zhores, and Medvedev, Roy. *A Question of Madness.* New York: Knopf, 1971.

Muravyov, Vladimir. *Ulochki-Shkatulochki. Moskovskie Dvory.* Moscow: Tverskaya, 13, 1998.

Nabokov, Vladimir. *Speak, Memory: An Autobiography Revisited.* New York: Perigee, 1966.

Nikishin, Aleksandr. "Pokhorony Akademika A. D. Sakharova." *Znamya,* no. 5, 1990.

Novikov, Igor D. *The River of Time.* Cambridge, England: Cambridge University Press, 1998.

Orlov, Yuri. *Dangerous Thoughts: Memoirs of a Russian Life.* New York: Morrow, 1991.

Orwell, George. *A Collection of Essays.* San Diego: Harcourt Brace Jovanovich, 1981.

Overbye, Dennis. *Lonely Hearts of the Cosmos: The Scientific Quest for the Secret of the Universe.* New York: HarperPerennial, 1992.

Overy, Richard. *Russia's War: Blood upon the Snow.* New York: TV Books, 1997.

Pasternak, Boris. *I Remember: Sketch for an Autobiography.* New York: Pantheon, 1959.

Paustovsky, Konstantin. *The Story of a Life.* New York: Pantheon, 1964.

Pestov, Stanislav. *Bomba: Tainy i Strasti Atomnoi Preispodnei.* St. Petersburg: Shans, 1995.

Pirozhkova, A. N. *At His Side: The Last Years of Isaac Babel.* South Royalton, Vermont: Steerforth Press, 1996.

Radzinsky, Edward. *The Last Tsar: The Life and Death of Nicholas II.* New York: Doubleday, 1992.

Remnick, David. "The Afterlife." *The New Yorker,* August 11, 1997.

———. *Lenin's Tomb: The Last Days of the Soviet Empire.* New York: Random House, 1993.

Rhodes, Richard. *Dark Sun: The Making of the Hydrogen Bomb.* New York: Touchstone, 1995.

———. *The Making of the Atomic Bomb.* New York: Touchstone, 1986.

Rothberg, Abraham. *The Heirs Of Stalin: Dissidence and the Soviet Regime, 1953–1970.* Ithaca, New York: Cornell University Press, 1972.

Rubenstein, Joshua. *Tangled Loyalties: The Life and Times of Ilya Ehrenburg.* New York: Basic Books, 1996.

Sagdeev, Roald Z. *The Making of a Soviet Scientist.* New York: John Wiley & Sons, 1994.

Sakharov, Andrei. *Alarm and Hope.* New York: Vintage, 1978.

———. "Lecture in Lyons: Science and Freedom." *Physics Today,* July 1999.

———. *Memoirs.* New York: Knopf, 1990.

———. *Moscow And Beyond: 1986 to 1989.* New York: Vintage, 1992.

———. *My Country and the World.* New York: Knopf, 1976.

———. *O Strane I Mire.* New York: Khronika, 1975.

———. *Reflections on Progress, Coexistence, and Intellectual Freedom.* New York: Norton, 1968.

———. *Sakharov Speaks.* New York: Knopf, 1974.

Saunders, George, ed. *Samizdat: Voices of the Soviet Opposition.* New York: Monad Press, 1974.

Sharansky, Natan. *Fear No Evil.* New York: Random House, 1988.

Shikhanovich, Yuri, ed. *Materialy Konferentsii, k 30-letiyu raboty A. D. Sakharova "Razmyshleniya o pogresse, mirnom sosushchestvovanii I intellektualnoi svobode."* Moscow: PRAVA CHELOVEKA, 1998.

Shklovsky, Iosif. *Five Billion Vodka Bottles to the Moon.* New York: Norton, 1991.

Sinyavsky, Andrei. *On Socialist Realism.* New York: Pantheon, 1960.

Smith, Hedrick. *The New Russians.* New York: Random House, 1990.

Snow, C. P. *The Physicists: A Generation That Changed the World.* Boston: Little, Brown, 1981.

———. *The Two Cultures and the Scientific Revolution.* New York: Cambridge University Press, 1959.

Solovyov, Vladimir, and Klepikova, Elena. *Boris Yeltsin: A Political Biography.* New York: G. P. Putnam's, 1992.

———. *Yuri Andropov: A Secret Passage into the Kremlin.* London: Robert Hale, 1983.

Solzhenitsyn, Alexander. *From under the Rubble.* Boston: Little, Brown, 1975.

———. *The Oak and the Calf: Sketches of Literary Life in the Soviet Union.* New York: Harper & Row, 1979.

Sparre, Victor. *The Flame in the Darkness.* London: Grosvenor Books, 1979.

Stites, Richard. *Russian Popular Culture: Entertainment and Society since 1900.* Cambridge, England: Cambridge University Press, 1992.

Strizhev, Aleksandr, ed. *Podvig Startsa Serafima.* Moscow: Palomnik, 1999.

Sudoplatov, Pavel, and Sudoplatov, Anatoli. *Special Tasks: The Memoirs of an Unwanted Witness—a Soviet Spymaster.* Boston: Little, Brown, 1994.

Szilard, Leo. *The Voice of the Dolphins and Other Stories.* New York: Touchstone, 1961.

Thomas, D. M. *Alexander Solzhenitsyn: A Century in His Life.* New York: St. Martin's Press, 1998.

Timberlake, Charles E., ed. *Religious and Secular Forces in Late Tsarist Russia.* Seattle: University of Washington Press, 1992.

Volkogonov, Dmitri. *Autopsy for an Empire: The Seven Leaders Who Built the Soviet Regime.* New York: The Free Press, 1998.

Wat, Aleksander. *My Century: The Odyssey of a Polish Intellectual.* Berkeley and Los Angeles: University of California Press, 1988.

Werth, Alexander. *Russia at War, 1941–1945.* New York: Carroll & Graf, 1964.

West, Anthony. *H. G. Wells: Aspects of a Life.* New York: Random House, 1984.

White, E. B. *Essays of E. B. White.* New York: Harper & Row, 1977.

White, Michael. *Isaac Newton: The Last Sorcerer.* Reading, Massachusetts: Perseus Books, 1997.

Yakir, Pyotr. *A Childhood in Prison.* New York: Coward, McCann & Geoghegan, 1973.

Yeltsin, Boris. *Against the Grain.* New York: Summit Books, 1990.

Zee, A. *Fearful Symmetry: The Search for Beauty in Modern Physics.* New York: Collier Books, 1986.

ACKNOWLEDGMENTS

Written in as much solitude as can ever be stolen from New York, this book was fed by many. Andrei Sakharov, an amazingly busy man as I discovered when writing this book, found time to speak with me in Boston and Moscow. In those days our primary subject of conversation was his memoirs, which I had translated into English, but inevitably our discussions strayed far past the nuances of tenses. Sakharov and Bonner also fed me well, accepting me as one of the family because their young granddaughter Anya was enamored of me. After dinner, despite the late hour, Sakharov would see me down to the street, hail me a gypsy cab, and slip the driver a five. My long, smoky conversations with Elena Bonner continued after Andrei Sakharov's death and still continue. Without her help it would have been impossible even to begin to grasp the nature of a person of greater intelligence and virtue than anyone I have ever encountered. She also provided details about the years of exile in Gorky that supplemented her own vivid writings on the subject. Bonner's daughter, Tatiana Yankelevich, both as director of the Andrei Sakharov Archives at Brandeis University and as a source of detail and insight, was simply invaluable. Efrem Yankelevich, who used to bring me Sakharov's memoirs in batches to translate, always with the same joke—Top Secret, burn before reading—sent me long e-mails from Israel that caught the flavor, weather, tenor of those now distant days of dissidence. Alexander Gribanov of the Sakharov Archives vetted the entire manuscript for accuracy, made creative and provocative suggestions, and alerted me to the existence of people and material that I might easily have otherwise overlooked. He was also a terrific source of the jokes that Russians love to tell and that, at times, somehow encapsulate a moment better than tomes and poems.

Andrei Sakharov's daughter Tanya was more than generous with her time and her honest insights into her father's psychology. She also supplied me with the letters that Sakharov wrote to his first wife, Klava, when they were separated at the end of the war. Sakharov's first marriage and first family would have been a mystery to me without her help. I am particularly grateful to her. Her daughter, Marina, provided a few key incidents and details of the sort that help change a chronology into a life. Sakharov's other two children—his second daughter, Lyubov, and his son, Dmitri, graciously met with me in Moscow.

Among Sakharov's scientific colleagues in Moscow, Evgeny Feinberg provided the most pleasurable and productive company. His memory is acute, as is his sense of discernment. He was particularly instrumental in giving me a feel for Sakharov, Tamm, Kurchatov, as men and as scientists, and for the atmosphere at FIAN, both the old and the new. He brought me to the new FIAN, showed me Sakharov's office, and introduced me to the dynamic Vitaly Ginzburg, who also shared his memories. Also in Moscow, Leonid Litinsky and Boris Bolotovsky proved patient helpers, sharing documents and reminiscences. Moscow, as always, was made a more homey place for me by the hospitality of Lyusa and Lyosha Persky.

I wish to thank Masha Gavrilova, formerly of the Sakharov Museum in Gorky, for taking a day of her life and showing me the apartment and the city where Sakharov spent seven years of his life. The current members of the staff were helpful in unearthing documents whose existence I never suspected and sustaining me with sandwiches and tea. In St. Petersburg, Klava's sister Zinaida spent a long winter's afternoon telling me stories of their youth in Ulyanovsk.

Two of Sakharov's colleagues, Leon Bell and Mates Agrest, both now living in the United States, afforded me useful insights into the young Andrei Sakharov, at the university and at the Installation. To Sakharov's friend Ed Kline I am indebted for glimpses of Sakharov in action, in particular, how the dynamic between him and Bonner actually operated. Former AP correspondent George Krimsky shared his memories of Sakharov, providing me as well with a tape of an interview—simply to hear Sakharov's voice again was every bit as important as the contents, themselves of great significance. My brother, Robert Lourie, physicist, "tutored" me in physics and also kindly vetted the manuscript for scientific errors, the responsibility for which, of course, remains with me. Discussions with the historian Martin Kenner reminded me of the relation of scale and detail in the telling of a life. First readers—Phil Pochoda, the initial editor of this book, and my cousin Jim Simons—are, in a sense, the most valuable of readers because, if their suggestions are heeded, they save you grief down the line. And there are people whose contribution to the writing of a book is less direct but no less important for that. I would like to single out Anne Isaak, whose easy generosity with her home in Sagaponack provided me with long hours away from the jackhammers of my city. Finally, I salute my wife, Jod, who finished a project of many years just as I was finishing this one and who for that reason best knows the struggle of a long involvement and the sudden blank freedom of completion.

INDEX